MW00651694

Genotype to Phenotype

Second edition

The HUMAN MOLECULAR GENETICS series

Series Advisors

D.N. Cooper, *Institute of Medical Genetics, University of Wales College of Medicine, Cardiff, UK*

S.E. Humphries, *Division of Cardiovascular Genetics, University College London Medical School, London, UK*

A. Wolffe, *Sangamo Biosciences Inc, Point Richmond Tech Center, Richmond, CA, USA*

Human Gene Mutation
Functional Analysis of the Human Genome
Environmental Mutagenesis
HLA and MHC: Genes, Molecules and Function
Human Genome Evolution
Gene Therapy
Molecular Endocrinology
Venous Thrombosis: from Genes to Clinical Medicine
Protein Dysfunction in Human Genetic Disease
Molecular Genetics of Early Human Development
Neurofibromatosis Type 1: from Genotype to Phenotype
Analysis of Triplet Repeat Disorders
Molecular Genetics of Hypertension
Human Gene Evolution
Analysis of Multifactorial Disease
Transcription Factors
Molecular Genetics of Cancer, Second edition
Genotype to Phenotype, second edition

Forthcoming titles
Prenatal Testing of Late Onset Neurogenetic Diseases

Genotype to Phenotype

Second edition

S. Malcolm
*Clinical and Molecular Genetics Unit, University College London,
Institute of Child Health, London WC1N 1EH, UK*

J. Goodship
*Institute of Human Genetics, University of Newcastle,
Newcastle Upon Tyne, UK*

**ACADEMIC
PRESS**

A Harcourt Science and
Technology Company

Published in the United States of America, its dependent territories and Canada by
arrangement with BIOS Scientific Publishers Ltd, 9 Newtec Place, Magdalen Road,
Oxford OX4 1RE, UK.

First published 2001

A CIP catalogue record for this book is available from the British Library.

ISBN 0–12–466257–9

Distributed exclusively throughout the United States, its dependent territories and
Canada by Academic Press, Inc., A Harcourt Science and Technology Company,
525 B Street, San Diego, CA 92101–4495. www.academicpress.com

Production Editor: Paul Barlass
Typeset by Saxon Graphics Ltd, Derby, UK.
Printed by Biddles Ltd, Guildford, UK.

Contents

Contributors

Bale, A.E. Department of Genetics, SHM I-321, Yale University School of Medicine, Box 208005, 333 Cedar Street, New Haven, CT 06520–8005, USA

Bonthron, D.T. Molecular Medicine Unit, University of Leeds, St James's Hospital, Leeds LS9 7TF, UK

Cambien, F. INSERM U525, Université Pierre et Marie Curie, Paris, France

Chinnery, P.F. Department of Neurology, University of Newcastle Upon Tyne, Newcastle Upon Tyne, UK

Christian, S.L. Department of Psychiatry, University of Chicago, 924E57th Street, Chicago, USA

Daly, A.K. Department of Pharmacological Sciences, University of Newcastle Upon Tyne, Framlington Place, Newcastle Upon Tyne, NE2 4HH, UK

Demant, P. Department of Pathology, Leiden University Medical Center, Leidenm The Netherlands

Fitzpatrick, D.R. MRC Human Genetics Unit, Western General Hospital, Edinburgh, EH4 2XU, UK

Gaspar, H.B. Molecular Immunology Unit, Institute of Child Health, University College London, and Great Ormond Street Hospital for Children NHS Trust, London WC1N 1EH, UK

Goodman, F.R. Molecular Medicine Unit, Institute of Child Health, 30 Guildford Street, London WC1N 1EH, UK

Goodship, J. School of Biochemistry and Genetics, 19 Claremont Place, Newcastle Upon Tyne NE2 4AA, UK

Humphries, S.E. Cardiovascular Genetics, British Heart Foundation Laboratories, Rayne Building, Royal Free and University College London Medical School, London WC1E 6JJ, UK

Kinnon, C. Molecular Immunology Unit, Institute of Child Health, University College London, London WC1N 1EH, UK

Ledbetter, D.H. Department of Human Genetics, University of Chicago, 920E58th Street, Chicago, USA

Lipoldova, M. Institute of Molecular Genetics, Academy of Sciences of the Czech Republic, Flemingovo nam. 2, 16637 Prague, Czech Republic

Malcolm, S. Clinical and Molecular Genetics Unit, Institute of Child Health, 30 Guilford Street, London WC1N 1EH, UK

Mitch, W.E. Renal Division, Woodruff Memorial Building, 1639 Pierce Drive, Emory University, Atlanta, GA 30322, USA

Montgomery, H. Cardiovascular Genetics, British Heart Foundation Laboratories, Rayne Building, Royal Free and University College London Medical School, London WC1E 6JJ, UK

Murray, J. Department of Paediatrics, University of Iowa, W229 General Hospital, Iowa City, Iowa, USA

Nagl, S.B. Bloomsbury Centre for Structural Biology, Department of Biochemistry, University College London, Gower Street, London, WC1E 6BT, UK

Price, S.R. Renal Division, Woodruff Memorial Building, 1639 Pierce Drive, Emory University, Atlanta, GA 30322, USA

Semina, E. Department of Paediatrics, University of Iowa, W229 General Hospital, Iowa City, Iowa, USA

Talmud, P.J. Cardiovascular Genetics, British Heart Foundation Laboratories, Rayne Building, Royal Free and University College London Medical School, London WC1E 6JJ, UK

Turnbull, D.M. Department of Neurology, University of Newcastle Upon Tyne, Newcastle Upon Tyne, UK

van Wezel, T. Division of Molecular Genetics, The Netherlands Cancer Institute, Plesmanlaan 121, 1066CX Amsterdam, The Netherlands

Wood, N.W. Department of Clinical Neurology, Institute of Neurology, Queen Square, London, UK

Worth, P.F. Department of Clinical Neurology, Institute of Neurology, Queen Square, London, UK

Abbreviations

ACE	angiotensin converting enzyme
ADCA	autosomal dominant cerebellar ataxias
AHO	Albright hereditary osteodystrophy
AMD	age-related macular regeneration
ANT	adenine nucleotide translocator
APC	adenomatous polyposis coli
APC	anaphase promoting complex
Apo	apolipoprotein
AS	Angelman syndrome
BAC	bacterial artificial chromosome
BCCs	basal cell carcinomas
BMP	bone morphogenetic protein
CAD	coronary artery disease
cDNA	complementary DNA
CDP	chondrodysplasia punctata
CES	cat eye syndrome
CFTR	cystic fibrosis transmembrane conductance regulator protein
CGD	chronic granulomatous disease
ci	cubitus interruptus
CM	chylomicrons
COX	cytochrome c oxidase
CPEO	chronic progressive external ophthalmoplegia
CRS	Cambridge reference sequence
CVID	combined variable immunodeficiency
cyt b	cytochrome b
DGS	DiGeorge syndrome
D-loop	displacement loop
DMRs	differentially methylated regions
DRPLA	dentatorubral pallidoluysian atrophy
DSBs	double stranded breaks
DUBs	deubiquitinating enzymes
EBV	Epstein-Barr virus
ENaC	epithelial sodium channel
ENU	N-ethyl-N-nitrosourea
ERG	ergosterol auxotrophic
ESE	exonic splice enhancers
EST	expressed sequence tags
FAP	familial adenomatous polyposis
FFA	free fatty acid
GLO	L-gulono-gamma lactone oxidase
GO	gene ontology

HD	Huntington's disease
HDL	high density lipoprotein
HFGS	hand-foot-genital syndrome
HNPP	hereditary neuropathy with liability to pressure palsies
HR	hazard ratio
HSPG	heparan sulfate proteoglycans
HSPs	heat shock proteins
HVR	hypervariable region of mtDNA within the noncoding D-loop
HWE	Hardy-Weinberg equilibrium
IDS	iduronate-2-sulfatase
KSS	Kearns-Sayre syndrome
LCRs	low copy repeats
LD	linkage disequilibrium
LDL	low density lipoprotein
LH	leuteinizing hormone
LHON	Leber's hereditary optic neuropathy
LPL	lipoprotein lipase
LVH	left ventricular hypertrophy
MELAS	mitochondrial encephalomyopathy with lactic acidosis and stroke-like episodes
MERRF	myoclonic epilepsy with ragged red fibers
MHC	major histocompatibility complex
MI	meconium ileus
MRI	magnetic resonance imaging
mtDNA	mitochondrial DNA
NBT	nitroblue tetrazolium
NF1	neurofibromatosis type 1
NI	intranuclear inclusions
NK	natural killer (cells)
NPH	familial juvenle nephronophthisis
PAC	P-1 derived artificial chromosome
PCCMT	prenylcysteine carboxymethyltransferase
PgP	P-glycoprotein
PHP	pseudohypoparathyroidism
PPTH	pseudopseudohypoparathyroidism
PTH	parathyroid hormone
PTHrP	parathormone-related peptide
PWS	Prader-Willi syndrome
QTL	Quantitative trait locus
RCS	recombinant congenic strains
RIS	recombinant inbred strains
RLGS	restriction landmark genomic scanning
RRF	ragged red fibers
RSV	Rous sarcoma virus
SBMA	Spinobulbar muscular atrophy
SCAs	spinocerebellar ataxias
SCIDs	severe combined immunodeficiences
SLAM	signalling lymphocyte activating molecule

SLOS	Smith-Lemli-Opitz syndrome
SMA	spinal muscular atrophy
SMS	Smith-Magenis syndrome
SNP	single nucleotide polymorphisms
SNRPN	small nuclear ribonucleoprotein N gene
SUMO	small ubiquitin related modifier
Tg	triglyceride
TGFB	transforming growth factor-beta
TPMT	thiopurine S-methyltransferase
TRAFs	TNF receptor associated factor
TSH	thyroid stimulating hormone
UBCs	ubiquitin-conjugating enzymes
UBLs	ubiqutin-like-modifier proteins
UCHs	ubiquitin C-terminal hydrolases
VCFS	velo-cardio facial syndrome
VHL	von Hipple-Lindau
VLDL	very low density lipoprotein
WAS	Wiskott-Aldrich syndrome
WBS	Williams-Beuren syndrome
XHM	X-linked hyper IgM syndrome
XLA	X-linked agammaglobulinemia
XLP	X-linked lymphoproliferative disease
XLT	X-linked thrombocytopenia

Preface

The Human Genome Project, which has so captured the public's imagination, is a classic example of non hypothesis driven research. It is merely the production of a catalogue and even the indexing is in its early stages. Within a relatively short period of time, the genome's potential for coding genes will be established as the momentum which has built up over the last few years and the investment put in has been so great. However, the sequence on its own will tell us very little about the function of the predicted genes. We may only get information that a gene is structurally related to another known gene which has tyrosine kinase activity, or has a section of sequence also found in a group of proteins which act as transcription factors or that it falls within a cluster of genes whose other members are metallo peptidases. The methods available to us to make these predictions, including web based tools such as the BLAST suite of programs, and some experimental techniques which can prove useful, such as yeast two hybrid technology and microarray assays are outlined in the chapter by Sylvia Nagl. The chapter explores the relationships between mutant phenotype, genetic interaction, physical interaction, sequence or structural similarity and expression patterns.

Over the last few years much of our understanding of the function of genes, at least in man and mouse, has been in reverse. Genes have been identified because they are mutated in genetic disorders and their role has been deduced from the phenotype of the disorder. Even more importantly, genes which work within pathways have been identified this way and interactions between proteins have been established. The detailed study of mutations has also been of vital importance. Not all mutations act by reducing or abolishing the activity of a gene (a fact which is likely to make the already difficult science of gene therapy even more difficult) but the exact mechanism of mutation will often throw considerable light on the role of a gene. Thus we have learnt much about the function of the fibroblast growth factor receptor gene families by the study of craniosynostosis, in which there is premature fusion of the cranial sutures. We now appreciate that missense amino acid changes which result in constitutive activation of the receptor activity and its subsequent intracellular signalling cause premature differentiation of the sutures and give rise to Crouzon and Pfeiffer syndromes. These are predominantly mutations which give rise to an unpaired cysteine residue which becomes available for inappropriate dimerisation. On the other hand specific mutations found in Apert syndrome act by changing the specificity of the Fibroblast Growth Factor Receptor 2 for the Fibroblast Growth Factor ligands.

This book aims to study the relationship between genotype and phenotype in several ways. A number of chapters deal with the issue of what classifies an amino acid change as a mutation or a polymorphism and what factors or special circumstances act to convert one into the other (see chapters by Cambien, Humphries, Malcolm and Daley). These factors may include the environment as in smoking, exposure to infectious disease or pharmacological agents. The chapter by Demant deals with methods of mapping loci which modify the action of major genes, particularly in cancer. Regrettably few of those loci have yet been converted into genes in our understanding, but progress is likely to be rapid in the next few years.

A number of examples are explored where our knowledge and comprehension of important pathways particularly in development, has arisen via reverse genetics. These include chapters by Gaspar and Kinnon (B and T cell differentiation), Goodman (homeobox genes), Semena and Murray (pitx2), Bale (hedgehog pathway) and Fitzpatrick (cholesterol).

A higher level of complexity is reached in the chapter by Bonthron where an apparently straightforward gene, the GNAS gene involved in G protein signalling, is shown not only to encode for three separate genes but to be imprinted i.e. expressed differentially from the maternal and paternal genomes. This introduces two concepts which further complicate matters. It is becoming apparent that man and other higher organisms may have fewer genes than expected and not so many more than simpler organisms such as flies and worms, or at least that may appear true at the nucleic acid level. However more complicated intron exon structures and creative uses of splicing may result in a more complex set of products.

Inherent instability in the human genome has also emerged from a study of its detailed structure. Christian and Ledbetter describe the profound, usually adverse, consequences for rearrangements of the genome which arise as a result of repeated elements within the genome. Both Goodman in the chapter on homeoboxes and polydactyly and Worth and Wood in their chapter on neurological disorders resulting from (CAG) repeats describe on a smaller scale how repeats within genes are also unstable and give rise to proteins with lethal properties.

The mitochondrial disorders follow a separate pattern of inheritance again because of the significance for mutational mechanisms of the presence of multiple copies in a single cell.

All of the above issues beg the question of how to define a gene in the twenty first century. If it is by phenotype then we will have several disorders i.e. phenotypes, resulting from one 'gene' through different mutational mechanisms. If it is by coding potential then we will have a number of repeated sequences which have a profound effect on phenotype by driving deletions and duplications although they have no significant coding potential, if it is defined as a transcription unit then several disparate proteins of unrelated function may arise from the same 'operon'. None of these definitions in the end matter. The deeper the study we make of genotype to phenotype correlations the more we will be amazed by the many and various methods available for going from one to the other.

S. Malcolm and J. Goodship

Genotype to phenotype: interpretation of the Human Genome Project

Sue Malcolm

The genome is done: you have seen it in the papers and heard it on the news. The year 2001 marked a milestone in our understanding of the human genome with the publication of two papers (International Human Genome Sequencing Consortium, 2001; Venter *et al*., 2001) as the culmination of a multi-million pound project. It is a giant step to move from there to the understanding of phenotype, and in particular how the single sequence which has been published can account for the wide range of phenotypic variation in appearance and susceptibility to disease actually found in humans.

1. The genome project

1.1 The sequence

The first two complete chromosomes to be sequenced were chromosome 22 in 1999 (Dunham *et al*., 1999) and chromosome 21 in 2000 (Hattori *et al*., 2000). They each contain 33 million base pairs. The published sequence does not correspond to any one individual as bacterial artificial chromosome (BAC), P-1 derived artificial chromosome (PAC), cosmid and fosmid libraries were used from multiple donor sources. The opportunity to donate DNA to the project was broadly advertised near the participating laboratories and volunteers of diverse backgrounds were accepted on a first-come, first-taken basis. Elaborate steps were taken to remove any way of identifying clones or sequences with a particular donor. Analysis of overlapping sequences on chromosome 21 revealed multiple nucleotide variations and small deletions and insertions, leading to an estimate of an average of one sequence difference for each 787 base pairs. Only a very small proportion of each chromosome remained unsequenced. Interestingly, on chromosome 22 one of these areas, on the proximal region of the long arm, corresponded to long, low copy number repeats which contribute to the instability

Genotype to Phenotype second edition, edited by S. Malcolm and J. Goodship.
© 2001 BIOS Scientific Publishers Ltd, Oxford.

associated with the DiGeorge or velocardiofacial critical region (Edelmann *et al.*, 1999) described in Chapter 9.

1.2 Finding the genes

Methods were developed for annotating the raw sequence and for searching for functional genes. A prediction of 225 genes was made for chromosome 21 (127 known genes and 98 predicted) and 545 genes for chromosome 22. Based on these figures a total gene count of between 30 000 and 40 000 was extrapolated for the whole genome. This is broadly in line with two other experimental determinations (Roest Crollius *et al.*, 2000; Ewing and Green, 2000) but considerably less than databases held by several biotech companies suggested (Liang *et al.*, 2000). The reason for the discrepancy is unknown, but may be connected with variable splice forms which have been registered more than once. Gene counts of around 30 000 were confirmed in the two papers analyzing the whole genome (International Human Genome Sequencing Consortium, 2001; Venter *et al.*, 2001). This is very few more, about twice as many, than in the worm and fly.

Both chromosome communities developed broadly similar hierarchical approaches to identifying genes. Basically, they combined gene prediction programs such as GENSCAN and GRAIL with sequence similarity searches using the BLAST suite of programs. The presence of CpG islands provided useful additional evidence. At the top came known human genes from the literature or public databases. Secondly came novel genes which could be correlated with known cDNAs or open reading frames from any organism. This category identified new members of human gene families as well as human homologs or orthologs of genes from yeast, *Caenorhabditis elegans*, *Drosophila*, mouse etc. After that the position became murkier with the next category containing novel genes which, in part, corresponded to a known protein domain such as a zinc finger. Finally, there was a class of novel anonymous genes defined solely by gene predictions including some quality guidelines, for example strongly predicted exons or adjacent predicted exons also being found spliced in Expressed Sequence Tags (EST). Both studies also found a surprisingly large number of pseudo genes (59 on chromosome 21 and 134 on chromosome 22) corresponding to 20% of the total. This should provide a cautionary note for amateur readers of the sequence.

To what extent does the definitive sequence provide the definitive catalog of all genes? Clearly extensive experimental studies will have to be carried out to confirm the nature of the predicted sequences, but there are frequent examples where experimental exploration of biological function leads to the discovery of an ever increasingly complex array of products from a single locus and it is sometimes purely a matter of semantics as to what should be defined as a/the gene. When a gene has the same coding region but has tissue specific splicing of 5' non coding exons from alternative promoters (Suter *et al.*, 1994), perhaps two genes could be defined named for their tissue specificity. A particularly complex example, the GNAS1 locus, is described in detail in Chapter 8. A clinically important example for our understanding of cancer is the set of tumor suppressor genes found in chromosomal region 9p21 which is the site of a major locus for predisposition to melanoma. Three genes have been identified in the region:

CDKN2A which encodes the p16 protein, *CDKN2B* which encodes p15 protein, and also *p14*[ARF] which is encoded by an alternative exon (1β) about 12kb upstream of *CDKN2A*. Exon 1β is spliced onto exon 2 of *CDKN2A* leading to an alternate reading frame (ARF) and no sequence homology to *CDKN2A* at the amino acid level (Randerson-Moor *et al.*, 2001; Stone *et al.*, 1995). Only half the melanoma families linked to 9p21–22 have detectable mutations in the *CDKN2A* gene but it appears to be the critical protein for tumor suppression of melanoma. A family has now been reported who have a phenotype of melanoma plus neural system tumors in which only the exon 1β is deleted. This will help to unravel the relative roles of *p14*[ARF] and *p16*.

Many of the most complex examples, perhaps only because they are also the most intensively investigated examples, arise from regions where genomic imprinting occurs. The small nuclear ribonucleoprotein N (*SNRPN*) gene was originally defined because of its role in splicing where spliceosomes contain small RNAs and associated polypeptides. The gene maps within the imprinted region on chromosome 15q11–13 implicated in the two neurodevelopmental disorders Prader–Willi syndrome and Angelman syndrome and is itself only expressed from the paternal chromosome. There are multiple alternatively spliced exons both 5′ and 3′ to the coding region of the gene (Buiting *et al.*, 1997; Dittrich *et al.*, 1996; Farber *et al.*, 1999) none of which alter the coding capability. Small deletions within the 5′ region have been found in Angelman syndrome and Prader–Willi syndrome patients who are unable to switch their imprint between generations (as shown by the methylation pattern; Buiting *et al.*, 1995; Dittrich *et al.*, 1996) leading to the definition of an imprinting center. These fall into two close but non overlapping clusters: those found in Angelman syndrome which stop the paternal to maternal switch (Buiting *et al.*, 1999) and those found in Prader–Willi syndrome which stop a maternal to paternal switch (Saitoh *et al.*, 1996). As there is to date no apparent connection between the spliceosome and the setting of a methylation imprint it is unclear how many 'genes' are coded for at the SNRPN locus.

Further glimpses into the complexity arising when sets of genes are regulated in a coordinated fashion, as they are at most imprinted loci, come from the discovery of spliced RNAs with no coding potential such as **I**mprinted in **P**rader–**W**illi (IPW) and H19 on chromosome 11p15 (Falls *et al.*, 1999)

1.3 Will the real DNA sequence please stand up

All humans, except monozygotic twins, have different DNA sequences and even monozygotic twins will have differences in somatic tissues, mitochondrial DNA and at certain loci such as the immunoglobulin genes. The published human sequence will be an average of all these because of the methods used in its deci- pherment, and we are unlikely ever to know whether this hypothetical individual would have had straight blonde hair and blue eyes, hypertension or would have developed cancer. In order to answer the above questions large scale systematic efforts at defining the differences between individuals have been undertaken. This has concentrated particularly on single nucleotide changes or polymor- phisms (SNPs) as these are most likely to have a functional role. A non-profit making SNP consortium was set up between 10 pharmaceutical companies, academic centers and the Wellcome Trust to identify and map a large number of SNPs and make these freely available (snp.cshl.org). These may occur in coding regions or intergenic DNA and may or may not lead to changes in amino acids. The significance of these changes and their frequency is discussed in Chapter 3.

2. Mutation vs. polymorphism

2.1 What is the difference?

As will already be clear, any substantial stretch of an individual's DNA which is sequenced is likely to contain differences, mainly heterozygous, from the 'standard' sequence. This variation certainly establishes individuality, but methods are still only poorly developed to establish which changes are responsible for pathological changes found in the individual. Traditionally, evidence has been derived from genetic methods and functional methods. The best genetic evidence arises when a *de novo* mutation in a gene is found in a sporadic case of a genetic disorder. This can be very powerful evidence for dominant disorders but it will not be possible to observe in recessive disorders where both parents will be carriers. Several mutations have been shown to occur at a sufficiently high level in certain populations that 100 so called 'normal' controls are likely to contain several individuals carrying the change. The gap junction protein Connexin 26 (CX26) provides a good example. Deletion of a guanine residue at cDNA position 35 (35delG), causes a frameshift of the coding sequence leading to premature chain termination at the 12th amino acid and has been found in numerous cases with recessive nonsyndromic deafness. Estivill *et al.* (1998) found mutations in the *CX26* gene in 49% of participants from Italy and Spain with a family history of recessive deafness and 37% of sporadic cases. The 35delG mutation accounted for 85% of *CX26* mutations. The carrier frequency of the 35delG mutation in the general population was 1 in 31 (95% CI, 1 in 19 – 1 in 87).

Another example comes from our knowledge of hemochromatosis which is probably the most common mutant gene in the Caucasian population. Iron overload in affected homozygous individuals can lead to liver cirrhosis and primary hepatocellular carcinoma. The faulty hemochromatosis gene (*HFE*) was isolated in 1996 (Feder *et al.*, 1996), the major evidence for it being the causative gene was that the majority of clinically confirmed cases are homozygously associated with a point mutation in *HFE* (the substitution of a tyrosine for cysteine at position 282, or C282Y). Both the frequency and non conservative nature of the amino acid change suggest this is a causative mutation. However, even in such an apparently clear cut case there are complications. Several reports have shown that a minority of individuals homozygous for C282Y do not fit normal diagnostic criteria (Tavill, 1999). Because the disease is fairly easily treatable, *HFE* screening has been proposed as a suitable subject for population screening but the reduced penetrance complicates this. A second sequence variant (H63D) has been found. On its own it is associated with only a slight increase in risk of disease, if any, but as a compound heterozygote together with C282Y it was found in 8/178 of the original cohort and its overall population frequency is 16.6%.

2.2 Changes involving splicing

Nucleotide changes within and around splice donor and acceptor sites have long been recognized as a major source of mutation in humans. More recently it has been suggested that some mutations within coding regions may in fact be exercising their effect through splicing by disrupting Exonic Splice Enhancers (ESEs; Blencowe, 2000; Cooper and Mattox, 1997). Accurate splicing is a complicated business requiring an array of small nuclear ribonucleoproteins and other factors which are components of the spliceosome. ESEs are present in constitutive and alternatively spliced exons and are required for efficient splicing of those exons. The ESEs in pre-mRNAs are recognized by serine/arginine rich (SR) proteins, a family of essential splicing factors that also regulate alternative splicing. ESEs contain a wide spectrum of sequences of approximately 6–8 nucleotides, but they are hard to detect because of their degeneracy. When missense mutations are identified in genomic DNA, particularly in a diagnostic laboratory, the usual interpretation is that the affected amino acid is crucial for the function of the protein if the change is non conservative or the residue changed is highly conserved in evolution. However, if these mutations disrupt an ESE the consequence may be incorrect splicing of the transcript. Nonsense mutations are well known to lead to altered splicing on occasion (NAS or nonsense associated altered splicing; Dietz *et al.*, 1993) and it may be that they are also acting by disrupting ESEs. Experimental evidence for this comes from studies in the breast cancer causing gene *BRCA1* (Liu *et al.*, 2001). A Glu1694Ter nonsense mutation in exon 18 has been found in a number of families with a breast cancer predisposition. It causes skipping of the entire exon (Mazoyer *et al.*, 1998). Although it is difficult to search directly for ESE disruption because of the sequence degeneracy the G→T change was predicted to disrupt a high scoring motif in a matrix analysis. Mini-gene constructs were tested for splicing in a HeLa cell nuclear extract. Exon jumping correlated with the disruption of high-scoring motifs rather than the presence of nonsense codons. It follows that the high-scoring motifs may also be disrupted by changes leading to missense mutations or silent third base changes and because of the degeneracy of the ESEs these events may be frequent.

The effects on phenoytpe may be varied and subtle. The molecular basis for the distinction between Duchenne and the clinically less severe Becker muscular dystrophy is generally explained by the frameshift hypothesis (Monaco *et al.*, 1988). Both disorders can result from deletions in the dystrophin gene but the outcome depends on the exact combination of exons removed. Deletions resulting in Duchenne have been shown to shift the translational open reading frame of triplet codons (because not all exons contain an exact multiple of three nucleotides) leading to in frame stop codons and a truncated protein. Deletions identified in Becker result from a combination of exons which maintain the open reading frame and predict a shorter but partially functional protein. In a minority of Duchenne cases nonsense mutations are observed, again predicting a truncated protein. A Japanese patient was reported with a nonsense mutation in exon 27 of the dystrophin gene but a phenotype of Becker. It was shown by functional studies using an *in vitro* splicing system that the mutation led to exon skipping of exon 27 by interruption of an exon specific enhancer (Shiga *et al.*, 1997).

Exon splice enhancers also provided the explanation for initially puzzling correlations of genotype and phenotype in spinal muscular atrophy (SMA). Two closely related survival motor neuron genes, predicted to encode identical proteins, map to the SMA locus. SMA is caused by mutations within the telomeric copy of the gene (*SMN1*) with 96% of patients showing a homozygous deletion of *SMN1* (Lefebvre *et al*., 1995). Not only is *SMN2* not able to compensate for loss of *SMN1* but homozygous absence of *SMN2* is found in 5% of control individuals with no phenotypic effect. *SMN1* produces full-length mRNA but *SMN2* expresses dramatically reduced full length mRNA and abundant levels of incorrectly spliced transcript lacking exon 7. The splicing of exon 7 is determined by a single nucleotide difference at position +6 in exon 7 (C in *SMN1* and T in *SMN2*) which disrupts an exon splice enhancer (Lorson and Androphy, 2000). A most exciting development for possible therapy has been the identification in *Drosophila* of a splicing factor, Htra2-β1, which promotes the inclusion of exon 7 into full length *SMN2* transcripts and works across species in human and mouse cells carrying an *SMN2* mini-gene (Hofmann *et al*., 2000)

3. The mutation defines the disorder

It is hardly surprising that the exact nature of a mutation leads to different phenotypic outcomes. A mutation at the DNA level may vary from complete deletion of a gene sequence, through small deletions or insertions resulting in a frame shift, to mutation of an amino acid to a termination codon (nonsense mutation), to amino acid changes leading to a reduction in gene activity or amino acid changes resulting in increased activity.

3.1 The ABCR gene associated with human retinopathies

Mutations in the gene encoding ABCR, a photoreceptor-specific ATP-binding cassette (ABC) transporter have been shown to be responsible, in a Mendelian pattern, for three separate clinically defined retinopathies. ABCR mutations were first found in autosomal recessive Stargadt disease, a juvenile-onset macular dystrophy associated with rapid central visual impairment (Allikmets *et al*., 1997; Azarian and Travis, 1997; Illing *et al*., 1997) but were subsequently also found in two other retinal dystrophies, autosomal recessive cone-rod dystrophy and autosomal recessive retinitis pigmentosa (Cremers *et al*., 1998). Overall, a remarkable heterogeneity of mutations has been found with more than 350 variant alleles observed (Allikmets, 2000) and as many of them are missense mutations it is difficult to determine their significance. However, the occurrence of a few mutations at a significant frequency (~10%) has made it possible to make some hypotheses of genotype–phenotype correlation (Maugeri *et al*., 1999). One mutation 2588 G→C was identified in 37.5% of a set of western European patients. The mutation is in linkage disequilibrium with a rare polymorphism (2828G→A) suggesting a founder effect. G2588 is part of the splice acceptor site of exon 17 and analysis of mRNA showed that the resulting mutant ABCR proteins either lack Gly863 or contain the missense mutation Gly863Ala. This is in itself likely to be a source of considerable variation. Maugeri *et al.* (1999) propose that this is likely to be a 'mild' mutation, leading to Stargadt's disease only in combination with a severe ABCR mutation. They support this with the observation that 5/8 Stargadt patients have a null, i.e. severe, mutation on the other chromosome and no Stargadt patient has been found with two copies of 2588 G→C, even though with a population carrier frequency of 1:35 western Europeans, homozygotes for this variant will not be uncommon. In contrast, the more severe autosomal recessive cone-rod dystrophy shows a different spectrum of mutations which can be interpreted as a combination of moderate and severe alleles (for example a complex double variant L541P: A1038V) leaving some residual activity. The most severe, autosomal recessive retinitis pigmentosa, results from a combination of two null alleles which will leave no residual activity (*see Table 1*). Although this is doubtless an over simplification, it is broadly backed up by functional studies. Sun *et al.* (2000) have developed a biochemical system which tests the effects of missense mutations by measuring ATP binding by photo-affinity labelling and retinal stimulated ATPase activity. This provides a direct method for determining the significance of missense changes. Mutations which introduce small deletions (i.e. null) produced greatly reduced amounts of protein and the two variants resulting from G2588→C both resulted in reduced yield and ATP binding.

Table 1. Genotype–phenotype correlation model of Maugeri *et al.* (1999): the severity of the disorder increases as the amount of residual ABCR activity decreases

Phenotype	Normal	Normal/AMD	Normal/AMD	STGD1	ArCRD	arRP
ABCR allele 1	Wild type	Moderate	Severe	Severe	Severe	Severe
ABCR allele 2	Wild type	Wild type	Wild type	Mild	Moderate	Severe

More controversially, but with emerging support, it seems that the continuum of phenotypes correlated with residual ABCR activity may extend to the genetically complex (i.e. non Mendelian) disorder Age Related Macular Degeneration (AMD) which affects nearly 30% of the population over 75 years of age. Case control studies have shown an association between AMD and heterozygous carriers of ABCR variants suggesting that heterozygous ABCR mutations may increase susceptibility to AMD. A recent meta-analysis of the two most common variants (D2177N and G1961E) showed a significant association for both of them with a four-fold increased risk in carriers of D2177N and seven-fold in G1961E (Allikmets, 2000). Both variants have been shown to affect the protein function in the in vitro system of Sun *et al.* (2000).

4. Interacting genes

4.1 Digenic inheritance

The genetic and allelic heterogeneity of inherited retinopathies has already been touched on in the above section. Mutations in rhodopsin, peripherin/RDS, cone-rod homeobox gene (CRX), aryl-hydrocarbon interacting protein-like 1 (AIPL1) and

RP1 have all been found (Sohocki *et al.*, 2001; http://www.sph.uth.tmc.edu/RetNet/) in patients with autosomal dominant retinitis pigmentosa. Further genes have been identified in autosomal recessive and X-linked forms of the disease. However, there can also be interactions between genes leading to digenic inheritance. Families have been identified with mutations in the unlinked photoreceptor-specific genes *ROM1* and peripherin/*RDS* in which only double heterozygotes develop retinitis pigmentosa (Dryja *et al.*, 1997; Kajiwara *et al.*, 1994). Individuals who inherit only one of the changes are essentially normal. *RDS* and *ROM1* are members of a photoreceptor-specific gene family which both localize to the rim of the rod disk. The *ROM1* gene product forms a heterodimer *in vivo* with the *RDS* gene product forming the sub-units of an oligomeric transmembrane protein complex. A molecular based rationale has been constructed for the observed digenic inheritance. In one family the peripherin/*RDS* missense mutation was L185P and expression studies showed that these mutant proteins cannot form the normal homotetramer complexes on their own but can still form the heterocomplexes with native *ROM1*. Only in individuals where there is also an apparently null mutation of *ROM1* is there insufficient complex for adequate function (Goldberg and Molday, 1996).

Doubtless other examples exist but the pedigrees in which they exist are rare and those in which they can be definitely established even rarer. A case has been made for digenic junctional epidermolysis bullosa with mutations in collagen XVII (COL17A1) and laminin 5 (LAMB3) both of which are components of the hemidesmosome-anchoring filament complex in the skin (Floeth and Bruckner-Tuderman, 1999). Another family with Waardenburg syndrome type 2 (WS2) in conjunction with ocular albinism apparently showed digenic inheritance in which the expression in heterozygotes for a tyrosinase mutant TYR R402Q is modified by a single copy of a mutant MITF (Micropthalmia-associated transcription factor) gene with one base pair deleted in exon 8 (Morell *et al.*, 1997). This would be particularly interesting because MITF is thought to be a transcription factor which regulates the tyrosinase gene.

4.2 Modifier genes

Although considerable progress has been made in defining modifier genes in mice (see Chapter 7) progress has inevitably been much slower in humans and they remain suspected but largely unproven.

The jump from rodent to human has been used in a study of cystic fibrosis (Zielenski *et al.*, 1999). Cystic fibrosis is caused by mutations in the *CFTR* gene, a transmembrane conductance regulator. Meconium ileus (MI) is a severe intestinal obstruction detected in a subset (15–20%) of cystic fibrosis patients at birth. None of the more than 800 different mutations which have been found in the *CFTR* gene are specific to the presence of MI but as there is a familial recurrence rate of 29% it suggests that there is a modifying genetic factor present. Mice experiments found a modifier locus on chromosome 7 which modulated the generally fatal intestinal disease in homozygous mice (Rozmahel *et al.*, 1996). As this region of mouse chromosome 7 is syntenic to part of human chromosome 19q13, the authors tested for a similar modifying effect in humans. One hundred and eighty five cystic fibrosis sibpairs were recruited and they were scored for the presence or absence of MI. They were then tested for sharing of alleles of polymorphic markers from chromosome 19. Sibs sharing a phenotype (concordant) would be expected to share markers from chromosome 19 whereas sibs with different phenotypes (discordant) would be expected to also have different alleles of the chromosome 19 markers. They found significant excess allele sharing in sibpairs who both had MI and lack of sharing in discordant sibpairs. Unfortunatley, it is difficult to move from there to identifiying a modifier gene. There is no obvious candidate in the 19q13 region which is over seven Megabases long and as no particular allele or haplotype was associated with the MI phenotype it is unlikely that there is a single mutation at the locus.

Two further examples of human modifiers have been mapped. Hirschsprung disease, congenital agangliosis, displays a complex pattern of inheritance. Approximately 50% of familial cases are heterozygous for mutations in the receptor tyrosine kinase *RET* but the penetrance of the mutations is 50–70% and is dependent on gender (Attie *et al.*, 1995). A set of families were exhaustively studied for mutations in the *RET* gene (Bolk *et al.*, 2000). Three categories were defined. Those with clear mutations in the *RET* gene, those with no mutations or only potentially polymorphic variations in the *RET* gene and a single family which did not map to the *RET* locus. The authors then searched for additional chromosomal regions shared in the families, and identified a locus on 9q31 which was shared by all the families where the disease did not map to the *RET* locus, but not shared in the families with a clear mutation. This can be interpreted as follows: a single coding region change in *RET* can be sufficient to cause Hirschsprung disease but there are other families with 'weaker' mutations which need an additional gene variant, from chromosome 9q31, to cause the disease. Neither the nature of the predisposing 'weak' mutations is known nor the nature of the modifying gene.

The second example comes from a study of recessive nonsyndromic deafness, for which there is already remarkble genetic heterogeneity with 30 loci mapped and six genes ranging in effect from gap junctions to myosins to components of the tectorial membrane (http://www.uia.ac.be/dnalab/hhh/). A vast (141 members) consanguineous Pakistani family was identified with eight individuals affected with profound, congenital, nonsyndromic, sensorineural hearing loss (Riazuddin *et al.*, 2000). A genome wide scan searching for homozygosity and linkage identified a locus on chromsome 4q31 with an impressive lod score of 8.10. However, further genotyping revealed that seven unaffected family members were also homozygous for the high risk haplotype and were, therefore, non-penetrant. A second scan was carried out to search for a dominant modifying locus mapping to each of the seven individuals. A modifier locus was mapped to a 5.5 centimorgan region on chromosome 1q24. The nature of this is completely unknown and it is harder to think of modifying mechanisms which compensate for a mutated gene rather than having a combined effect which pushes the phenotype over into expression of a disorder.

There will certainly be other examples of modifiers in humans but the above examples show that only in a minority of cases will the pedigrees reveal them, either because of the relatively high frequency of the disorder or because of an exceptionally large pedigree. Establishing their nature is likely to be even more difficult.

5. Conclusions

Publication of the entire human genome sequence is the beginning of a journey rather than the end. Methods for finding genes among the three billion base pairs are inadequate and the situation is further complicated by complex alternative splicing. Our methods for determining the significance of sequence variants, of which there are many, are also woefully inadequate and we need to improve greatly our understanding of regulatory regions and splicing mechanisms as well as the three dimensional structure of predicted proteins.

References

Allikmets, R., Singh, N., Sun, H. *et al.* (1997) A photoreceptor cell-specific ATP-binding transporter gene (ABCR) is mutated in recessive Stargardt macular dystrophy. *Nat. Genet.* **15**: 236–246.

Allikmets, R. (2000) Simple and complex ABCR: Genetic predispostion to retinal disease. *Am. J. Hum. Genet.* **67**: 793–799.

Attie, M., Till, A., Pelet, J., Amiel, P., Edery, L., Boutrand, A., Munnich, A. and Lyonnet, S. (1995) Mutation of the endothelin-receptor B gene in Waardenburg-Hirschsprung disease. *Hum. Mol. Genet.* **4**: 2407–2409.

Azarian, S.M. and Travis, G.H. (1997) The photoreceptor rim protein is an ABC transporter encoded by the gene for recessive Stargardt's disease (ABCR). *FEBS Lett.* **409**: 247–252.

Blencowe, B.J. (2000) Exonic splicing enhancers: mechanism of action, diversity and role in human genetic disease. *TIBS* **25**: 106–110.

Bolk, S., Pelet, A., Hofstra, R.M.W., Angrist, M., Salomon, R., Croaker, D., Buys, C.H.C.M., Lyonnet, S. and Chakravarti, S. (2000) A human model for multigenic inheritance: Phenotypic expression in Hirschsprung disease requires both the *RET* gene and a new 9q31 locus. *Proc. Natl Acad. Sci. USA* **97**: 268–273.

Buiting, K., Saitoh, S., Gross, D., Dittrich, B., Schwartz, S., Nicolls, R.D. and Horsthemke, B. (1995) Inherited microdeletions in the Angelman and Prader-Willi syndromes define an imprinting centre on human chromosome 15. *Nat. Genet.* **9**: 395–400.

Buiting, K., Dittrich, B., Endele, S. and Horsthemke, B. (1997) Identification of novel exons 3′ to the human SNRPN gene. *Genomics* **40**: 132–137.

Buiting, K., Lich, C., Cottrell, S., Barnicoat, A. and Horsthemke, B. (1999) A 5-kb imprinting center deletion in a family with Angelman syndrome reduces the shortest region of deletion overlap to 880 bp. *Hum. Genet.* **105**: 665–666.

Cooper., T.A. and Mattox, W. (1997) The regulation of splice-site selection and its role in human disease. *Am. J. Hum. Genet.* **61**: 259–266.

Cremers, F.P., van de Pol, D.J., van Driel, M. *et al.* (1998) Autosomal recessive retinitis pigmentosa and cone-rod dystrophy caused by splice site mutations in the Stargardt's disease gene ABCR. *Hum. Mol. Genet.* **7**: 355–362.

Dietz, H.C., Valle, D., Francomano, C.A., Kendzior, R.J., Pyeritz, R.E. and Cutting, G.R. (1993) The skipping of constitutive exons in vivo induced by nonsense mutations. *Science* **259**: 680–683.

Dittrich, B., Buiting, K., Korn, B. *et al.* (1996) Imprint switching on human chromosome 15 may involve alternative transcripts of the *SNRPN* gene. *Nat. Genet.* **14**: 163–170.

Dryja, T.P., Hahn, L.B., Kajiwara, K. and Berson, E.L. (1997) Dominant and digenic mutations in the peripherin/RDS and ROM1 genes in retinitis pigmentosa. *Invest. Opthalmol. Vis. Sci.* **38**: 1972–1982.

Dunham, I., Shimizu, N., Roe, B.A. *et al.* (1999) The DNA sequence of human chromosome 22. *Nature* **402**: 489–495.

Edelmann, L., Pandita, R.K. and Morrow, B.E. (1999) Low-copy repeats mediate the common 3-Mb deletion in patients with velo-cardio-facial syndrome. *Am. J. Hum. Genet.* **64**: 1076–1086

Estivill, X., Fortina, P., Surrey, S. *et al.* (1998) Connexin-26 mutations in sporadic and inherited sensorineural deafness. *Lancet* **351**: 394–398.

Ewing, B. and Green, P. (2000) Analysis of expressed sequence tags indicates 35,000 human genes. *Nat. Genet.* **25**: 232–234.

Q6Falls, J.G., Pulford, D.J., Wylie, A.A. and Jirtle, R.L. (1999) Genomic imprinting: implications for human disease. *Am. J. Pathol.* **154**: 635–647.

Farber, C., Dittrich, B., Buiting, K. and Horsthemke, B. (1999) The chromosome 15 imprinting centre (IC) region has undergon multiple duplication events and contains an upstream exon of SNRPN that is deleted in all Angelman syndrome patients with an IC microdeletion. *Hum. Mol. Genet.* **8**: 337–343.

Feder, J.N., Gnirke, A., Thomas, W. *et al.* (1996) A novel HMC class I-like gene is mutated in patients with hereditory haemodiromatosis. *Nature* **13**: 399–408.

Floeth, M. and Bruckner-Tuderman, L. (1999) Digenic junctional epidermolysis bullosa: mutations in *COL17A1* and *LAMB3* genes. *Am. J. Hum. Genet.* **65**: 1530–1537.

Goldberg, A.F.X. and Molday, R.S. (1996) Defective subunit assembly underlies a digenic form of retinitis pigmentoas linked to mutations in peripherin/rds and rom-1. *Proc. Natl Acad. Sci. USA* **93**: 13726–13730.

Hattori, M., Fujiyama, A., Taylor, T.D. *et al.* (2000) The DNA sequence of human chromosome 21. The chromosome 21 mapping and sequencing consortium. *Nature* **405**: 311–319.

Hofmann, Y., Lorson, C.L., Stamm, S., Androphy, E.J. and Wirth, B. (2000) Htra2-β1 stimulates an exonic splicing inhancer and can restore full-length SMN expression to survival motor neuron 2 (SMN2). *Proc. Natl Acad. Sci. USA* **97**: 9618–9623.

Illing, M., Molday, L.L. and Molday, R.S. (1997) The 220kDa rim protein of retinal rod outer segments is a member of the ABC transporter superfamily. *J. Biol. Chem.* **272**: 10303–10310.

International Human Genome Sequencing Consortium (2001) Initial sequencing and analysis of the human genome. *Nature* **409**: 860–921.

Kajiwara, K., Berson, E.L. and Dryja, T.P. (1994) Digenic retinitis pigmentosa due to mutations at the unlinked peripherin/RDS and ROM1 loci. *Science* **264**: 1604–1608.

Lefebvre, F., Burglen, L., Reboullet, S., Clermont, O., Munnich, A., Dreffus, G. and Zeviani, M. (1995) Identification and characterisation of a spinal-muscular-atrophy-determining gene. *Cell* **80**: 155–165.

Liang, F., Holt, I., Pertea, G., Karamycheva, S., Salzberg, S.L. and Quackenbush, J. (2000) Gene index analysis of the human genome estimates approximately 120,000 genes. *Nat. Genet.* **25**: 239–240.

Liu, H.-X., Cartegni, L., Zhang, M.Q. and Krainer, A.R. (2001) A mechanism for exon skipping caused by nonsense or missense mutations in *BRCA1* and other genes. *Nat. Genet.* **27**: 55–58.

Lorson, C.L. and Androphy, E.J. (2000) An exonic enhancer is required for inclusion of an essential exon in the SMA-determining gene SMN. *Hum. Mol. Genet.* **22**: 259–265.

Maugeri, A., van Driel, M.A., van de Pol, D.J. *et al.* (1999) The 2588 G→C mutation in the *ABCR* gene is a mild frequent founder mutation in the Western Europena population and allows the classification of ABCR mutations in patients with Stargardt disease. *Am. J. Hum. Genet.* **64**: 1024–1035.

Mazoyer, S., Puget, N., Perrin-Vidoz, L., Lynch, H.T., Serova-Sinilnikova, O.M. and Lenoir, G.M. (1998) A BRCA1 nonsense mutation causes exon skipping. *Am. J. Hum. Genet.* **62**: 713–715.

Moeller, N., Moore, T., Morikang, E. *et al.* (1996) A novel MHC class I-like gene is mutated in patients with hereditary haemochromatosis. *Nat. Genet.* **13**: 399–408.

Monaco, A.P., Bertelson, C.J., Liechti-Gallati, S., Moser, H. and Kunkel, L.M. (1988) An explanation for the phenotypic differences between patients bearing partial deletions of the DMD locus. *Genomics* **2**: 90–95.

Morell, R., Spritz, R.A., Ho, L., Pierpoint, J., Guo, W., Friedman, T.B. and Asher, J.H. (1997) Apparent digenic inheritance of Waardenburg syndrome type 2 (WS2) and autosomal recessive ocular albinism (AROA). *Hum. Mol. Genet.* **6**: 659–664.

Randerson-Moor, J.A., Harland, M., Williams, S. *et al.* (2001) A germline deletion of p14ARF but not CDKN2A in a melanoma-neural system tumors syndromefamily. *Hum. Molec. Genet.* **10**: 55–62.

Roest Crollius, H., Jaillon, O., Bernot, A. *et al.* (2000) Estimate of human gene number provided by genome-wide analysis using Tetraodon nigroviridis DNA sequence. *Nat. Genet.* **25**: 235–238.

Riazuddin, S., Castelein, C.M., Ahmed. Z.M. *et al.* (2000) Dominant modifier DFNM1 suppresses recessive deafness DFNB26. *Nat. Genet.* **26**: 431–434.

Rozmahel, R., Wilschanski, M., Matin, A. *et al.* (1996) Modulation of disease severity in cystic fibrosis transmembrane conductance regulator deficient mice by a secondary genetic factor. *Nat. Genet.* **12**: 280–287.

Saitoh, S., Buiting, K., Rogan, P.K. *et al.* (1996) Minimal definition of the imprinting center and fixation of chromosome 15q11-q13 epigenotype by imprinting mutations. *Proc. Natl Acad. Sci. USA* **93**: 7811–7815.

Shiga, N., Takeshima, Y., Sakamoto, H., Inoue, K., Yokota, Y., Yokoyama, M. and Matsuo, M. (1997) Disruption of the splicing enhancer sequence within exon 27 of the dystrophin gene by a nonsense mutation induces partial skipping of the exon and is responsibe for Becker musculardystrophy. *J. Clin. Inv.* **100**: 2204–2210.

Sohocki, M.M., Daiger, S.P., Browne, S.J. *et al.* (2001) Prevalence of mutations causing retinitis pigmentosa and other inherited retinopathies. *Hum. Mutation* **17**: 42–51.

Stone, S., Jiang, P., Dayananth, P., Tavtigian, S.V., Katcher, H., Parry, D., Peters, G. and Kamb, A. (1995) Complex structure and regulation of the P16 (MTS1) locus. *Cancer Res.* **55**: 2988–2994.

Sun, H., Smallwood, P.M. and Nathans, J. (2000) Biochemical defects in ABCR protein variants associated with human retinopathies. *Nat. Genet.* **26**: 242–246.

Suter, U., Snipes, G.J., Schoener-Scott, R., Welcher, A.A., Pareek, S., Lupski, J.R., Murphy, R.A., Shooter, E.M. and Patel, P.I. (1994) Regulation of tissue-specific expression of alternative peripheral myelin protein-22 (*PMP22*) gene transcripts by two promoters. *J. Biol. Chem.* **269**: 25795–25808.

Tavill, A.S. (1999) Clinical implications of the Hemochromatosis gene. *New. Engl. J. Med.* **341**: 755–757 (Editorial).

Venter, J.C. *et al.* (2001) The sequence of the human genome. *Science* **291**: 1304–1351.

Zielenski, J., Corey, M., Rozmahel, R. *et al.* (1999) Detection of a cystic fibrosis modifier locus for meconium ileus on human chromosome 19q13. *Nat. Genet.* **22**: 128–129.

2

From protein sequence to structure and function

Sylvia B. Nagl

1. Introduction

Roughly 25 years ago, *in vitro* comparisons between DNA sequences of retroviral genes implicated in various cancers and vertebrate genes that regulate cell growth and differentiation, produced the startling insight that oncogenes are mutated versions of these regulatory proteins. The discovery of the cellular version of the *src* oncogene by DNA annealing was one of the first in a wave of seminal work establishing this link (Hunter, 1989). Complementary DNA (cDNA) unique to the oncogenic *src* region of the Rous sarcoma virus (RSV) genomic RNA was shown to hybridize to the DNA of all tested vertebrate animals, with efficiency of formation and stability of the duplexes directly related to the evolutionary distance between the DNA source and chickens, the natural host of RSV (Varmus, 1989). Subsequent work revealed that the cellular sequences related to the RSV *src* gene were exons of a large cellular gene that is extremely well conserved from arthropods to humans and encodes a protein tyrosine kinase assuming an onco-genic phenotype when constitutively activated by mutation (Hunter, 1989). These early experiments marked the start of an era of unprecedented progress in cancer research. They serve as a poignant example illustrating the power of function prediction by similarity.

Today, an exponentially increasing amount of electronic sequence data is being generated by genome projects worldwide, and gene function assignment by sequence similarity has largely moved from the wet lab to the computer. Contemporary approaches to function prediction from sequence employ a combi-nation of highly efficient sequencing techniques and sophisticated electronic searches against millions of gene and protein sequences. On a wider scale, a drive toward the integration of computational and experimental techniques is profoundly changing the ways in which biomedical research is done.

Bioinformatics is rapidly expanding its methodological repertoire for the prediction of the biochemical functions and cellular roles of disease-associated gene products. Essentially a cross-disciplinary field, it includes aspects of mole-cular biology, mathematics, computer science and software engineering (Baldi

Genotype to Phenotype second edition, edited by S. Malcolm and J. Goodship.
© 2001 BIOS Scientific Publishers Ltd, Oxford.

and Brunak, 1998; Baxevanis and Ouellette, 1998; Gusfield, 1997; Searls, 1998). By unlocking access to biological databases and applying powerful predictive tools, it can help to guide, focus, and at times crucially speed-up research. At present, bioinformatics is predominantly employed to support the experimental biologist at two distinct stages of a research project: at the beginning, in formulating testable hypotheses and models, as this type of theoretical analysis can be carried out well in advance of any significant bench effort, and, at the end, in the interpretation of experimental results. At both stages, its rewards can be enormous.

However, to realize the fullest potential of the emerging biomedical science of the 21st century, experimental and computational methods will need to become integrated more closely in the discovery process (Searls, 2000). This integration becomes possible when a computational model is not simply used at isolated points in the discovery process, e.g. to pass on candidate genes to an independent bench validation, but when it can be linked to a biological model, so that theoretical predictions can be directly tested in the laboratory, and the results be fed back for creating refinements in the theoretical model. The refined model in turn becomes the starting point for suggesting further experiments, and new results then lead to further model refinement. Such an integrated approach can be envisaged not only to substantially speed up the discovery process, but also to enhance our understanding of the biological system under study. An analogy can be drawn between future roles of bioinformatics in biomedical science and computational physics and its now integral role in advancing this branch of the natural sciences (Searls, 1998).

Already, bioinformatics has become an indispensable partner to experimental molecular biology, and the purpose of this chapter is to offer a map through this fast-changing terrain from a biologist's perspective. Specifically, it provides a guide to current computational techniques for function assignment from protein sequence and three-dimensional structure.

2. Predicting gene function

2.1 What is 'function'?

One of the most elusive and multi-faceted concepts in molecular biology is 'function'. For example, a single protein's function may be simultaneously described as 'protein phosphatase' (from the perspective of biochemistry), or 'cell division control gene' (from cell biology), or 'radiation resistance gene' (from genetics; Sander, 2000). Therefore, before we proceed to discuss gene function prediction using bioinformatics techniques, it is beneficial to pause and consider what it is that we wish to discover, and to consider some of the complexities we will face in the process. In order to do so, let us approach this problem from a perspective focusing on the underlying biology.

Fundamentally, proteins represent polygenic traits (*Figure 1*; Klose, 2000). Proteins exhibit different phenes and phenotypes, and these may result, for example, from mutations in the primary sequence, from variations in the amount of protein synthesized, from differences in degradation rates, or from the degree of

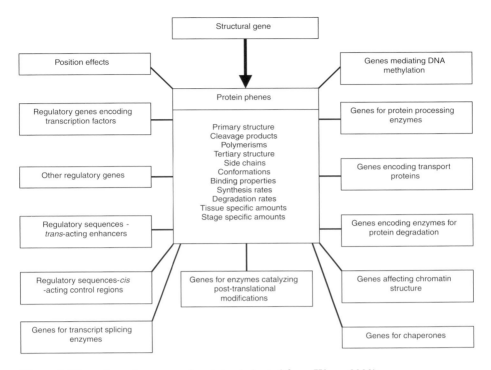

Figure 1. The polygenic nature of proteins (adapted from Klose, 2000).

post-translational modifications. Furthermore, a single gene or protein can have multiple forms and functions that are context-dependent, e.g. only relevant at a certain cellular localization, developmental stage, or cell-cycle point. Consequently, its functional characteristics cannot be understood without reference to the effects of other gene products on modulating its functions. For an understanding of the roles certain genes play in multifactorial diseases, it is therefore crucial to know how the different phenes and phenotypes of the gene products depend on a network of other genes (Klose, 2000). A mutation in a structural gene may affect several phenes of its product simultaneously (e.g. charge, molecular weight, binding properties), whereas mutations in regulatory sequences will most likely lead to quantitative changes. Changes in phenotype are also often linked to altered phosphorylation and glycosylation states and thus depend on the concentration and action of certain other enzymes. Genetic linkage studies can be employed to detect other genes affecting the phenes and phenotypes of a given protein. The new technologies of functional genomics can elucidate global states in gene activity providing the context leading to the manifestation of a specific protein phenotype at the cell, tissue or organism-wide level (see section on Future directions).

A special class of multi-functional proteins, termed 'moonlighting' proteins, also serves to highlight the often highly context-dependent nature of protein function (Jefferey, 1999). The function of a moonlighting protein can vary as a consequence of changes in cellular localization, cell type, oligomeric state, or the cellular concentration of a ligand, substrate, cofactor or product (*Figure 2*). Their multi-functionality can thus be seen as distinct from that observed in proteins that are the result of gene fusions, homologous but non-identical proteins, splice variants, proteins whose post-translational modifications can vary and proteins that have a single function but can operate in different locations or utilize different substrates (Jefferey, 1999). Crystallins which play a structural role in the eye lens and are identical to cytoplasmic proteins acting as heat shock proteins and dehydrogenases are among the growing list of moonlighting proteins.

In contrast, some genes do not appear to have a recognizable phenotype or function at all, as has been observed in some gene knockout experiments. The reason for an apparent lack of phenotype could either lie with the fact that the

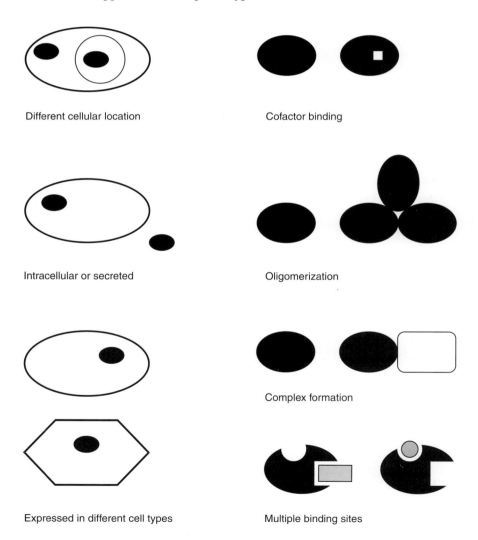

Different cellular location

Cofactor binding

Intracellular or secreted

Oligomerization

Complex formation

Expressed in different cell types

Multiple binding sites

Figure 2. The function of a moonlighting protein is context-dependent (adapted from Jefferey, 1999).

gene performs a very circumscribed, and hard to assay, function under precise environmental or developmental conditions only, or could be related to inbuilt redundancies and convergent pathways that make the cell's functioning error-resistant by providing surrogates for missing components (Boguski, 1999).

Finally, it is important to remember that 'function' is relative to the experimental method used to determine it. An *in vitro* assay for catalytic activity in an enzyme will characterize function in biochemical terms which is still the most common notion of a protein's function. In contrast, a 2-hybrid system experiment (Rudert *et al.*, 2000) will yield a radically different description of function, such as 'the function of protein A is to interact with protein B', and an expression microarray assay (Brown and Botstein, 1999; Schena *et al.*, 1998) may result in the conclusion that 'the expression of gene C is correlated with the expression of gene D'.

We seem to have reached a cross-roads where we cannot simply ignore the complexities of multi-functionality and context-dependence in favor of a simpler notion based on classical biochemical function alone, but at the same time, we do not have a framework yet that would comfortably accommodate these phenomena. Clearly, new concepts and categories are needed that can effectively represent the relationships between conserved function due to evolutionary relationships and divergent function due to context dependence. Whilst standard analysis and prediction methods will remain valid, the now familiar notion of static single-gene function will need to be expanded. Boguski (1999) suggested that we may come to change our question from 'what is the function of this protein?' to 'what roles does this sequence play in one or more biological processes that are operational under these conditions?'

A new scheme, called Gene Ontology (GO), aims to provide a more context-sensitive functional classification scheme for gene and protein annotation (Ashburner *et al.*, 2000). An ontology is a conceptual device that attempts to describe all entities within an area of reality and all relationships between those entities. GO employs a structured, precisely defined, common controlled vocabulary for describing the roles of genes and gene products in any eukaryotic organism. The GO structure reflects the current representation of biological knowledge as well as serving as a guide to the organization of new data. As it was developed for a generic eukaryotic cell, it does not encompass specialized cell types, organs or body parts. There are three ontologies within GO: molecular function, biological process, and cellular component (*Table 1*). These are all *independent* attributes of genes, gene products or gene-product groups. The relationships of a gene product (protein) or gene product group to these three categories are one-to-many, reflecting the fact that a particular protein may function in several biological processes, contain domains that carry out diverse molecular functions, and participate in multiple interactions with other proteins, organelles or locations in the cell (Ashburner *et al.*, 2000). The approach to the classification of gene function adopted by the GO Consortium provides a very useful frame of reference for a discussion of the strengths and limitations of current bioinformatics techniques used in the assignment of function. The complementary role of bioinformatics vis-à-vis experimental approaches in protein function assignment is highlighted in *Table 2* which correlates types of biological evidence to experimental methods based on the GO evidence codes (Ashburner *et al.*, 2000).

Table 1. The three functional categories of the Gene Ontology

Molecular function[a]	The biochemical activity (including specific binding to ligands or structures) of a gene product. This definition also applies to the capability that a gene product (or gene product complex) carries as a potential. It describes only what is done without specifying where or when the event actually occurs. Examples of broad functional terms are 'enzyme', 'transporter' or 'ligand'. Examples of narrower functional terms are 'adenylate cyclase' or 'Toll receptor ligand'.
Biological process	A biological objective to which the gene or gene product contributes. A process is accomplished via one or more ordered assemblies of molecular functions. Processes often involve a chemical or physical transformation. Examples of broad (high-level) biological process terms are 'cell growth and maintenance' or 'signal transduction'. Examples of more specific (lower level) process terms are 'translation', 'pyrimidine metabolism' or 'cAMP biosynthesis'.
Cellular component	The place in the cell where a gene product is active. These terms reflect our understanding of eukaryotic cell structure. This category includes such terms as 'ribosome' or 'proteasome', specifying where multiple gene products would be found. It also includes terms such as 'nuclear membrane' or 'Golgi apparatus'.

[a] Definitions are quoted directly from Ashburner *et al.* (2000).
URL of the GO website: http://www.geneontology.org

2.2 Computational function prediction

Bioinformatics seeks to elucidate the relationships between biological sequence, three-dimensional structure and its accompanying functions, and then to use this knowledge for predictive purposes. The guiding principles for the study of sequence-structure-function relationships, and the predictive techniques resulting from this analysis, are derived from the processes of molecular evolution (in particular, gene duplication, evolution of multi-gene families; Li, 1997; Page and Holmes, 1998). Most generally, protein function is inferred from the known functions of homologous proteins; i.e. proteins predicted to share a common ancestor with the query protein based on significant sequence or structural similarity. However, we still lack a quantitative measure of functional relatedness, a fact which imposes serious limitations on transferring functional annotations between proteins with different degrees of homology.

Computational methods are employed in the prediction of protein function from sequence or structure, or in the prediction of structure from sequence. For homologous proteins with easily recognizable sequence similarity, this type of prediction is based on the 'similar sequence-similar structure-similar function' paradigm. This rule assumes a one-to-one relationship between its components, and a linear path leading from sequence to a unique three-dimensional structure, and from structure to a unique function (*Figure 3a*). It is important to be aware that, whilst this usually holds true for closely related proteins, more distant relatives belonging to multi-gene families present in the same genome exhibit a complex pattern of shared and distinct characteristics in terms of molecular

Table 2. Evidence categories for function assignment within the Gene Ontology scheme

Evidence category	Methods	GO categories
Mutant phenotype	Any gene mutation/knockout Overexpression/ectopic expression of wild-type or mutant genes Anti-sense experiments Specific protein inhibitors	Mol. function Biol. process
Genetic interaction	Traditional genetic interactions (suppressors, synthetic lethals, etc.) Functional complementation Rescue experiments Any mutation experiment involving more than one gene product, done in a non-wild-type background	Mol. function Biol. process
Physical interaction	2-hybrid interactions Co-purification Co-immunoprecipitation Ion/protein binding experiments	Mol. function Cell. component

Sequence or structural similarity	**Bioinformatics:**
	    sequence similarity
	    (homolog of/most closely related)
	    recognized domains
	    structural similarity
	Southern blots
		Mol. function
		Biol. process
		Cell. component

Direct assay	Enzyme assays *In vitro* reconstitution (e.g. transcription) Immunofluorescence (for cellular component) Cell fractionation (for cellular component) Binding assay (for cellular component)	Mol. function Biol. process Cell. component
Expression pattern	Transcript levels (Northern blots, microarray data) Protein levels (Western blots, proteomics)	Biol. process Cell. component

function, ligand specificity, gene regulation, protein–protein interactions, tissue specificity, cellular location (cellular component), developmental phase of activity, biological process, etc. Their functional classification following the GO scheme may differ substantially. Across different genomes, one can expect to find an entire spectrum of functional conservation in evolutionarily related genes, ranging from those with even greater degrees of context-dependent functional divergence (in particular, between paralogs) to others with highly conserved function, the latter most likely being orthologous proteins involved in core biological processes.

A high degree of gene conservation across organisms, correlated with conservation of function, has been well established by the sequencing of 32 entire prokaryotic genomes (six archeal and 26 bacterial genomes, as of August 2000) and

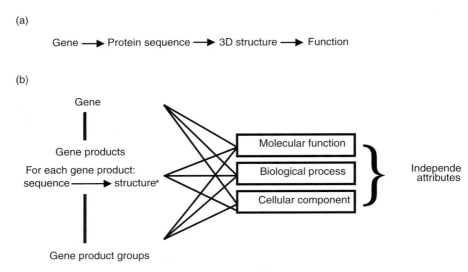

Figure 3. Two conceptual frameworks for sequence-structure-function relationships. (a) the 'similar sequence-similar structure-similar function paradigm', (b) relationships between entities in the Gene Ontology scheme. →, one-to-one mapping; _____, one-to-many mapping. *Sequence-structure is mostly understood as a one-to-one mapping, but it can be context dependent. For instance, in cases of proteins that adopt a unique structure only open upon binding to ligand.

the complete genomes of three eukaryotic model organisms (the budding yeast *Saccharomyces cerevisiae*, completed in 1996; the nematode worm *Caenorhabditis elegans*, completed in 1998; and the fruitfly *Drosophila melanogaster*, completed in early 2000). The comparison between two complete eukaryotic genomes, budding yeast and nematode, suggested orthologous relationships between a substantical number of genes in the two organisms. About 27% of the yeast genome (~5700 genes) encode proteins with significant similarity to ~12% of nematode genes (~18 000 genes) (Chervitz *et al.*, 1999). Furthermore, the same set of yeast genes also has putative orthologs in the *Drosophila* genome (Rubin *et al.*, 2000). Most of these shared proteins perform conserved roles in processes common to all eukaryotes, such as metabolism, gene transcription, and DNA replication.

When one is dealing with distantly related proteins, it is important to keep in mind that the 'sequence-structure-function' paradigm does not address context-dependence of function and multi-functionality. Predictions made from within this paradigm therefore can not reveal these relationships, frustratingly, we are thus able to conceptualize this functional complexity in terms of the GO one-to-many mappings (*Figure 3b*), but to date still lack the appropriate predictive tools. However, encouragingly, provided that a homologous relationship between query and target protein can reasonably be inferred based on significant sequence similarity over most of the aligned sequence, fairly detailed assignments regarding molecular function and cellular component can be expected to be made, but biological process may vary. For pairs of protein domains that share the same fold, precise function appears to be conserved down to ~40% sequence identity, whereas broad functional class is conserved to ~25% identity (Wilson *et al.*, 2000). Yet, given a significant local match to one or more protein domains, attribution of the entire protein to any of the three GO ontologies will be less reliable. Ideally, function prediction by inferred homology ought to be confirmed by appropriate experimental analysis (*Table 2*).

Predicting function from sequence. The most widely used approach to function identification from amino acid sequence aims at the recognition of significant sequence similarity between a given query and a protein or protein family whose function is already known. So, in practice, the first step toward function identification routinely involves searching sequence databases with a query sequence using tools such as BLAST (Altschul *et al.*, 1990), PSI-BLAST (Altschul *et al.*, 1997) or FASTA (Pearson and Lipman, 1988).

In the best case, a statistically significant match to a well characterized protein over the entire length of the query sequence will be returned. However, more typically, such a search will reveal local similarities between the query sequence and a number of diverse proteins (Attwood, 2000). These local matches often signal the presence of shared homologous domains. Larger proteins are modular in nature, and their structural units, protein domains, can be covalently linked to generate multi-domain proteins with extensive or minimal contact between domains. Some of these structural domains are thermodynamically stable and fold independently within the context of the whole protein. Domains are not only structurally, but also functionally discrete units, i.e. specific molecular functions are carried out by specific domains. It is thought that families of homologous domains arise from gene duplication events. Family members are structurally and functionally conserved and recombined in complex ways during evolution. Novelty in protein function often arises as a result of the gain or loss of domains, or by re-shuffling existing domains along the linear amino acid sequence. Domains can be thought of as the 'units of evolution', and, therefore, both structural and functional similarity between proteins needs to be approached at the domain level.

In sequence database searching, local matches to domains present in multi-domain proteins need to be evaluated with caution. The main issues to be addressed are the correct identification of the domain(s) that corresponds to the query and the biological context in which the domain(s) occurs in both query and matched sequence. It is important to be aware that domain function can vary significantly depending on context, and automatic transfer of functional annotation can be misleading (*Figure 4*). The relationship between homologous domains is analogous to the relationship between genes belonging to multi-gene families: Database search algorithms cannot distinguish between a matched ortholog (the functionally conserved version of a gene in another species) and a matched paralog (a homolog with a different but related function in the same organism; Attwood, 2000; *Figure 5*). As this distinction cannot be derived automatically with the available algorithms, the researcher needs to evaluate the returned matches on a case by case basis. The higher the level of sequence conservation, the more likely it is that the query and the matched sequence are orthologous.

If a sequence search identified an entire protein, or a protein domain, of known three-dimensional structure with a level of identity of 35% or better to the query sequence, the structure of the query protein can be predicted by homology

Figure 4. Domain structure of proteins containing a histidine kinase domain. These four examples from the Pfam database (domain numberPF00512) illustrate the modular nature of proteins. The same domain can be found together with a variety of other domains in different proteins. (a) phytochrome C from *Sorghum bicolor* (TrEMBL code P93528), (b) sensory transduction histidine kinase from *Methanobacterium thermoautotrophicum* (TrEMBL code O26988), (c) virulence regulator protein from *Pseudomonas solanacearum* (TrEMBL code Q52582), (d) mitochondrial pyruvate dehydrogenase kinase isoform 1 (PDK1) from rat. The histidine kinase domain is shown in black.

modelling (Marti-Renom *et al.*, 2000). The predicted structure may contribute valuable detail on the protein's biochemical functions, active site, interaction surfaces etc. Using the matched protein as the structural template, a comparative modelling engine such as MODELLER (Sali and Blundell, 1993), COMPOSER (Sutcliffe *et al.*, 1987) or SWISS-MODEL (Guex and Peitsch, 1997) allows automatic modelling and model evaluation. At present, only 15–25% of all sequences have a homologous protein of known structure (Elofsson and Sonnhammer, 1999), but with structural genomics efforts now underway around the world, this percentage will increase dramatically in the near future (Burley *et al.*, 1999; Skolnick *et al.*, 2000). If the goals of structural genomics are realized, i.e. if all possible folds are experimentally determined and any protein sequence comes within modelling distance of a known protein structure, homology modelling most likely will quickly be added to the list of standard sequence analysis tools.

Current global and local sequence alignment methods are reliable for identifying sequence relatedness when the level of sequence identity is above 30%. However, these tools cannot consistently detect structural and functional similarities when the level of sequence similarity drops below the 'twilight zone' of 25–30% identity

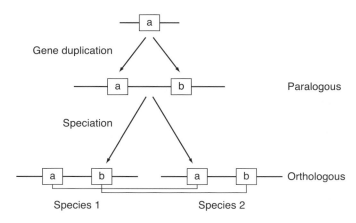

Figure 5. Paralogous and orthologous relationships between genes and proteins.

(Hobohm and Sander, 1995). Therefore, purely sequence-based comparisons fail to identify many of the evolutionary relationships that emerge once the structures of proteins are compared. Significantly, the BLAST algorithm finds only 10% of known relationships in the Protein Data Bank of three-dimensional structures (PDB; Bernstein *et al.*, 1977), and although much more sensitive, the iterative PSI-BLAST method also misses many of these relationships (Altschul *et al.*, 1990; Altschul *et al.*, 1997; Brenner *et al.*, 1998).

Predicting function from structure. As the number of known three-dimensional protein structures has increased dramatically over the last 5 years, with over 12 000 structures now deposited in the PDB, structural data have become an increasingly important information source for function prediction. Over the years there have been various estimates as to how many unique protein folds might exist in nature, with values ranging from 1000 to 100 000. The elucidation of evolutionary relationships among structures which are hidden at the sequence level supports the hypothesis that the number of protein families is finite, and relatively small, and closer to the minimum estimate of 1000 folds (Chothia, 1993). However, as most of our knowledge about protein structure comes from a subset of proteins that are water-soluble, small and that crystallize easily (Dodge *et al.*, 1998; Holm and Sander, 1999), it is presently unclear how representative the PDB data set is of the entire universe of protein structures. This experimental constraint has resulted in a bias against 'flexible' proteins that only adopt a unique structure upon binding to a receptor or ligand, and membrane-bound proteins, for example. As membrane-bound structures represent over 20% of eukaryotic genomes, this exclusion is obviously significant. Nevertheless, this does not diminish the usefulness of the structures that are available, as each individual structure can provide much information about the specific mechanisms giving rise to a particular molecular function carried out by the protein.

In order to understand the universe of protein structures as a whole, and to explore the evolutionary relationships between structure and function, a rational 'molecular taxonomy' is needed. Over the last 10 years, several groups have developed structural classification schemes, usually based on a combination of sequence and structure comparison methods (e.g. SSAP, Taylor and Orengo, 1989; DALI, Holm and Sander, 1993). Lists of structural neighbors have been constructed (VAST, Hogue *et al.*, 1996; DALI, Holm and Sander, 1996; DIAL, Sowdhamini *et al.*, 1996), and several more complete domain based classification schemes, providing phenetic description of structure together with phylogenetic relationships, have been generated (e.g., SCOP, Murzin *et al.*, 1995; HOMSTRAD, Mizuguchi *et al.*, 1998; CATH, Orengo *et al.*, 1997).

From these studies and classification efforts, it is now clear that protein structure is much better conserved than sequence; and many examples of proteins with similar folds but no detectable sequence similarity have been found. Several possible causes for this phenomenon have been suggested, ranging from very distant homology, to convergent evolution and even random similarity (Murzin, 1998). Regardless of its origins, such structural similarity can give important clues to a protein's functions.

In order to detect signals of these remote relationships in protein sequences, new techniques have been developed that extend the range of representation of similarity between proteins. 'Pattern and profile' methods enhance sequence information by a mutation pattern at a given position along the sequence that is derived from an alignment of homologous proteins. Different analytical approaches have been employed to create a range of representations (termed regular expressions, rules, profiles, signatures, fingerprints, blocks, Hidden Markov Models, etc.) with different diagnostic strengths and weaknesses, and different areas of optimum application (Attwood, 2000). Patterns diagnostic of a particular protein family (typically ~10–20 amino acids in length) tend to correspond to the core structural and functional elements (motifs) of the proteins. These methods can therefore be applied to the problem of structure prediction by limiting the dataset of proteins searched to those with known structure (Koonin *et al.*, 2000). Even more specific motif representations can be created by combining conserved sequence patterns with information about structural context (Yu *et al.*, 1998). Pattern and profile methods can also be used for the purpose of recognizing remote similarities between proteins for which none of the structures are known. Using various representation methods, databases of diagnostic protein family patterns have been created and can be searched with a query sequence (Attwood, 2000; *Table 3*). This may reveal the presence of one or several regions corresponding to particular functions in the query sequence. Pattern and profile database searches therefore constitute an important resource for function prediction.

In 'threading' methods sequence information is enhanced or replaced by residue interaction preferences (Bowie *et al.*, 1991; Godzik *et al.*, 1992; Jaroszewski *et al.*, 1998; Jones *et al.*, 1992; Murzin, 1999; Russell *et al.*, 1996). These methods aim at

Table 3. Resources on the World Wide Web

Homology searching
Searching for homologous sequences:
BLAST	http://www.ncbi.nlm.nih.gov/BLAST/
	http://www.sanger.ac.uk/DataSearch/
FASTA	http://bioweb.pasteur.fr/seqanal/interfaces/fasta-simple.html

Searching for homologous proteins for which a 3D structure is known:
NRL-3D	http://pir.georgetown.edu/pirwww/dbinfo/nrl3d.html
SAS	http://www.biochem.ucl.ac.uk/bsm/sas/

Searching for homologous domains:
PRINTS	http://bioinf.man.ac.uk/fingerPRINTScan/
Pfam	http://www.sanger.ac.uk/Software/Pfam/
PRODOM	http://protein.toulouse.inra.fr/prodom.html
BLOCKS	http://www.blocks.fhcrc.org/
SMART	http://smart.embl-heidelberg.de/
PROFILESCAN	http://www.isrec.isb-sib.ch/software/PFSCAN_form.html
EMOTIF	http://dna.Stanford.EDU/identify/
MEME	http://meme.sdsc.edu/meme/website/

Homology modelling
SWISS-MODEL	http://www.expasy.ch/swissmod/SWISS-MODEL.html

3D active site template searches
PROCAT	http://www.biochem.ucl.ac.uk/bsm/PROCAT/PROCAT.html

---

identifying the structural 'signature' of a sequence and at finding structurally similar proteins in cases of very distant relatedness, or even in the absence of homology. Although threading methods can be successful in recognizing remotely related proteins, the associated structures often differ extensively. A particularly common problem is poor alignment between the query sequence and the template structure. Whilst model building algorithms do not currently address this problem, tools for model refinement have been developed that may be able to reduce the backbone root mean square deviation (r.m.s.d.) from native from 10 to 4–6 Å (Kolinski *et al.*, 1999). In addition to threading methods which are limited to the identification of known folds, *ab initio* approaches may to used to predict novel folds. Although still limited by their inability to handle beta proteins and larger proteins, for small proteins at least, results to date indicate that the quality of *ab initio* models is comparable to those obtained by threading (Orengo *et al.*, 1999).

Once the structure of a protein has been either solved or predicted (by homology modelling, pattern/profile analysis, threading or *ab initio* methods), how can its functions be determined? Recent analysis of protein sequences and structures showed that functional information can be automatically derived from structural information only to a limited extent (Kasuya and Thornton, 1999). It was found, for example, that two functions were associated with seven folds each. However, some folds can have as many as 16 associated functions (Hegyi and Gerstein, 1999). These findings demonstrate that knowledge of the overall structure or domain family is not sufficient to reliably assign function at the biochemical level.

A protein's biochemical function can be predicted by scanning its structure for a match to the geometry and chemical composition of a known active site. This approach can be applied not only to experimentally determined structures, but also to low-resolution structures obtained by available prediction methods (Skolnick *et al.*, 2000, and references therein). For example, PROCAT is a library of three-dimensional motifs that specify the relative positions in space of atoms involved in functional sites (*Figure 6*; Wallace *et al.*, 1996). To validate the method, a motif for the serine proteases was developed that was able to identify all known serine protease and triacylglycerol lipases in a set of protein structures. This type of three-dimensional motif exploits the high level of detail regarding placement of specific side chain atoms that is contained in high resolution structures. However, it cannot be usefully applied to the identification of functional sites in models predicted by threading or *ab initio* approaches. To extend the scope of three-dimensional motifs, well conserved active sites can be excised from crystal structures and used to represent the active site geometry at a level that is suitable for high-resolution structures and inexact models (Kasuya and Thornton, 1999; Sanchez and Sali, 1998; Skolnick *et al.*, 2000). Skolnick and colleagues have developed a structural motif library, populated by structural motifs called fuzzy functional forms, that can be used for the identification of functional sites in threading and *ab initio* models (Fetrow and Skolnick, 1998; Skolnick *et al.*, 2000).

Function identification by three-dimensional motifs transcends the limitations of homology-based sequence analysis methods. What is more, these methods are able to detect functional sites where structural similarity is not the result of descent from a shared ancestral protein, but has come about by an as yet not well

(a)   (b)

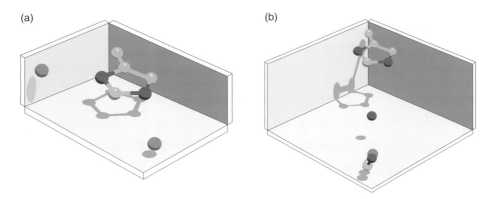

Figure 6. Two examples of PROCAT 3D active site templates. (a) the consensus template for the Ser-His-Asp catalytic triad (carboxypeptidase C, EC number 3.4.16.5), (b) glutamine-tRNA ligase (EC number 6.1.1.18).

understood evolutionary mechanism that re-creates similar sites in different folds and protein families (Fischer *et al.*, 1994).

3. Future directions: toward a 'global biology'

Most biological functions result from interactions among many components giving rise to collective properties. In recognition of this fact, the new field of functional genomics seeks to develop a powerful new perspective on the holistic operation of biological systems (Hieter and Boguski, 1997). The development of new methods that allow a multiplicity of genes and proteins to be studied simultaneously is progressing rapidly (*Table 4*), and new analysis algorithms that take account of the parallel-processing properties of tissue or organism-wide interaction

Table 4. Levels of gene-function analysis in functional genomics

Level of analysis	Definition	Methods[a]
Genome	Complete set of genes of an organism or organelle	DNA sequencing
Transcriptome	Complete set of mRNA transcripts[b]	Microarrays High-throughput Northern blotting
Proteome	Complete set of proteins[b]	Two-dimensional gel electrophoresis Peptide mass fingerprinting 2-hybrid analysis
Metabolome	Complete set of low-molecular weight metabolites[b]	Infra-red spectroscopy Mass spectrometry Nuclear magnetic resonance spectrometry

[a] for references, see Oliver (2000)
[b] present in a cell, tissue or organ at a given time

networks are also being developed (e.g. Eisen *et al.*, 1998). In future, we can expect to arrive at a fundamentally new conception of protein function brought about by the integration of sequence, structure and interaction network analysis at an unparalleled level of complexity.

References

Altschul, S. F., Gish, W., Miller, W., Myers, E.W. and Lipman, D.J. (1990) Basic local alignment search tool. *J. Mol. Biol.* **215**: 403–410.

Altschul, S.F., Madden, T.L., Schaffer, A.A., Zhang, J., Zhang, Z., Miller, W. and Lipman, D.J. (1997) Gapped BLAST and PSI-BLAST: a new generation of protein database search programs. *Nucleic Acids Res.* **25**: 3389–3402.

Ashburner, M., Ball, C.A., Blake, J.A. *et al.* (2000) Gene ontology: tool for the unification of biology. *Nature Genet.* **25**: 25–29.

Attwood, T.K. (2000) The quest to deduce protein function from sequence: the role of pattern databases. *Int. J. Biochem. Cell Biol.* **32**: 139–155.

Baldi, P. and Brunak, S. (1998) *Bioinformatics: The Machine Learning Approach.* MIT Press, Cambridge, MA.

Baxevanis, A. and Ouellette, B.F.F. (1998) *Bioinformatics: A Practical Guide to the Analysis of Genes and Proteins.* John Wiley & Sons, New York.

Bernstein, F.C., Koetzle, T.F., Williams, G.J., Meyer, E.F. Jr, Brice, M.D., Rodgers, J.R., Kennard, O., Shimanouchi, T. and Tasumi, M. (1977) The Protein Data Bank. A computer-based archival file for macromolecular structures. *J. Mol. Biol.* **112**: 535–542.

Boguski, M.S. (1999) Biosequence exegesis. *Science* **286**: 453–455.

Bowie, J.U., Luethy, R. and Eisenberg, D. (1991) A method to identify protein sequences that fold into a known three dimensional structure. *Science* **253**: 164–170.

Brenner, S.E., Chothia, C. and Hubbard, T.J. (1998) Assessing sequence comparison methods with reliable structurally identified distant evolutionary relationships. *Proc. Natl Acad. Sci. USA* **26**: 6073–6078.

Brown, P.O. and Botstein, D. (1999) Exploring the new world of the genome with DNA microarrays. *Nature Genet.* **21**(1 Suppl): 33–37.

Burley SK, Almo, S.C., Bonanno, J.B. *et al.* (1999) Structural genomics: beyond the human genome project. *Nature Genet.* **23**: 151–157.

Chervitz, S.A., Hester, E.T., Ball, C.A. *et al.* (1999) Using the Saccharomyces Genome Database (SGD) for analysis of protein similarities and structure. *Nucleic Acids Res.* **27**: 74–78.

Chothia, C. (1993) One thousand families for the molecular biologist. *Nature* **357**: 543–544.

Dodge, C., Schneider, R. and Sander, C. (1998) The HSSP database of protein structure-sequence alignments and familiy profiles. *Nucleic Acids Res.* **26**: 313–315.

Eisen, M.B., Spellman, P.T., Brown, P.O. and Botstein, D. (1998) Cluster analysis and display of genome-wide expression patterns. *Proc. Natl Acad. Sci. USA* **95**: 14863–14868.

Elofsson, A. and Sonnhammer, E.L. (1999) A comparison of sequence and structure protein domain families as a basis for structural genomics. *Bioinformatics* **15**: 480–500.

Fetrow, J.S. and Skolnick, J. (1998) Method for prediction of protein function from sequence using the sequence-to-structure-to-function paradigm with application to glutaredoxins/thioredoxins and T1 ribonucleases. *J. Mol. Biol.* **281**: 949–968.

Fischer, D., Wolfson, H., Lin, S.L. and Nussinov, R. (1994) Three-dimensional, sequence order-independent structural comparison of a serine protease against the crystallographic database reveals active site similarities: potential implications to evolution and to protein folding. *Protein Sci.* **3**: 769–778.

Godzik, A., Skolnick, J. and Kolinski, A. (1992) A topology fingerprint approach to the inverse folding problem. *J. Mol. Biol.* **227**: 227–238.

Guex, N. and Peitsch, M.C. (1997) SWISS-MODEL and Swiss-Pdb Viewer: An environment for comparative protein modelling. *Electrophoresis* **18**: 2714–2723.

Gusfield, D. (1997) *Algorithms on Strings, Trees, and Sequences.* Cambridge University Press, Cambridge.

Hegyi, H. and Gerstein, M. (1999) The relationship between protein structure and function: a comprehensive survey with application to the yeast genome. *J. Mol. Biol.* **288**: 147–164.

Hieter, P. and Boguski, M.S. (1997) Functional genomics: It's all how you read it. *Science* **278**: 601–602.

Hobohm, U. and Sander, C. (1995) A sequence property approach to searching protein databases. *J. Mol. Biol.* **251**: 390–399.

Hogue, C.W., Ohkawa, H. and Bryant, S.H. (1996) A dynamic look at structures: WWW-Entrez and the Molecular Modelling Database. *Trends Biochem. Sci.* **21**: 226–229.

Holm, L. and Sander, C. (1993) Protein structure comparison by alignment of distance matrices. *J. Mol. Biol.* **233**: 123–138.

Holm, L. and Sander, C. (1996) Mapping the protein universe. *Science* **273**: 595–602.

Holm, L. and Sander, C. (1999) Protein folds and families: sequence and structure alignments. *Nucleic Acids Res.* **27**: 244–247.

Hunter, T. (1989) Oncogene products in the cytoplasm: The protein kinases. In: Weinberg, R.A. (ed.), *Oncogenes and the Molecular Origins of Cancer.* Cold Spring Harbor Laboratory Press, Cold Spring Harbor NY, pp. 147–174.

Jaroszewski, L., Rychlewski, L., Zhang, B. and Godzik, A. (1998) Fold prediction by a hierarchy of sequence and threading methods. *Protein Sci.* **7**: 1431–1440.

Jefferey, C.J. (1999) Moonlighting proteins. *Trends Biochem. Sci.* **24**: 8–11.

Jones, D.T., Taylor, W.R. and Thornton, J.M. (1992) A new approach to protein fold recognition. *Nature* **358**: 86–89.

Kasuya, A. and Thornton, J.M. (1999) Three-dimensional structure analysis of Prosite patterns. *J. Mol. Biol.* **286**: 1673–1691.

Klose, J. (2000) Genotypes and Phenotypes. In: Dunn, M.J. (ed), *From Genome to Proteome: Advances in the Practice and Application of Proteomics.* Wiley-VCH, New York, pp. 63–72.

Kolinski, A., Rotkiewicz, P., Ilkowski, I. and Skolnick, J. (1999) A method for the improvement of threading based protein models. *Proteins* **37**: 592–610.

Koonin, E.V., Wolf, Y.I. and Aravind, L. (2000) Protein fold recognition using sequence profiles and its application in structural genomics. *Advances Protein Chem.* **54**: 245–275.

Li, W.-H. (1997) *Molecular Evolution.* Sinauer Associates. Sunderland, MA.

Marti-Renom, M.A., Stuart, A.C., Fiser, A., Sanchez, R., Melo, F. and Sali, A. (2000) Comparative protein structure modelling of genes and genomes. *Rev. Biophys. Biomol. Struct.* **29**: 291–325.

Mizuguchi, K., Deane, C.M., Blundell, T.L. and Overington, J.P. (1998) HOMSTRAD a database of protein structure alignments for homologous families. *Protein Sci.* **7**: 2469–2471.

Murzin, A.G. (1998) How far divergent evolution goes in proteins. *Curr. Opin. Struct. Biol.* **8**: 380–387.

Murzin, A.G. (1999) Structure classification-based assessment of CASP3 predictions for the fold recognition targets. *Proteins* **37(S3)**: 88–103.

Murzin, A.G., Brenner, S.E., Hubbard, T. and Chothia, C. (1995) SCOP: a structural classification of the protein database for the investigation of sequence and structures. *J. Mol. Biol.* **247**: 536–540.

Oliver, S. (2000) Guilt-by-association goes global. *Nature* **403**: 601–603.

Orengo, C.A., Michie, A.D., Jones, S., Jones, D.T., Swindells, M.B. and Thornton, J.M. (1997) CATH – a hierarchic classification of protein structures. *Structure* **5**: 1093–1108.

Orengo, C.A., Bray, J.E., Hubbard, T., LoConte, L. and Sillitoe, I. (1999) Analysis and assessment of ab initio three-dimensional prediction, secondary structure, and contact prediction. *Proteins* **37(S3)**: 149–170.

Page, R.D.M. and Holmes, E.C. (1998) *Molecular Evolution: A Phylogenetic Approach.* Blackwell Science. Oxford.

Pearson, W.R. and Lipman, D.J. (1988) Improved tools for biological sequence comparison. *Proc. Natl Acad. Sci. USA* **85**: 2444–2448.

Rubin, G.M., Yandell, M.D., Wortman, J.R. *et al.* (2000) Comparative genomics of the eukaryotes. *Science* **287**: 2204–2215.

Rudert, F., Ge, L. and Ilag, L.L. (2000) Functional genomics with protein-protein interactions. *Biotechnol. Ann. Rev.* **5**: 45–86.

Russell, R.B., Copley, R.R. and Barton, G.J. (1996) Protein fold recognition by mapping predicted secondary structures. *J. Mol. Biol.* **259**: 349–365.

Sali, A. and Blundell, T.L. (1993) Comparative protein modelling by satisfaction of spatial constraints. *J. Mol. Biol.* **234**: 779–815.

Sanchez, R. and Sali, A. (1998) Large-scale protein structure modeling of the Saccharomyces cerevisiae genome. *Proc. Natl Acad. Sci. USA* **95**: 13597–13602.

Sander, C. (2000) Genomic medicine and the future of health care. *Science* **287**: 1977–1978.

Schena, M., Heller, R.A., Theriault, T.P., Konrad, K., Lachenmeier, E. and Davis, R.W. (1998) Microarrays: biotechnology's discovery platform for functional genomics. *Trends Biotechnol.* **16**: 301–306.

Searls, D.B. (1998) Grand challenges in computational biology. In: Salzberg, S.L., Searls, D.B. and Kasif, S. (eds), *Computational Methods in Molecular Biology*. Elsevier, New York, pp. 3–10.

Searls, D.B. (2000) Using bioinformatics in gene and drug discovery. *Drug Discovery Today* **5**: 135–143.

Skolnick, J., Fetrow, J.S. and Kolinski, A. (2000) Structural genomics and its importance for gene function analysis. *Nature Biotechnol.* **18**: 283–287.

Sowdhamini, R., Rufino, S.D. and Blundell, T.L. (1996) A database of globular protein structural domains: clustering of representative family members into similar folds. *Folding and Design* **1**: 209–220.

Sutcliffe, M.J., Haneef, I., Carney, D. and Blundell, T.L. (1987) Knowledge based modelling of homologous proteins, Part I: Three-dimensional frameworks derived from the simultaneous superposition of multiple structures. *Protein Eng.* **1**: 377–384.

Taylor, W.R. and Orengo, C.A. (1989) Protein structure alignment. *J. Mol. Biol.* **208**: 208–229.

Varmus, H. (1989) An Historical Overview of Oncogenes. In: Weinberg, R.A. (ed), *Oncogens and the Molecular Origins of Cancer*. Cold Spring Harbor Laboratory Press, Cold Spring Harbor NY, pp. 3–44.

Wallace, A.C., Laskowski, R.A. and Thornton, J.M. (1996) Derivation of 3D coordinate templates for searching structural databases: application to Ser-His-Asp catalytic triads in the serine proteinases and lipases. *Protein Sci.* **5**: 1001–1013.

Wilson, C.A., Kreychman, J. and Gerstein, M. (2000) Assessing annotation transfer for genomics: Quantifying the relations between protein sequence, structure and function through traditional and probabilistic scores. *J. Mol. Biol.* **297**: 233–249.

Yu, L., White, J.V. and Smith, T.F. (1998) A homology identification method that combines protein sequence and structural information. *Protein Sci.* **7**: 2499–2510.

3

Genes in population

François Cambien

This chapter is an introduction to notions of population genetics that may be useful to set up and interpret genotype–phenotype association studies. There are excellent introductions to population and molecular evolutionary genetics, providing a broader account of the subject to which the reader may refer (Hartl and Clark, 1997). The study of genetic variability is a common basis for evolutionary genetics, population genetics and genetic epidemiology. However, all biology is affected by genetic variability since biological systems owe their origin, fine tuning and coordination to genetic variability and reciprocally, they connect genotypes among themselves and with the external world. For Lewontin (2000) 'the greatest methodological challenge that population genetics now faces is to connect the observations between outcome of evolutionary processes to the tradition of experimental functional biology'. Such a view singularly departs from the impression gained from contemporary literature and debates that the challenge is on the contrary to generate and manage ever increasing amounts of data. It may be that the difficulties we experience presently in our attempts to understand how genetics contributes to quantitative traits and common multifactorial diseases, are more related to our insufficient consideration of the phenotype than to a lack of genetic data. Certainly, there are great promises in the availability of all human gene sequences in the near future. This introduction to genes in populations, despite being mainly focused on the structure, generation, maintenance and evolution of genetic variability in humans, is written with this idea in mind.

1. Population structure and dynamics

For research purposes, a population is a set of individuals considered appropriate for investigating scientific hypotheses. In human genetics, depending on the question asked, a relevant study group may range from a single family to the whole population of a country. Most of the time, the actual collection of individuals investigated is a subset of subjects appropriately representing the population from which it has been selected. Representativeness of the study sample is essential to allow inference to the whole underlying population and is best obtained by random sampling.

In population genetics, an important characteristic of a population is that males and females assemble in a random way to produce offspring. Random

Genotype to Phenotype second edition, edited by S. Malcolm and J. Goodship.

mating is often required because the validity of major statistics relies on it. The Hardy–Weinberg principle specifies that for a locus with two alleles, in a large randomly mating population with no selection, mutation and migration, observed genotype frequencies should not deviate significantly from theoretical expectations (p^2:$2pq$:q^2), predicted from the observed allele frequencies (p: q). A consequence of Hardy–Weinberg equilibrium (HWE) is that allele and genotype frequencies will remain constant from generation to generation This important property of Mendelian segregation insures the conservation of genetic variability. Departure from HWE may be observed at several loci in subdivided populations with different allele frequencies, where random mating occurs preferentially within the subpopulations. The population structure resulting from non-random mating may be quite complicated and hard to detect, and for reasons of convenience, most population genetics theories and interpretations of data ignore or minimize its influence. In genotype–phenotype association studies population stratification may generate spurious associations when both the prevalence of the phenotype and allele frequencies differ among population strata. To deal with this problem, a set of unlinked genetic markers may be used to explore the presence of population structure and if the latter exists, it may be taken into account for investigating candidate loci (Pritchard *et al.*, 2000). The present structure of a population reflects the past dynamics of its component subpopulations, and integrates the consequences of stagnations, expansions, dispersion, aggregation, migration, drastic reduction in size and extinction. The genetic pool of the human species is in large part the outcome of this complicated demographic history.

2. Genetic variability and evolution

2.1 The generation of genetic variability

Genetic variation, coupled with differential survival and reproductive fitness, has provided the basis for the evolution of all biological structures and functions through its contribution to the variability of the phenotype at all its levels from molecules to organisms. Two main categories of processes generate variation: (1) mutations create novel sequences through substitution, addition or subtraction of one or several nucleotides; (2) recombination creates rearrangements of already existing genetic material through exchange or conversion between chromosomes (*Figure 1*). Over long periods, variation accumulates, generating haplotypes, i.e. alleles defined by combinations of variants at different loci closely linked on a chromosome. A mutation occurs on a single haplotype, and is therefore exclusively associated with this haplotype, in complete linkage disequilibrium, until recombination events occur that will induce a progressive decay of linkage disequilibrium. Pairs of alleles or haplotypes at homologous positions define the genotype of an individual at a locus. In space and time, the genotype may be viewed as generating a signal which after modification by epigenetic and environmental factors, and a non negligible addition of stochastic noise translates into the phenotype. An adequate definition of evolution would therefore integrate changes in both the genotypic and phenotypic spaces (Lewontin, 2000).

Original sequence $G_{15}..T_{213}..A_{563}..C_{987}..G_{1356}$

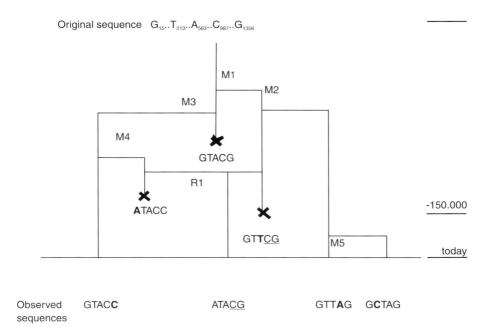

| Observed | GTAC**C** | | ATAC<u>CG</u> | GTT**AG** | GC**TAG** |

Figure 1. Evolution of haplotypes. A hypothetical genealogy for four haplotypes observed today. Originally, one million years ago, the ancestral DNA sequence had G, T, A, C and G at positions 15, 213, 563, 987 and 1356 respectively. Five mutational events (M1–M5) and one recombinational event (R1) occurred during the evolution of the sequence generating six new alleles, however only four out of seven alleles survived to the present time: GTACC, ATACG, GTTAG and GCTAG. M5 occurred after the human population dispersed all over the world from Africa (estimated 150 000 years ago), as a consequence GCTAG should not be found in all human populations. Note that reconstructing the phylogeny from the observed haplotypes would be impossible.

2.2 Mutation

Most variants identified on a sequence are the consequence of a single mutation that occurred in the past in one germinal cell, and led to the generation of a new allele that has survived until the present time. Not all new variation is transmitted to the following generations, even if it is potentially highly beneficial for the organism. Actually, the probability that a new allele survives may be quite low, especially in large populations, as it is inversely proportional to the population size. As for neutral mutations, the mutation rate is equal to the rate of allelic substitution per generation (Kimura, 1968), an estimate of the neutral mutation rate per generation may be obtained by comparing non functional DNA sequences such as pseudogenes, i.e. non functional homologs of functional genes, in species whose generation times and divergence (approximately six millions years for human and chimpanzee) are known. Estimates derived from a comparison of processed pseudogenes in human and chimpanzee suggest that the average mutation rate per generation may be around 2.5×10^{-8} per nucleotide (Nachman, 2000). A change in the sequence of a gene if not neutral (functionally equivalent) is more likely to degrade than to improve its function; the overall impact of muta-tions must therefore be deleterious. As slightly deleterious variants have a longer survival in the population, they may in the long run cause a reduction of fitness as large or greater than those with marked effect (Crow, 1997). Based on rates of synonymous (no change in the amino acid) and non synonymous (change in amino acid) substitutions in 46 protein-coding sequences, and on estimates of the fraction of the genome that is under selective constraint, the occurrence of more than four amino acid altering mutations per diploid genome, per generation in human since the separation from the chimpanzee lineage has been inferred (Eyre-Walker and Keightley, 1999). It was estimated from the same study that 38% of these mutations had been eliminated, and that more than 1.6 new deleterious mutations affecting protein-coding sequence per diploid genome, per generation had occurred (Eyre-Walker *et al.*, 1999). Even higher rates of so called deleterious mutation in humans have been reported (Nachman, 2000). It has been stressed that the rate of deleterious mutation in humans is so high, that if selection was acting on each deleterious variant independently it would not be compatible with the survival of the species. Synergistic epistasis among deleterious mutations leading to a larger decrease in relative fitness (a greater chance of being eliminated from the population) for each additional mutation has been proposed to explain this paradox (Crow, 1997; Eyre-Walker and Keightley, 1999; Kondrashov, 1995), but this may not be the whole story.

2.3 Recombination

Exchange of material between homologous regions of chromosomes is responsible for the decay of linkage disequilibrium between linked sites. Recombination rate may be defined as the probability that a randomly selected gamete produced by a double heterozygote is a recombinant. Recombination rate can be measured in humans by considering markers over long distances of several centimorgans. Over such distances, crossing-over is responsible for the progressive decrease of linkage disequilibrium from generation to generation; but for shorter regions, for example smaller than 1000 bp, conversion i.e. the transfer of short tracks of sequence from one chromosome to the other without crossing-over, may be of greater importance for shuffling the variant sites between chromosomes (Andolfatto and Nordborg, 1998; Wiehe *et al.*, 2000). Recombination generates a mosaic of sequences, with different evolutionary history, this explains why different evolutionary trees may be inferred from the analysis of different parts of a gene sequence having experienced recombination. Recombination may also occur between different genomic regions having sufficient homology. This process plays an important role in the generation of new genes and alleles. For example, pseudogenes are sometimes involved in the generation of new alleles (Haino *et al.*, 1994).

Detecting and accounting for the presence of recombination may thus be essential when analysing DNA sequences. Recombination is suggested by the observed pattern of polymorphism found on a gene. For example, from a detailed analysis of polymorphisms within the *ACE* gene (Rieder, 1999) a recombination breakpoint could be mapped between two polymorphisms located at less than 300 bp from each other. This allowed the definition of sub-regions of the gene within which the phylogeny of haplotypes was easier to infer and the location of a gene region where the functional polymorphism associated with the ACE phenotype is more likely to be found. *Figure 2* provides a simple example of inferred recombination between non synonymous variants within the P-selectin gene (Herrmann *et al.*, 1998). Phylogenies or networks derived from haplotypes in which recombinations occurred are characterized by the presence of apparent homoplasies, as for the P-selectin, where the same mutation is present on more than one branch of the haplotypes network (*Figure 2*). It may be difficult, if not impossible, in the presence of homoplasy (i.e. the same nucleotide in two molecular sequences which is not the consequence of common ancestry) to differentiate a recurrent mutation from a recombination event; although overall, the later is more common than the former.

Recent attempts have been made to derive the frequency of recombination events from homoplasies found in gene trees constructed under the hypothesis of no recombination. In the Lipoprotein lipase gene for example 29 statistically significant recombination events were deduced using this approach (Templeton *et al.*, 2000).

2.4 Evolution

'Evolution as a process of cumulative change depends on a proper balance of the conditions which at each level of organization – gene, chromosome, cell, indi-

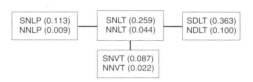

Figure 2. Haplotypes generated by non synonymous polymorphisms of the P-selectin gene. Haplotypes (and their frequencies), from left to right: positions 290, 562, 599, 715 In Europeans the P-selectin gene sequence carries at least 13 common polymorphisms in coding and regulatory regions (Herrmann *et al.*, 1998). Four of the polymorphisms Ser290Asn (S290N), Asn562Asp (N562D), Leu599Val (L599V) and Thr715Pro (T715P) are associated with non synonymous codon changes. The minor allele frequencies at these loci being 0.171, 0.462, 0.114 and 0.125, respectively. The Ser290Asn polymorphism, which is located in the 7th exon of the gene is only in weak linkage disequilibrium with the three other polymorphisms, located in exons 11, 12 and 13, respectively, which on the contrary are themselves in tight pairwise linkage disequilibrium. Haplotype frequencies combining alleles at the four sites were estimated using 1264 chromosomes from samples of subjects recruited in France and Belfast, all of European origin (ECTIM Study). Only eight haplotypes out of 16 possible were present. As shown in the diagram above, NLT is connected by a single mutational event to NLP, NVT and DLT. However, both codons at position 290 are found on the four haplotypes defined by the polymorphisms at positions 562, 590 and 715. This suggests the presence of a recombination hot spot between the sites encoding codons 290 and 562. In the presence of recombination, the network of haplotypes suggests that the same mutation (S290N) may have occurred on different haplotypes (homoplasy), which in the present case is the signature of recombination or conversion between haplotypes.

vidual, local race – make for genetic homogeneity or genetic heterogeneity of the species' (Wright, 1931; quoted in Wright, 1972)

Evolution is conventionally defined in a restricted way limited to the genotypic side, as a change of allele frequencies in a population over time. It is the result of two types of process: one, depending on mutation, recombination and gene flow (resulting from migration), contributes to the appearance of new alleles and increases genetic variability; the other, combining genetic drift and natural selection, reduces variability and ultimately may result in the disappearance of some alleles and the fixation of others. Addition of alleles in a gene pool is largely random, whereas subtraction of alleles is in part non random. Alleles that determine a more reproductively fit individual being favored relative to those which do not affect or reduce reproductive fitness. Since Darwin, for a large fraction of biologists, natural selection which corresponds to this non random process, is considered to have been the major force driving the evolution of species. Alternatively, according to the neutral theory of evolution developed by Kimura (1989), most evolutionary changes at the molecular level are caused by random fixation, due to sampling drift in finite populations, of selectively equivalent or nearly equivalent variants under continued mutation pressure. The neutral theory implies that genetic variability within species is maintained by the balance between mutational input and random extinction, rather than by balancing selection. Whether survival of the fittest or of the luckiest drives evolution is still a matter of intense debate. However, as the examples provided below show, there is good empirical support for both views. The entire organism being the unit of selection, the whole genome, not single genes, is affected by selection. Depending on the nature of selection, its pressure is very heterogeneously distributed across genes, this explains why different structures and functions in an organism may have had a very different tempo of evolution. Being able to characterize the constraints that are responsible for such striking heterogeneity is of great interest in evolutionary genetics.

2.5 Demonstrating selection in humans

'The problem is how every amino acid substitution can make a physiological difference that is ultimately translated into an average difference in viability and fertility' (Lewontin, 2000).

The consequences of natural selection on evolution, as initially postulated by Darwin, depend on very small differences of fitness between organisms within a species. As a consequence Darwinian type natural selection is quite difficult to demonstrate in natural or experimental populations, except in organisms having a rapid reproductive cycle. However, even if weak selection is important, it will become apparent when discussing the selective force of infectious diseases, that strong selection may also have had a considerable impact on human evolution.

As already mentioned, selection pressure is not evenly distributed across the genome. Traces of its past influence may be detected by suitable comparisons of DNA or protein sequences within or between populations or species. An interesting example, taken from a large possible number, is provided by the rapid evolution, in a number of species, of traits related to male reproduction, leading to a very high level of divergence in male reproductive proteins between closely related species. This phenomenom is particularly marked in primates and humans. Wyckoff *et al.* (2000) used three different statistical approaches to assess whether the rapid evolution of several reproductive proteins in male humans and primates is more likely to be the consequence of positive selection than of other mechanisms. First, they observed that the rate of non synonymous substitutions in these genes was higher than expected, whereas this was not the case for synonymous and non coding substitutions. Second, they observed that the ratio of non synonymous over silent nucleotide changes was significantly higher for the between-species divergence than for the within-species polymorphisms. The third approach used accounted for the fitness consequences of non synonymous nucleotide substitutions based on physico-chemical similarities between residues. For the protein studied, positive selection was only observed in conservative classes of amino acid substitutions. This observation fits the Darwinian perspective for which evolution is more likely to favor small than large changes. In their systematic analysis of the polymorphism of 106 genes, Cargill *et al.* (1999) assessed the force of selection by comparing non synonymous single nucleotide polymorphisms according to whether they are conservative or not, using an empirical amino acid substitution matrix which assigns scores to amino acids exchanges according to their chance of occurrence (Henikoff and Henikoff, 1992); their results showed a reduced frequency of non conservative coding polymorphisms, suggesting a 50% survival relative to conservative polymorphisms.

2.6 The major importance played by Infectious diseases in shaping the human gene pool

Wars, conflicts, genocides, famines, natural disasters, infectious diseases and many other plagues have greatly affected the human gene pool in a way which may not be entirely blind with regard to the genotype of the victims and survivors. This is particularly well established for infectious diseases which appear to have exerted a considerable pressure on the gene pool and resulted in the selection of alleles favoring survival.

Hemoglobinopathies and malaria. Malaria has been and is still a major selective force in human evolution. There may be several reasons for that (Luzzato, 2000): (1) *Plasmodium falciparum* infection is highly lethal at an early age; (2) the disease has been endemic for thousands of years in some tropical and subtropical areas, allowing the selection of genetic characteristics conferring resistance; (3) the disease is complex and several mechanisms of resistance at different levels of the pathophysiological pathway are conceivable. *P. falciparum* infection has influenced the selection of red blood cell hemoglobins, enzymes and membrane proteins. Although the hypothesis that abnormalities of the red blood cell may have conferred protection against malaria was formulated more than 50 years ago by Haldane, the epidemiological and *in vitro* demonstration that it is effectively the case has not been easy and is still not obvious for some defects. A striking feature of the gene variants that may have been favored by malaria is their very large number and localized pattern of distribution over the world, suggesting a very recent origin for most of them.

Hemoglobinopathies are the most common monogenic diseases in human, hundreds of β globin gene variants have been reported which are found only in particular populations. Some forms however, such as the Glu→Val substitution of the 6th codon of the β globin gene, which is responsible for sickle cell anemia when present in homozygous form, are quite common and have a worldwide distribution in regions where malaria is or has been endemic, or where populations originating from such regions have migrated. In some areas of Africa the frequency of the sickle cell gene is as high as 40%. α and β thalassemias are not due to structural abnormalities but impaired production of hemoglobin. Both forms of thalassemia are very frequent in many parts of the tropics and the disease in some regions, highly infested by malaria, confers such a strong protection that it may explain the near fixation of the thalassemia gene in some populations. For example, in the Tharu population living in the Terai region of Nepal the α thalassemia homozygous state decreases the risk from malaria by about 10-fold (Modiano et al., 1991). The large number of different thalassemia alleles may account for a large part of the variable severity of the disease. In addition, the existence of compound heterozygotes, the simultaneous presence of α and β variants and the role of modifier genes and of the environment lead to a fairly complicated and heterogeneous pattern of genotype–phenotype relationship. (Weatherall, 1999).

Another target of selection by malaria has been the enzyme Glucose-6-phosphate dehydrogenase (G6PD). G6PD is a rate limiting enzyme in the pentose pathway. In the red blood cell, through the generation of NAPDH, G6PD is a major defence against oxidative stress. Over 300 variants of the *G6PD* gene, which is located on the X chromosome, are known, approximately one third having a frequency >1% (Ruwende and Hill, 1998). G6PD deficiency, the commonest enzymopathy in humans, affects over 400 million people worldwide and is mainly distributed in tropical and subtropical regions. In some regions, particular deficient forms are highly prevalent and almost exclusive; this is the case in Africa where three major G6PD alleles exist : G6PD B is the commonest variant, G6PD A is associated with an enzyme activity of 80% and has a high prevalence (0.15–0.40) and G6PD A- is responsible for a severe reduction of enzyme activity (12%) and has a very variable regional distribution (0.00–0.25). Acute Hemolytic anemia is the commonest manifestation of G6PD deficiency. The striking geographical co-occurrence of G6PD deficiency and malaria has suggested that deficient forms of the G6PD gene could have been favored by natural selection. In support of this hypothesis, *in vitro* studies have shown an impaired growth of *P falciparum* in enzyme-deficient erythrocytes (Roth et al., 1983) and case-control studies strongly suggest that in Africa, the deficient G6PD A- allele confers resistance to severe malaria (Ruwende et al., 1995).

HIV-1 and coreceptors. The *CCR5* gene encodes a chemokine receptor that is essential for the cellular entry of some HIV-1 strains. The CCR5-d32 variant leads to truncation and loss of the receptor on lymphoid cells. Homozygotes for the deletion have a very high but not complete resistance to HIV-1 infection, while heterozygotes have a delayed onset of the acquired immunodeficiency syndrome (AIDS; Michael, 1999). The frequency of the CCR5-d32 variant, estimated in 4166 individuals from 38 populations worldwide, has revealed a strong north–south gradient in Europe, from 14% in Sweden to 4% in Greece and a virtual absence of the allele in Africans, East Asians and American Indians (Stephens *et al.*, 1999). This suggested that the mutation originated in a member of the ancestral European population. To investigate this hypothesis further, the age of the haplotype bearing the CCR5-d32 variant was determined, using a set of polymorphisms of the gene to construct haplotypes and derive a haplotype tree (Stephens *et al.*, 1999). The result is remarkable in suggesting that the mutation may have occurred very recently, about 700 years ago. For the authors, the most likely explanation for the high frequency and such recent occurrence of the variant is that it may have conferred a high protection to one or several infectious agents that decimated the European population at the end of the middle ages. Similar conclusions were reached in another study of 18 European populations (Libert *et al.*, 1998). In fact, different CCR5 haplotypes, comprising the CCR5-d32 variant or promoter polymorphisms, and combination of haplotypes may be associated with variable HIV-1 disease modifying effect in African Americans and Caucasians (Gonzalez *et al.*, 1999). Strong interaction between pairs of haplotypes with different evolutionary histories, have been observed implying that the genotype rather than the haplotype is important with regard to the disease-modifying effect of HIV-1.

These examples of gene selection induced by infectious diseases, especially those related to hemoglobinopathies, show that a considerable number of variants having generally low frequency may have been selected by the disease. If this particular mode of genetic determination extended to most common complex disorders, i.e. multiple loci with a plethora of risk alleles being involved (Weiss and Terwilliger, 2000), we would face a major difficulty and could wisely question the scientific relevance of enumerating all responsible variants. However, malaria is presently a highly active selective force and most of the malaria-selected variants are young. If the infection disappeared or could be prevented, balancing selection would not be maintained and the malaria selected variants would drift to disappearance if the reduced fitness conferred by the hemoglobinopathy was not counterbalanced in some way. It must also be remembered that hemoglobinopathies are major monogenic disorders and there is little support for the view that common complex disorders are similarly monogenic. An alternative to the 'plethora' model to account for the genetic component of common multifactorial diseases is that disease susceptibility depends on a relatively small number of genes with frequent alleles having weak or moderate effect (Cambien, 1997).

As suggested by a number of examples, the gene selection induced by infectious diseases is likely to extend beyond malaria and AIDS. The response to selection may affect a limited number of 'strategic' genes encoding host proteins involved in the pathogenesis of infectious diseases. As exemplified by the CCR5-d32 polymorphism, some of these genes may exert an effect which is not limited to a single disease. The imprint of past selection on these genes may be difficult to detect, especially if the infectious agent has disappeared or has a much smaller impact now than in the past. Obviously, the suggested link of the CCR5 polymorphism with epidemics in the middle ages could not have been formulated before its association with AIDS attracted the attention of biologists. Compared with the slow evolutionary pace that is generally assumed by Darwinian selection, the rapidity of allele frequency changes that may occur under the pressure of an infectious agent is amazing. It would therefore not be surprising if a number of alleles had reached fixation as a consequence of differential exposure to infectious agents.

2.7 Other selective forces shaping human gene variability

'Genes and combinations of genes which were at one time an asset may in the face of environmental change become a liability' (Neel, 1962).

Apart from infectious diseases, numerous environmental factors may have affected the human allele pool through their influence on survival and reproductive fitness. As for infectious diseases, the consequences of this selection are considered very important to explain the present pattern of susceptibility or resistance to disease in human populations.

As a first example we may consider the protection conferred by skin pigmentation against sun light. The melanocortin 1 receptor (MC1R) is a key element in the regulation of melanogenesis. In European populations the *MC1R* gene is highly polymorphic. Three particular variants associated with reduced or abolished function have been reported to be associated with red-hair when present as compound heterozygotes or homozygotes (Valverde *et al.*, 1995). Heterozygosity for these polymorphisms has also been shown to determine sun sensitivity in people without red hair (Healy *et al.*, 2000) as well as predisposition to risk of cutaneous malignant melanoma (Palmer *et al.*, 2000). A detailed analysis of the polymorphism of the *MC1R* gene in several populations in Africa, Asia and Europe (Harding *et al.*, 2000), displayed a very heterogeneous distribution of non synonymous variants. Indeed, no such variant could be found in Africans, whereas in the European and Asians, ten non synonymous variants were detected. This suggested a strong constraint in Africa against any degradation of the MC1R function that might lead to a diversion from eumelanin production, whereas such constraint has been less strong or absent in European and Asian populations.

The present adaptive paradigm is to link the high prevalence of atherosclerosis, hypertension, thrombosis, obesity, diabetes, dyslipidemia, and several other metabolic and degenerative disorders in many human populations to genetic variants that were positively selected in the past to fulfil an essential function, but are now deleterious in the particular environments in which we live (Neel, 1962). A typical adaptive scenario is composed of two parts: (1) a specific environmental constraint is applied to a population for a sufficiently long period of time to allow the selection of genetic features that are well adapted to this constraint. The length of the selection period is inversely proportional to the strength of the constraint, for example malaria is a strong constraint and in populations first exposed to it, gene frequencies may evolve quite rapidly. Conversely, other types of environmental constraints, such as low availability of salt may have had a weaker and more gradual effect on gene selection that had to be maintained for tens of thousands of years to become manifest; (2) the constraint is relieved or modified so that the genetic features that were formerly beneficial become ill-adapted, resulting in dysfunction and disease. For example, in populations not exposed to malaria, the disadvantage of bearing a hemoglobinopathy gene will not be counterbalanced by the protection it confers against malaria. In the case of hypertension the low salt constraint when relieved leaves an over-active salt retaining mechanism that may favor the rise of blood pressure.

Relaxation of environmental constraints may have resulted in the loss of many genes in the lineage leading to humans. As a striking example, most primates are deficient in the enzyme L-gulono-gamma-lactone oxidase (GLO), which is present in all other mammalian species (except the guinea pig) and is required for the synthesis of ascorbic acid. This deficit when not compensated by an appropriate dietary intake of vitamin C leads to a number of abnormalities culminating in scurvy. The most plausible explanation for the disappearance of the functional *GLO* gene from our ancestors' genome is that their dietary intake of vitamin C was sufficiently high that no endogenous production was required. In the absence of constraint, mutations could therefore accumulate on the *GLO* gene leading to its progressive or abrupt inactivation. The inactive *GLO* gene is still present in the primate and human genome as a pseudogene (Nishikimi, 1994). Later, after climate change or migration of our ancestors to regions where vitamin C from the environment was less abundant, the deficit became patent and selection may have favored individuals able to compensate for this deficit. Note that the foraging ability that this compensation implies may have led to the selection of genes that had an important role in the evolution of higher functions. The loss of most odorant receptors in humans (Rouquier, 1998) is another example of gene inactivation resulting from environmental constraint relaxation, or substitution of one ability by others. How common these natural gene knockouts have been, and what their impact has been on human evolution, is a subject of great interest on which systematic genome sequence comparisons within and between species will provide new insights.

An adaptive interpretation of disease derives its logic from the different time scales on which genetic and environmental changes take place. The rapid and profound modifications that have affected the climate on earth during the last hundreds of thousands of years may have constituted a major stress for the human genome, possibly accounting for its evolution (Hewitt, 2000). However, while the general adaptive framework of interpretation is plausible and often considered self-evident, it is by no means the unique possible explanation to account for relationships between genetic patterns of variation and common traits.

Genes that predispose to obesity, diabetes or other common traits for which adaptive explanations are commonly supplied, may actually have been selected because they favored an unsuspected or unknown function. This is made possible by the pleiotropic effect of genes and the involvement and interaction of functions at very different levels of organization, from metabolic or signalling pathways to behavioral traits. Obesity may be a good example. The often quoted hypothesis of selection of a thrifty genotype (Neel, 1962) to account for the high prevalence of this trait in several societies implicates selection of alleles resulting in improved fat storage. However, obese individuals in some primitive societies may not fulfil the same social function as the non-obese. They may, for example, be more likely to stay in the village accomplishing sedentary tasks than roaming the territory for food or fight. This social specialization may become a cultural feature of the society and have a number of consequences that possibly affect reproductive fitness, and account for the selection and high frequency of alleles predisposing to obesity. Speculations of this kind may be proposed for the maintenance of most traits that are considered disadvantageous according to our current Western view.

The qualitative or quantitative features of a biological system say nothing about its adaptation to past or present circumstances and a fortiori about why it reached its present level of functionality. This argument bears some analogy with the major criticism addressed by Gould and Lewontin (1979) to the adaptationist programme of evolutionary genetics, which says that 'the immediate utility of an organic structure often says nothing at all about the reasons of its being'. Whether metabolic or socialization features drive human adaptation is an interesting and far reaching issue that the limited scope of this presentation does not allow to be further developed.

2.8 Random drift

Random genetic drift refers to the time-related change of allele frequencies in a population as a consequence of chance alone, independent of natural selection. The segregation of parental genes implies that allele frequencies can increase or decrease from generation to generation. In large populations, the change will not be important but the smaller the population the larger it will be. Random genetic drift has no direction, accumulates with time and has no tendency to force allele frequencies to return to their ancestral level. Drift causes a reduction of genetic variation within a population and an increase of genetic variation among populations. In sufficiently small populations, mildly deleterious alleles may even drift to fixation. Drift may be quite important when a small group migrates a long distance away from the original population, or becomes isolated within a population, or when a population undergoes a drastic reduction in size, a bottleneck, as a consequence of some catastrophic event. Because of sampling error, a newly formed small group of individuals will not carry most of the rare alleles that were present in the original population but will carry some of them at an increased frequency. This explains why genetic diseases, especially recessive ones, may be more prevalent in human groups having experienced an important drift, than in large stable populations. Migration has been an important factor counteracting the consequences of random genetic drift on the genetic structure of small populations. For example, migration is considered responsible for the pattern of polymorphism frequencies across Europe which shows a gradient that originates in the Middle East and is directed to the northwest (Cavalli-Sforza and Piazza, 1993).

Is genetic drift important for evolution? The debate is still open. We have already mentioned the neutral theory of Kimura which places great emphasis on random drift (Kimura, 1989). It is difficult here not to mention the competing views of evolution proposed in the 1930s by Fisher and Wright. Both founders of population genetics viewed evolution in the context of Mendelian genetics. But according to Fisher the most important evolutionary changes took place in large populations through selection of nearly independent loci having small phenotypic effects, the role of chance being negligible. However, Wright thought that the fitness of an organism was largely determined by the combined interacting effect of its genes : 'I recognized that an organism must never be looked upon as a mere mosaic of 'unit characters', each determined by a single gene, but rather as a network of interacting systems' (Wright, 1978). Wright also thought that species were often subdivided into small local populations, allowing wide stochastic variability at a large number of loci. Wright's views fit well with what we presently know of human evolution, genetic polymorphism and biological organization and it is conceivable that random genetic drift drives the evolution of complexity in higher organisms (Zuckerkandl, 1997).

2.9 The age of polymorphisms and the human phylogeny

When polymorphic alleles are found in several populations, this may be an indication that the mutation occurred in a common ancestor of the populations, i.e. before they diverged. However, this supposes a strong isolation of the populations, as migration could explain the diffusion of an allele from one population where it originated to other populations. Greater differences over a large number of loci, possibly reflect a more distant common ancestry; this offers a rationale to construct distance matrices among populations, and generate phylogenic trees. It is commonly the case that more alleles and higher levels of heterozygosity are observed in African than in European and Asian populations. This is compatible with the ancestral human population originating from Africa. The results of studies using genetic markers from mitochondrial DNA (Vigilant *et al.*, 1991), chromosome Y (Hammer *et al.*, 1998) and autosomes (Armour *et al.*, 1996) from a large number of ethnic groups, consistently suggest a recent primary subdivision between African and non-African populations. As some polymorphisms are frequent in non-African populations but absent in subsaharian populations, it has been suggested that there was a relatively long intermediate period before the dispersal of human groups all over the world took place (Mitchell *et al.*, 1999).

2.10 Systematic studies of gene polymorphisms

Recent systematic studies have provided a wealth of information on the characteristics of gene polymorphisms within and between populations. Nucleotide diversity defined as the expected number of differences per nucleotide site, on a pair of chromosomes taken at random from the population, is reported for different categories of polymorphisms in *Table 1*. The nucleotide diversity measured in our own study (Tiret *et al.*, unpublished data) indicates that two randomly chosen coding sequences would differ approximately every 3300 bp. The results in *Table 1* confirm the greater nucleotide diversity in Africans than in Europeans and in

Table 1. Nucleotide diversity for different categories of polymorphisms in systematic analyses of gene polymorphisms

Study	No. of genes	No. of SNPs	Population	Coding	Synony-mous	Non synonymous	Non coding
Halushka (1999)	75	726	Europeans	0.00045	0.00090	0.00031	0.00054
Halushka (1999)	75	726	Africans	0.00063	0.00129	0.00042	0.00068
Cargill (1999)	106	560	Europeans	0.00056	0.00098	0.00037	0.00049
Onishi (2000)	41	187	Japanese	0.00024	0.00012	0.00012	0.00043
Tiret[1]	50	228	Europeans	0.00030	0.00056	0.00023	0.00043

[1] Unpublished extension of previously published results (Cambien, 1999)

Europeans than in Japanese. Furthermore, except in the Japanese study, non synonymous changes are less common than synonymous changes in coding sequences; this observation is consistent with the effect of natural selection. It is also interesting that in Europeans and Africans, nucleotide diversity in non coding sequences (mostly 3′ and 5′ UTR and short portions of introns) is lower than at synonymous sites. An observation that is compatible with functional changes associated with variability in non coding regions. Considerable differences of nucleotide diversity in different genes were observed in the four studies.

2.11 Linkage disequilibrium and haplotypes

Linkage disequilibrium (LD) reflects the non random association of alleles at two or more loci. LD may be generated by mutation, natural selection, random genetic drift, and population admixture and it decreases at a rate proportional to the recombination fraction between the loci and the number of generations. As already mentioned, each new mutation is in complete LD with the haplotype on which it occurred, until some recombination event associates it with another haplotype. If n different diallelic polymorphisms are found on a DNA sequence, the minimum number of haplotypes that have existed during the evolution of this region is $n+1$. However, the maximum number of haplotypes deriving from a set of n diallelic polymorphisms is 2^n and would correspond to a situation where recombination has shuffled all polymorphisms. In such case, significant LD over the region may still exist if haplotype frequencies differ from the frequencies expected if alleles at the different loci were combined randomly.

2.12 Linkage disequilibrium in some regions of the genome is very strong

A good example is provided by the major histocompatibility complex (MHC) region. The complete sequence and gene map of the human MHC has been reported (MHC sequencing consortium, 1999). The region is the most gene-dense in the human genome, with 224 gene loci, a high proportion of them being immune-related genes. A particularly striking aspect of the genes in the MHC region is their high degree of polymorphism and considerable LD in some subregions. The high degree of polymorphism appears to be the result of selection pressure imposed on the immune system by infectious agents. Important genetic determinants for a number of diseases including type I diabetes, psoriasis, rheumatoid arthritis, celiac disease and hemochromatosis have been located within the MHC region. The extended LD makes it difficult to localize with precision which genes may be responsible. As an illustration we may consider the genetics of hemochromatosis. Associations between HLA haplotypes and idiopathic hemochromatosis were reported more than 2 decades ago (Simon *et al.*, 1976). The gene responsible for the disease, called *HFE*, was cloned in 1996 (Feder *et al.*, 1996); it encodes a MHC class I-like protein which bears missense mutations, Cys282Tyr and His63Asp, that are responsible for most cases of hereditary hemochromatosis (Feder *et al.*, 1996). The *HFE* gene is located in a region of high LD. The recombination rate, estimated from 784 informative meioses in the

CEPH pedigrees, in the 4Mb between HLA-A and HFE was 0.19% instead of 1% expected (Malfroy *et al.*, 1997). High LD in the MHC generates a large number of associations between markers and diseases to which local genes contribute and complicates the identification of the susceptibility genes. However, there is an advantage to this high LD because a large number of markers will be informative for the disease loci. Identifying regions of the genome with high LD may therefore lead to the rapid identification of genotype–phenotype associations.

2.13 Whole genome LD

In theory, the LD generated by random drift and population admixture should be homogeneously distributed all over the genome. Conversely, the LD generated by mutation or selection should be localized, with regions of reduced LD in the presence of mutation hot spots and regions of increased LD around loci that are under selective pressure, as in the MHC region. To characterize the degree of LD over the whole human genome, Huttley *et al.* (1999) analyzed the genotype data from 5048 microsatellite polymorphisms on 54 independent haplotypes derived from a set of seven European families. The analysis revealed a highly heterogeneous LD across the genome. By comparing LD distribution from single chromosomes to that of the rest of the genome, the null hypothesis of spatial homogeneity of LD could be rejected. Nine regions presenting a degree of LD equal or greater than that of the HLA region were identified. The non uniform pattern of LD in the human genome is consistent with the operation of natural selection (Huttley *et al.*, 1999). However, detecting localized traces of selection, is not incompatible with LD being the consequence of random drift elsewhere in the genome. The relative importance of selection and drift on LD may vary according to the resolution of the analysis (regions of genes, genes, regions of chromosomes etc.).

In the genome wide LD study discussed above, one of the genes found in a region exhibiting a LD greater than that found in the HLA region was *BRCA1*. BRCA1 is involved in the maintenance of genomic integrity and mutations in the *BRCA1* gene confer an increased risk of breast cancer. This led Huttley *et al.* (2000) to hypothesize that positive selection could exist at that locus in the lineage leading to humans. To test this hypothesis, differences in the ratio of replacement to silent substitutions in the *BRCA1* gene in humans and non human primates were compared. In humans and chimpanzees, values for this ratio of 3.1 and 2.6, respectively were much higher than the average 0.68 value found in the other primates. This suggests that the *BRCA1* sequence in humans and chimpanzees may have been under selection pressure. There is no evidence however that the influence of natural selection at the *BRCA1* locus is responsible for the strong regional LD identified on chromosome 17q21-q22 (Huttley *et al.*, 1999); indeed there are 27 other known genes in the region and probably a number of still unknown genes. We may suppose that if several genes contributed to the high LD in some regions of the genome, as it is the case in the HLA region, this would possibly lead to haplotypes with highly differentiated functions and pleiotropic effects that might account for important phenotypic differences.

2.14 Complete association

A remarkable characteristic of the human genome, probably reflecting the tumul-
tuous history of our species, is the very high frequency of complete or nearly
complete association between alleles at nearby loci. Complete association involving
n biallelic sites is reflected by the strong predominance of only two haplotypes.
When genotypes of two biallelic polymorphisms are crossed, nine combinations
may be observed. Nearly complete association is characterized by the predominance of genotypes located on the diagonal of the 3*3 table. Such pattern may
extend to a larger number of polymorphisms. We evaluated the frequency of nearly
complete associations between common polymorphisms, in 35 genes for which we
had identified at least two polymorphisms (Tiret, unpublished data). Nearly
complete association between at least two polymorphisms was observed for 18 of
these genes; 18 of these associations involved two polymorphisms, five involved
three polymorphisms, two involved four polymorphisms and two involved six
polymorphisms. An example of two nearly complete associations involving three
polymorphisms is shown in *Figure 3* for the beta fibrinogen gene.

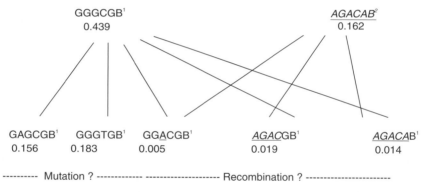

Figure 3. Inferred haplotype structure of the beta fibrinogen gene and example of complete
association. To identify polymorphisms on the beta fibrinogen gene, a systematic search was
performed using DNA from 11 smokers having high levels of plasma fibrinogen (Behague
et al., 1996). Ten polymorphisms were detected and genotyped in a population-based case
control study of myocardial infarction (2628 alleles). Three polymorphisms located at the 5′
end of the gene (G-1420A, C148T, G+1532/in1T) were almost completely concordant and
the same was true for three polymorphisms at the 3′ end of the gene (Bcl1[B¹/B²], Y345Y
and +192/in6[ins/del]). Statistical analysis inferred the presence of 20 haplotypes, of which
seven had a frequency greater than 5/1000 and accounted for 97.7% of all haplotypes.

In the above network of haplotypes proposed for the beta fibrinogen gene, the three
haplotypes GGGCGB¹, GAGCGB¹ and GGGTGB¹ are connected by a single mutational
event. Haplotypes GGGCGB¹ and *AGACAB²*, differ at four out of six positions suggesting
that there common ancestor may be quite ancient. Recombination between these two
haplotypes may have generated the three remaining haplotypes.

The G-455A and Bcl1[B¹/B²] polymorphisms exist in the Japanese population,
although at a lower frequency than in individuals of European descent (Iso *et al.*, 1995)
but no systematic study of the beta fibrinogen gene polymorphisms in African and other
populations has been performed yet. It is therefore impossible to determine which
haplotypes are present in all human populations.

Extreme genetic drift and extinction of lineage, leading to the disappearance or quasi disappearance of a large number of haplotypes may explain the high frequency of complete associations between polymorphisms located close to each other on the human genome. Haplotypes defined by completely associated polymorphisms are likely to be quite ancient, however, they may accumulate more recent variation as shown for the beta fibrinogen gene polymorphism (*Figure 3*).

It is important to know the age of an haplotype. Theory predicts that for a given polymorphism, under neutral evolution, the probability that an allele is the oldest is equal to its frequency (Watterson and Guess, 1977). There are many examples showing young haplotypes that are common in a particular population. Conversely very ancient haplotypes may be rare. An interesting example of such a situation is provided by the ApoE polymorphism. From a detailed reconstruction of the evolutionary history of ApoE allelic divergence (Fullerton *et al.*, 2000), it was inferred that the most ancient haplotype of the *ApoE* gene, still present in humans and also found in chimpanzees, was present in only two individuals out of 96. Assigning an approximate age to haplotypes can be carried out by comparing molecular data collected from different ethnic groups and non human primates. It is possible to establish whether a particular haplotype is specific to one, several or all human groups and if it is shared by some or all primate species. The time since the separation between human ethnic groups and between humans and other primates being known from independent sources, it is therefore possible to infer an approximate age to the haplotypes.

2.15 *The exploitation of linkage disequilibrium in genotype–phenotype association studies*

For a number of years and with undeniable success, sets of markers have been used in family linkage studies to explore the whole genome in search of a region where genes causing monogenic diseases are located. Approximately 400 highly polymorphic markers such as microsatellites or 3–4 times as many frequent single nucleotide polymorphisms randomly distributed over the genome, are considered sufficient to detect the cotransmission of a marker allele and a genetic disease from one generation to the next. Recently, with the prospect that new technologies such as DNA-chips will become usable on a large scale at acceptable cost, whole genome association studies have been proposed as an appropriate strategy to explore the genetic component of common disorders. The use of genetic markers in association studies is justified by the expectation that the LD between the marker(s) and an unknown functional variant affecting the phenotype of interest, will generate a significant association between the marker(s) and the phenotype in a study of reasonable sample size. The ideal marker is the functional variant itself, this justifies the integral exploration of gene polymorphism that is discussed below; however, the quality of a random marker is also an issue as the use of such markers has been proposed for whole genome association studies.

The question of the number of polymorphisms needed for an exhaustive exploration of the genome has received different answers. A recent estimate based on extensive computer simulations, suggested that 500 000 may be required (Kruglyak, 1999). However, extrapolations based on empirical data led to the proposition that 30 000 well-chosen markers would be sufficient for an examination of the whole genome (Collins *et al.*, 1999). This last proposal would correspond to approximately one marker every four genes, a density which appears quite insufficient for exploring the genetics of common diseases.

More practically, a consortium of private and academic laboratories was set-up with an aim to provide 300 000 SNPs to the scientific community. It may be instructive to consider what we can expect from a 300 000 whole genome random SNP map, using chromosome 22, the sequence of which is known. Chromosome 22, representing 1.8% of the whole genome sequence, would harbor 5400 markers; given that genes (including introns) occupy one third of the sequence and exons only 3%, the number of markers located in genes and exons would be approximately 1800 and 162, respectively. Six hundred and seventy nine genes were found on chromosome 22 with a total number of exons greater than 3500. Very few markers would therefore be located in coding sequences. In fact, an analysis of the data actually produced during the sequencing effort found 12 000 SNPs of which 1.6% (192) fell in the 3% of coding region (Dawson *et al.*, 2001), although 35% fell within 5 kb of an exon. These data may at first sight indicate that a 300 000 markers random map would provide little information on important regions of the genome except for those regions where LD is particularly strong and provided the markers are highly informative for the functional variants sought. We believe that sufficient data is available now demonstrating a lack of LD or low LD on a number of genes (Cambien *et al.*, 1999), to suggest that a random marker strategy, even with 1 000 000 markers, cannot be adopted as a general approach for the study of genotype–phenotype association.

LD being very heterogeneous across the genome, it would appear logical to adapt the density of the markers used for a whole genome search accordingly; more markers would be used in regions of low LD and less markers in regions of high LD. This methodological shift may be a first step toward a gene focused approach which we believe is the most appropriate for resolving the genetic component of multifactorial disorders (Cambien, 1999).

The gene focused approach implies identifying all common polymorphisms in coding, and regulatory sequences of genes (Cambien *et al.*, 1999). These polymorphisms would constitute markers, but they would also include most functional polymorphisms whose identification is the ultimate goal of association studies. From our work on 36 candidate genes for cardiovascular disorders we extrapolated that there might be approximately 400 000 relatively common SNPs within regulatory and coding sequences of human genes. Halushka *et al.* (1999) estimated that there might be 500 000 SNPs in non coding regions of genes, and 400 000 coding SNPs, half being synonymous. Cargill (1999) extrapolated that the number of coding SNPs could be between 240 000 and 400 000. Our approach based on the gene scan of 40 alleles from individuals of European origin, is biased toward the detection of frequent, ancient alleles. This is justified by a working model specifying that most of the genetic contribution to common multifactorial disorders (in term of attributable risk) is the consequence of gene forms that are common, generally of ancient origin and shared by most human populations. If however the 'plethora' of alleles model was correct (Weiss and Terwilliger, 2000) the variability of each candidate gene should be investigated in each population (possibly reducing to single families) where the disease has a high prevalence. It would indeed be very unlikely to find these variants by examining the type of standard sample of individuals that is used to identify common polymorphisms.

Choosing putatively functional regions of genes for study is based on *a priori* assumptions; however, some authors argue that functionality is not predictable and that reconstructing the complete evolutionary history of a gene is indispensable for the purpose of investigating its implication in disease and requires that all its polymorphisms are known (Fullerton *et al.*, 2000; Templeton *et al.*, 2000). These arguments are debatable and before being advocated, this maximalist strategy, which for a typical gene would require the characterization of tens to hundreds of polymorphisms, will have to demonstrate its merits over simpler and more hierarchical approaches focused on a priori defined putative functional regions.

2.16 The advantage of using haplotypes

Knowing the haplotype structure of a gene and not only the genotypes of single polymorphisms may be quite useful. For example when considering two polymorphisms, ignoring the haplotype structure may mean that we ignore whether variants *a* and *b* are on the same or different chromosomes in heterozygotes. Haplotypes are also important because they can be functional. In such cases, two or more variants on a gene may confer a particular functionality to the encoded protein when they occur together on the same allele. Combination of variants in regulatory regions may interact by affecting the binding of regulatory proteins (Terry *et al.*, 2000); whereas in coding regions, variants may confer a particular structure to the encoded protein that affects its function. Finally, knowing haplotypes may allow one to reconstruct the phylogeny of a gene or a network of haplotypes, which is essential for evolutionary and population genetic studies, but may also be of interest in genetic epidemiology, because haplogroups may be derived that provide a rational principle to categorize phenotypes. However, there is no guarantee that a set of haplotypes that is consistently related to a phenotype will correspond to a haplogroup or set of haplogroups derived from the phylogeny of a gene. Unfortunately when investigating a set of polymorphisms in a sample of independent individuals, except in trivial situations, haplotypes are not readily deducible from the genotypes. A sequential procedure (Clark, 1990; Clark *et al.*, 1998, Fullerton *et al.*, 2000) may help to resolve haplotypes from multiple polymorphisms of a gene. The method first identifies unambiguous haplotypes in homozygotes and heterozygotes at a single site, then use allele-specific PCR to assign phase at appropriately chosen pairs of sites and when ambiguity remains for genotypes that cannot be assigned to a single haplotype. The number of unresolved haplotypes at the end of the procedure will depend on several factors including the number of individuals studied, the number of polymorphisms and the degree of LD among them. Different methods have been proposed to estimate haplotype frequencies from genotypes at several loci in a genomic region, without knowledge of phase. An investigation, using simulated data, of the efficiency of the Expectation-Maximization (EM) algorithm to recover true allele frequencies (Fallin and Schork, 2000) has shown that for large sample size, even in the presence of non optimal conditions, the haplotype frequency estimates appeared remarkably accurate. Developments are underway which we hope will allow to investigate haplotype–phenotype relationships in associations studies.

3. Conclusion

The classic view that evolution proceeds slowly as a consequence of selection acting on small heritable changes, is challenged by empirical data suggesting that strong selection and genetic drift are important determinants of gene frequencies changes in human populations. A number of contingent environmental conditions, among which infectious diseases may have been the most powerful, have exerted variable pressures on the human genome and favored the selection of alleles interfering with disease physiopathology. Strong selection at one locus, especially when it occurs in small fragmented populations may also influence unlinked loci through random drift. Conversely, random drift, as is observed in the presence of a founder effect, may introduce a gene in a population that will have a strong selective effect subsequently. In many circumstances, both random drift and selection, as well as their interaction, have driven the change of the human gene pool, and trying to evaluate their respective contribution may be as hopeless as trying to separate the effect of gene and environment on a multifactorial phenotype.

Strong selection, as it acts on a limited space and time may favor population specific variants as we see for hemoglobinopathies. However, older alleles, shared by all human populations may account for most human inter-individual genetic variability. Even if they have a moderate phenotypic effect at the individual level, these alleles because of their high frequency may have an important impact on the population. Furthermore, they may combine together and possibly interact among each other and with environmental factors to affect phenotypes. A reasonable hypothesis is that such alleles may be responsible for most of the genetic component of common multifactorial diseases.

A major project in the next few years will be to produce a catalog of all common forms of human genes. An appropriate use of this information should lead to a better understanding of human evolution. However, the major challenge will be to relate this variability to the structure and function of biological systems. Unravelling this relationship will be essential to understand and possibly counteract the genetic contribution to common multifactorial diseases.

References

Andolfatto, P. and Nordborg, M. (1998) The effect of gene conversion on intralocus associations. *Genetics* **148**: 1397–1399.

Armour, J.A., Anttinen, T., May, C.A. *et al.* (1996) Minisatellite diversity supports a recent African origin for modern humans. *Nat. Genet.* **13**: 154–160.

Behague, I., Poirier, O., Nicaud, V. *et al.* (1996) Beta fibrinogen gene polymorphisms are associated with plasma fibrinogen and coronary artery disease in patients with myocardial infarction. *Circulation* **93**: 440–449.

Cambien, F., Poirier, O., Mallet, C. and Tiret, L. (1997) Coronary heart disease and genetics in epidemiologist's view. *Mol. Med. Today* **3**: 197–203.

Cambien, F., Poirier, O., Nicaud, V. *et al.* (1999) Sequence diversity in 36 candidate genes for cardiovascular disorders. *Am. J. Hum. Genet.* **65**: 183–191.

Cargill, M., Altshuler, D., Ireland, J. *et al.* (1999) Characterization of single-nucleotide polymorphisms in coding regions of human genes. *Nat. Genet.* **22**: 231–238.

Cavalli-Sforza, L.L. and Piazza, A. (1993) Human genomic diversity in Europe: a summary of recent research and prospects for the future. *Eur. J. Hum. Genet.* **1**: 3–18.

Clark, A.G., Weiss, K.M., Nickerson, D.A. *et al.* (1998) Haplotype structure and population genetic inferences from nucleotide-sequence variation in human lipoprotein lipase. *Am. J. Hum. Genet.* **63**: 595–612.

Clark, A.G. (1990) Inference of haplotypes from PCR-amplified samples of diploid populations. *Mol. Biol. Evol.* **7**: 111–122.

Collins, A., Lonjou, C. and Morton, N.E. (1999) Genetic epidemiology of single-nucleotide polymorphisms. *Proc. Natl Acad. Sci. USA* **96**: 15173–15177.

Crow, J.F. (1997) The high spontaneous mutation rate: Is it a health risk? *Proc. Natl Acad. Sci. USA* **94**: 8380–8386.

Dawson, E., Chen, Y., Hunt, S. *et al.* (2001) A SNP Resource for Human Chromosome 22: Extracting Dense Clusters of SNPs From the Genomic Sequence *Genome Res.* **11**: 170–178.

Dunham, I., Shimizu, N., Roe, B.A. *et al.* (1999) The DNA sequence of human chromosome 22. *Nature* **402**: 489–495.

Eyre-Walker, A. and Keightley, P.D. (1999) High genomic deleterious mutation rates in hominids. *Nature* **397**: 344–347.

Fallin, D. and Schork, N.J. (2000) Accuracy of haplotype frequency estimation for biallelic loci, via the Expectation-Maximization algorithm for unphased diploid genotype data. *Am. J. Hum. Genet.* **67**: 947–959.

Feder, J.N., Gnirke, A., Thomas, W. *et al.* (1996) A novel MHC class-IIike gene is mutated in patients with hereditary hemochromatosis. *Nat. Genet.* **13**: 399–408.

Fullerton, S.M., Clark, A.G. and Weiss, K.M. (2000) Apolipoprotein E variation at the sequence haplotype level: Implications for the origin and maintenance of a major human polymorphism. *Am. J. Hum. Genet.* **67**: 881–900.

Gonzales, E., Bamshad, M., Sato, N. *et al.* (1999) Race-specific HIV-1 disease-modifying effects associated with CCR5 haplotypes. *Proc. Natl Acad. Sci. USA* **96**: 12004–12009.

Gould, S.J. and Lewontin, R.C. (1979) The spandrels of San Marco and the Panglossian paraolign: a critique of the adaptationist programme. *Proc. R. Soc. Lond. B. Biol. Sci.* **205**: 581–598.

Haino, M., Hayashida, H., Miyata, T *et al.* (1994) Comparison and evolution of human immunoglobin VH segments located in the 3′ 0.8 megabase region. Evidence for unidirectional transfer of segmental gene sequences. *J. Biol. Chem.* **269**: 2619–2626.

Halushka, M.K., Fan, J.B., Bentley, K. *et al.* (1999) Patterns of single nucleotide polymorphisms in candidate genes for blood pressure homeostasis. *Nat. Genet.* **22**: 239–247.

Hammer, M.F., Karafet, T., Rasanayagam, A., Wood, E.T., Altheide, T.K., Jenkins, T., Griffiths, R.C., Templeton, A.R. and Zegura, S.L. (1998) Out of Africa and back again: nested cladistic analysis of human Y chromosome variation. *Mol. Biol. Evol.* **15**: 427–441.

Harding, R.M., Healy, E., Ray, A.J. *et al.* (2000) Evidence for variable selective pressure at MC1R. *Am. J. Hum. Genet.* **66**: 1351–1361.

Hartl, D.L. and Clark, A.G. (1997) *Principles of Population Genetics*. Sinauer, Sunderland, MA.

Healy, E., Flannagan, N., Ray, A., Todd, C., Jackson, I.J., Matthews, J.N., Birch-Machin, M.A. and Rees, J.L. (2000) Melanocortin-1-receptor gene and sun sensitivity in individuals without red hair. *Lancet* **355**: 1072–1073.

Henikoff, S. and Henikoff, J.G. (1992) Amino acid substitution matrices from protein blocks. *Proc. Natl Acad. Sci. USA*; **89**: 10915–10919.

Herrmann, S.M., Ricard, S., Nicaud, V., Mallet, C., Evans, A., Ruidavets, J.B., Arveiler, D., Luc, G. and Cambien, F. (1998) The P-selectin gene is highly polymorphic: reduced frequency of the Pro715 allele carriers in patients with myocardial infarction. *Hum. Mol. Genet.* **7**: 1277–1284.

Hewitt, G. (2000) The genetic legacy of the quaternary ice ages. *Nature* **405**: 907–913.

Huttley, G.A., Easteal, S., Southey, M.C., Tesoriero, A., Giles, G.G., McCredie, M., Hopper, J.L. and Venter, D.J. (2000) Adaptive evolution of the tumor suppressor BRCA1 in humans and chimpanzees. *Nat. Genet.* **25**: 410–413.

Huttley, G.A., Smith, M.W., Carrington, M. and O'Brien, S.J. (1999) A scan for linkage disequilibrium across the human genome. *Genetics* **152**: 1711–1722.

Iso, H., Folsom, A.R., Winkelmann, J.C. *et al.* (1995) Polymorphisms of the beta fibrinogen gene and plasma fibrinogen concentration in Caucasian and Japanese population samples. *Thromb. Haemost.* **73**: 106–111.

Kimura, M. (1989) The neutral theory of molecular evolution and the world view of the neutralists. *Genome* **31**: 24–31.

Kimura, M. (1968) Evolutionary rate at the molecular level. *Nature* **217**: 624–626.

Kondrashov, A.S. (1995) Contamination of the genome by very slightly deleterious mutations: why have we not died 100 times over? *J. Theor. Biol.* **175**: 583–594.

Kruglyak, L. (1999) Prospect for whole-genome linkage mapping of common diseases. *Nature Genet.* **22**: 139–144.

Lewontin, R. (2000) The problem of population genetics. In: Singh, R.S., Krimbas, C.B. (eds), *Evolutionary Genetics – from molecules to morphology*. Cambridge University Press, Cambridge.

Libert, F., Cochaux, P., Beckman, G. *et al.* (1998) The deltaccr5 mutation conferring protection against HIV-1 in Caucasian populations has a single and recent origin in Northeastern Europe. *Hum. Mol. Genet.* **7**: 399–406.

Luzzato, L. (2000) Genetic factors in malaria resistance. In: Boulyjenkov, V., Berg, K., Christen, Y. (eds), *Genes and resistance to disease*. Springer-Verlag Berlin Heidelberg New York, pp. 105–120.

Malfroy, L., Roth, M.P., Carrington, M., Borot, N., Volz, A., Ziegler, A. and Coppin, H. (1997) Heterogeneity in rates of recombination in the 6-MB region telomeric to the human major histocompatibility complex. *Genomics* **43**: 226–231.

MHC Sequencing Consortium (1999) Complete sequence and gene map of the human major histocompatibility complex. *Nature* **401**: 921–923.

Michael, N.L. (1999) Host genetic influences on HIV-1 pathogenesis. *Curr. Opin. Immunol.* **11**: 466–474.

Mitchell, R.J., Howlett, S., White, N.G. *et al.* (1999) Deletion polymorphism in the human *COL1A2* gene: genetic evidence of a non-African population whose descendants spread to all continents. *Hum. Biol.* **71**: 901–914.

Modiano, G., Morpurgo, G., Terrenato, L. *et al.* (1991) Protection against malaria morbidity: near fixation of the α thalassemia gene in a Nepalese population. *Am. J. Hum. Genet.* **48**: 390–397.

Nachman, M.W. and Crowell, S.L. (2000) Estimate of the mutation rate per nucleotide in humans. *Genetics* **156**: 297–304.

Neel, J.V. (1962) Diabetes Mellitus: a 'thrifty' genotype rendered detrimental by 'progress'? *Am. J. Hum. Genet.* **14**: 353–362.

Nishikimi, M., Fukuyama, R., Minoshima, S., Shimizu, N. and Yagi, K. (1994) Cloning and chromosomal mapping of the human nonfunctional gene for L-gulono-gamma-lactone oxidase, the enzyme for L-ascorbic acid biosynthesis missing in man. *J. Biol. Chem.* **269**: 13685–13688.

Palmer, J.S., Duffy, D.L., Box, N.F., Aitken, J.F., O'Gorman, L.E., Green, A.C., Hayward, N.K., Martin, N.G. and Sturm, R.A. (2000) Melanocortin-1 receptor polymorphisms and risk of melanoma: is the association explained solely by pigmentation phenotype? *Am. J. Hum. Genet.* **66**: 176–186.

Pritchard, J.K., Stephens, M., Rosenberg, N.A. and Donelly, P. (2000) Association mapping in structured populations. *Am. J. Hum. Genet.* **67**: 170–181.

Rieder, M.J., Taylor, S.L., Clark, A.G. and Nickerson, D.A. (1999) Sequence variation in the human angiotensin converting enzyme. *Nat. Genet.* **22**: 59–62.

Roth, E.F., Raventos-Suarez, C., Rinaldi, A. and Nagel, R.L. (1983) Glucose-6-Phosphate dehydrogenase deficiency inhibits *in vitro* growth of *Plamodium Falciparum* Malaria. *Proc. Natl Acad. Sci. USA* **80**: 298–299.

Rouquier, S., Taviaux, S., Trask, B.J., Brand-Arpon, V., van den Engh, G., Demaille, J. and Giorgi, D. (1998) Distribution of olfactory receptor genes in the human genome. *Nat. Genet.* **18**: 243–250.

Ruwende, C., Khoo, S.C., Snow, R.W. *et al.* (1995) Natural protection of hemi- and heterozygotes for G6PD deficiency in Africa by resistance to severe Malaria. *Nature* **376**: 246–249.

Ruwende, C. and Hill A. (1998) Glucose-6-Phosphate dehydrogenase deficiency and malaria. *J. Mol. Med.* **76**: 81–88.

Simon, M., Bourel, M., Fauchet, R. and Genetet, B. (1976) Association of HLA A3 and HLA B14 with idiopathic hemochromatosis. *Gut* **17**: 332–334.

Stephens, J.C., Reich, D.E., Goldstein, D.B. *et al.* (1999) Dating the origin of the CCR5-Delta32 AIDS-resistance allele by the coalescence of haplotypes. *Am. J. Hum. Genet.* **62**: 1507–1515.

Templeton, A.R., Clark, A.G., Weiss, K.M., Nickerson, D.A., Boerwinkle E. and Sing C.F. (2000) Recombinational and mutational hotspots within the human lipoprotein lipase gene. *Am. J. Hum. Genet.* **66**: 69–83.

Terry, C.F, Loukaci V. and Green, F.R. (2000) Cooperative influence of genetic polymorphisms on interleukin 6 transcriptional regulation. *J. Biol. Chem.* **275**: 18138–18144.

Valverde, P., Healy, E., Jackson, I., Rees, J.L. and Thody, A.J. (1995) Variants of the melanocyte-stimulating hormone receptor gene are associated with red hair and fair skin in humans. *Nat. Genet.* **11**: 328–330.

Vigilant, L., Stoneking, M., Harpending, H., Hawkes, K. and Wilson, A.C. (1991) African populations and the evolution of human mitochondrial DNA. *Science* **253**: 1503–1507.

Watterson, G.A. and Guess, H.A. (1977) Is the most frequent allele the oldest? *Theor. Population Biol.* **11**: 141–160.

Wright, S. (1978) The relation of livestock breeding to theories of evolution. *J. Anim. Sci.* **46**: 1192.

Weatherhall, D. (1999) From genotype to phenotype: genetics and medical practice in the new millennium. *Phil. Trans. R. Soc. Lond. B* **354**: 1995–2010.

Weiss, K.M. and Terwilliger, J.D. (2000) How many diseases does it take to map a gene with SNPs ? *Nat. Genet.* **26**: 151–157.

Wiehe, T., Mountain, J., Parham, P. and Slatkin, M. (2000) Distinguishing recombination and intragenic gene conversion by linkage disequilibrium patterns. *Genet. Res.* **75**: 61–73.

Wyckoff, G.J., Wang, W. and Wu, C.I. (2000) Rapid evolution of male reproductive genes in the descent of man. *Nature* **403**: 304–309.

Zuckerkandl, E. (1997) Neutral and non neutral mutations: the creative mix-evolution of complexity in gene interaction systems. *J. Mol. Evol.* **44 (suppl 1)**: S2–S8.

4

Gene–environment interaction: lipoprotein lipase and smoking and risk of CAD and the ACE and exercise-induced left ventricular hypertrophy as examples

Steve E. Humphries, Philippa J. Talmud and Hugh Montgomery

1. Introduction

In common with all past genetic analyses of human disease, genetic research into coronary artery heart disease (CAD) has been based on two major premises. The first was that the identification of disease-causing mutations would lead to the development of DNA-based tests to identify those at risk. To date there has been little progress in developing such a 'battery' of genetic tests for CAD (much to the chagrin of several commercial companies), but a detailed understanding of the way common genetic variations of modest effect interact with environmental factors, suggests that this goal was always going to be difficult and technically challenging to achieve.

As we predict that the impact of single common mutations on CAD development will be modest (increasing Relative Risk (RR) by 20–40% at most), the main issue of clinical relevance is whether the conferred risk of such a mutation is very much higher in some population subgroups. To be clinically useful in a risk algorithm, we might require for any factor to have a RR of 2 or greater; which is, for example, the RR estimated to be associated with smoking habit in middle-aged

Genotype to Phenotype second edition, edited by S. Malcolm and J. Goodship.
© 2001 BIOS Scientific Publishers Ltd, Oxford.

men (Doll and Hill, 1966). Such subgroups might be those carrying a second important mutation in another gene (i.e. those with a gene–gene interaction), and such individuals might be identified using conventional genetic strategies. Alternatively, one might identify individuals exposed to a given environment which amplifies the risk associated with that gene (i.e. gene–environment interaction). This chapter focuses on common genetic variants (at polymorphic frequency i.e. >1% carrier frequency) which are associated with significant excess risk only when the individual is exposed to a 'high-risk' environment. The examples given show the way forward in developing potentially useful DNA tests over the next few years.

The second major goal of molecular genetic research has been to use molecular genetics to understand the pathological processes that are involved in determining CAD, and this aspect of the research has already made considerable contributions over the last 10 years. The detection of association between a mutation in a particular enzyme and plasma levels of an 'intermediate phenotype' for CAD risk (e.g. plasma levels of lipids, clotting factors or homocysteine) identifies the enzyme (and the pathway that the enzyme is involved in) as a pharmacological target. Although this chapter focuses on gene–environment interactions that have been identified in the field of CAD, much of the general approach is relevant to the analysis of any complex multi-factorial disorder of late onset, such as diabetes or hypertension, and probably disorders such as Parkinson's disease, Alzheimer's disease or even cancer. It highlights how understanding of the mechanisms by which these mutations interact with environmental factors sheds light on the pathological processes of disease and may lead to novel therapeutic or interventional strategies to reduce an individual's risk of disease.

2. Risk factors for CAD

There are several physiological systems that are involved in maintaining cardiovascular health, where disease can be predicted to develop due to a failure to maintain homeostasis. The first might be the various systems through which plasma and intracellular lipid metabolism is regulated. These are likely targets due to the influence of metabolic profile on cardiovascular risk, and also because of the direct atherogenic effects of some lipid products such as oxidized low density lipoprotein (LDL). The second is the coagulation cascade, whose components may directly drive atherogenesis, but which also influence plaque growth and acute luminal occlusion in response to plaque ulceration or rupture. Third is the endothelial lining of the vessel wall with its role in preventing thrombosis, maintaining normal wall structure, and influencing vessel tone. One might also consider the normal homeostatic systems which regulate short-term blood pressure and intravascular volume (and hence cardiac output). However, the systems which regulate the responses to altered cardiovascular work are probably more important, as these lead to alterations in myocardial structure and contractile function, and changes particularly in left ventricular size and muscle composition in response to long term increase in load, for example as a result of hypertension. If taken to pathological extremes this can lead to left ventricular hypertrophy (LVH), and eventually to heart failure and death (Levy *et al.*, 1990). It can therefore be predicted that any environmental challenge which stresses any of these systems may be a 'risk-environment' for CAD.

As a result of epidemiological and clinical studies over the last 20 years many CAD 'intermediate phenotypes', have been identified, with high risk being associated with elevated plasma levels of lipids (Castelli *et al.*, 1986; Hokanson and Austin, 1996), fibrinogen (Meade *et al.*, 1986), and homocysteine (Ma *et al.*, 1996), and with low levels of high density lipoprotein (HDL)-cholesterol (Wilson *et al.*, 1996). In addition it is now recognized that the inflammatory system occupies a key role in the atherosclerotic process (Ross, 1999), which may explain why elevated levels of markers of the acute phase response system such as C-reactive protein (CRP; Ridker *et al.*, 2000a), as well as cytokines such as interleukin-6 (IL-6), are also elevated in CAD subjects (Ridker *et al.*, 2000b). Based on this understanding, it can similarly be predicted that an environmental challenge which results in change in any of these phenotypes will be a 'risk-environment for CAD'. Such challenges include increasing age, male gender, dietary intake of fats or vitamins, use of cigarettes and alcohol, presence of hypertension, diabetes and obesity, and as shown in *Table 1* these have all been implicated in potential gene–environment interactions of clinical relevance. Of these, smoking is particularly relevant, and will be examined further.

Smoking is known to roughly double life-time risk of CAD (Doll and Hill, 1966), and is thought to increase cardiovascular risk by several different mechanisms. The products of tobacco combustion directly damage vascular endothelium, leading to increased secretion of adhesion molecules which enhance binding of platelets and

Table 1. Examples of CAD risk traits which are affected by environment challenges, and reported examples of specific candidate gene–environment interaction in determining an individuals response

Plasma CAD risk trait	Environment stressor	Candidate genes	Reference
Cholesterol	Dietary fat	*APOE/CETP/LPL*	Ordovas and Schaefer (2000); Wallace *et al.* (2000)
Triglyceride	Obesity/lack of exercise/smoking	*LPL/APOC3*	Fisher *et al.* (1995); Waterworth *et al.* (2000)
HDL	Smoking/gender/ alcohol	*APOA1/CETP*	Sigurdsson *et al.* (1992), Fumeron *et al.* (1995); Gudnason *et al.* (1997)
Fibrinogen	Injury/infection/ surgery/smoking	*FIBB*	Humphries *et al.* (1997); Montgomery *et al.* (1996); Gardemann *et al.* (1997)
CRP/IL-6	Injury/infection/ surgery/smoking	*IL6*	Fishman *et al.* (1998); Brull *et al.* (2001)
Homocysteine	Low folate diet	*MTHFR*	Ma *et al.* (1996); Dekou *et al.* (2001)
LV mass	Hypertension/ exercise	*ACE*	Montgomery *et al.* (1997); Myerson *et al.* (2001)

monocytes to the vessel wall, thus promoting thrombosis and atherosclerosis (Allen *et al.*, 1988; Blann, 1992; Lowe, 1993). Smoking disturbs lipoprotein metabolism by increasing insulin resistance and lipid intolerance, and is implicated in the production of small dense LDL. By stimulating catecholamines, smoking upregulates hormone sensitive lipase, increasing circulating free fatty acid (FFA) levels (Eliasson *et al.*, 1997), thus causing atherogenic dyslipidaemia. In addition, smoking-induced lung damage may lead to an interleukin-6-mediated inflammatory response, causing hepatic up-regulation of fibrinogen expression (Dalmon *et al.*, 1993) and increased risk of thrombosis (Meade *et al.*, 1986). Smokers have lower levels of antioxidants such as ascorbate and tocopherol and thus smoking may favour the oxidation of LDL (Fickle *et al.*, 1996).

3. Homeostasis and use of stressing the genotype to identify functional variants

In the context of a cell (or even an organ or organism), genes can be considered simply to code for the synthesis of proteins, which allow the maintenance of intracellular homeostasis in the face of extracellular or environmental changes. At the basic level the cell requires oxygen, energy and various chemicals in order to reproduce and, in a Darwinian sense, to pass on its genes to the next generation. Naturally-occurring genetic variation means that some individuals are better able to maintain homeostasis than others in the face of the same environmental challenge. In the western culture people now experience many challenges to maintaining cardiovascular health, such as a high-fat diet, high levels of smoking and alcohol intake, and reduced physical exercise leading to obesity. Even in the face of these environmental 'insults', some individuals maintain cardiovascular health into old age, while others, with a different genetic make-up, fail to maintain homeostasis (e.g. plasma levels of cholesterol or fibrinogen within an optimal range), and thus develop atherosclerosis. Identifying the genes involved in maintaining cardiovascular homeostasis in the face of environmental challenge should thus lead to progress in understanding pathophysiology and etiology, as well as in genetic risk prediction, as mutations in such genes are likely to be strongly predisposing to or protecting from heart disease.

As an example of this, fibrinogen is a major acute phase reactant and levels rise rapidly after infection or injury. As elevated fibrinogen levels are an independent risk factor for CAD (Meade *et al.*, 1986), an individual who has a genetic predisposition to making particularly large responses, may be at greater thrombotic risk than an individual with a genotype predisposing to a modest response. For any individual, the measured level of fibrinogen in the blood is thus due to that individual's genetically-determined ability to maintain homeostasis in response to the environment that has been, or is currently being, experienced. If genetic variation can be identified that distinguishes the 'plastic' from the 'stable' genotype, such information, in conjunction with information about current level may add significantly to future risk prediction.

This can be examined by a 'genotype-stress' study, which has already been shown by us and other groups to be an excellent way of magnifying modest genotype effects on traits. A beta-fibrinogen gene promoter polymorphism (–455G>A) has been shown to be consistently associated with a modest effect on baseline plasma fibrinogen levels, with carriers of the –455GA allele (frequency roughly 0.20: 95% CI: 0.18–0.22) in European populations (reviewed in Humphries *et al.*, 1997) having on average 0.28 g/l higher fibrinogen levels than those with the genotype GG (weighted average in healthy men, see Humphries *et al.*, 1997). The molecular mechanism of the effect has not been fully elucidated, but is likely to be due to a greater rate of transcription of the beta-fibrinogen gene A-allele compared to the –455G allele, leading to higher levels of secretion of the mature protein from the liver (van t'Hooft *et al.*, 1999). The β-gene promoter has been well studied, and as well as elements for basal transcription, contains an interleukin 6 (IL-6) responsive element (Dalmon *et al.*, 1993). The –455G>A is in almost complete allelic association with a –148T>C change which is near the IL-6 element and may itself be the functional sequence change which is marked by the –455 site although other sequence changes in the promoter have been reported that may also be important (van t'Hooft *et al.*, 1999).

We examined the effects of acute intensive exercise on plasma fibrinogen levels and the relationship of these responses to the fibrinogen G-455A genotype (Montgomery *et al.*, 1996). One hunded and fifty six male British Army recruits were studied at the start of their 10 week basic training (which emphasizes physical fitness) and between 0.5 and 5 days after a major 2 day strenuous military exercise. Within 12 hours of the completion of the exercise, fibrinogen levels were 14.5% higher than those at baseline, suggesting that the acute response had already begun, and levels were significantly higher on days 1–3 after the exercise, suggesting an 'acute phase' response to strenuous exercise). The fibrinogen-raising effect associated with the A allele was seen at baseline and after 5 days, but in both cases the effect was small and non significant in this small sample. However, the degree of rise at 2–3 days after the exercise, was strongly related to the presence of the A allele. In the A allele carriers levels had risen by 37% compared to only 27% in GG subjects, while in men homozygous for the A allele (representing 4% of the population) fibrinogen levels had risen by more than 100% compared to their 'untrained' levels. Although the group was small, the differences were statistically significant ($p=0.01$), and the size of the effect was sufficient to be of biological significance (i.e. risk of thrombosis), if occurring in an individual with atherosclerosis compared to these fit young men where no such side effects were noted.

Similar differences in the fibrinogen response to the injury of coronary surgery have been reported (Cotton *et al.*, 2000; Gardemann *et al.*, 1997), with again the –455A allele carriers having a greater response than GG subjects. These finding support an association of the A allele with greater fibrinogen level 'responsiveness' after a cytokine inducing event, and suggest that the stress model may be a good analytical tool to analyze gene–environment interaction. Other stress situations that have been used in CAD genetic research include a fatty meal (Gerdes *et al.*, 1997) or a high fat diet (Wallace *et al.*, 2000), with some individuals having a modest response and others a very large response. These controlled stress-induced gene–environment interactions are likely also to be the key to understanding the 'causes' determining other multifactorial disorders.

4. How mutations identify rate-limiting steps in biochemical pathways

The model for how association studies allow rate-limiting steps in a biochemical pathway to be explored is shown in *Figure 1*. In this simplified pathway, a dietary component is absorbed from the gut and is metabolized, for example in the liver, by a series of enzymes coded for by genes A–C. The end product is secreted from the liver and can be measured in plasma, and high levels are noted in epidemiological studies of CAD to be associated with increased risk. From a therapeutic point of view it may be most efficient to target the rate-limiting step in the pathway for lowering drug therapy. Common variants in all of the genes have been identified using standard molecular screening techniques, and some of these may be directly of functional consequence while others could be in allelic association with functional variants that have yet to be found. An initial frequency comparison in subjects with high vs. low plasma levels of the trait or, better, an association study determining trait levels in population-based subjects with different genotypes at all loci, would reveal largest differences with the gene coding for the rate-limiting step – in this case gene B whose level in the cell is the lowest. Complete knock-outs of gene A or C would also have major effects on trait level, but would be rare (probably pediatric) metabolic defects and would not be likely to be common in the general population (because of negative selection pressure). As well as identifying gene B as a pharmacological target, it may also be the basis for a CAD predisposition genetic test, especially (and maybe only) in the presence of an environment containing high levels of the dietary component.

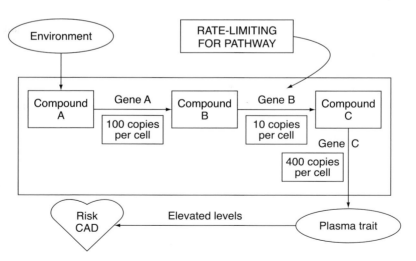

Figure 1. Model showing how genetic variation studies can help identify 'rate-limiting' steps in biochemical pathways. Variants in gene B that result in for example ±30% difference in levels are likely to effect plasma trait levels and therefore risk of CAD, while similar effects in genes A or C will not have significant effects on plasma trait level since they are not rate-limiting.

4.1 Examples of gene–environment interaction –1 the LPL Gene

As an example of gene–environment interaction an example of a common mutation in the *LPL* gene is particularly informative (see review by Fisher *et al.*, 1997). The plasma level of any CAD intermediate phenotypes such as plasma Tg will be due to the balance between the rate of production (in this case very low density lipoprotein (VLDL) secretion from the liver) and rate of removal. Lipoprotein lipase (LPL) plays a central role in lipid metabolism, hydrolyzing triglyceride (Tg)-rich particles in muscle, adipose tissue and macrophages and generating FFA and glycerol for energy utilization and storage (Olivecrona and Olivecrona, 1995), as well as its 'bridging' role as a ligand in lipoprotein–cell surface interactions and receptor-mediated uptake of lipoproteins (Beisiegel *et al.*, 1991; Mulder *et al.*, 1993). Any mutation that results in a partial deficiency of LPL would thus be predicted to result in a modest increase in plasma Tg levels, with the increase being proportional to the degree of deficiency caused by the mutation.

To date, two *common* mutations have been identified in the LPL gene (Fisher *et al.*, 1997), one that causes the substitution of the Aspartic acid residue at codon 9 for Asparagine (D9N) and the second that alters Asparagine at 291 to a serine (N291S). We have shown by *in vitro* mutagenesis and expression in COS cells that the first of these causes a modest 15–20% decrease in secreted LPL activity (Mailly *et al.*, 1995; Zhang *et al.*, 1996), probably because a larger proportion of the LPL-N9 is retained in the cells, and this has now been confirmed and extended in stably-transfected cell lines (Fisher *et al.*, 2001). It is known that LPL is only active as a 'head-to-tail' dimer (see refs in Fisher *et al.*, 1997), LPL-S291 constructs produce an LPL protein with significantly decreased dimer stability, which results in the levels of secreted LPL activity reduced by up to 50% (Zhang *et al.*, 1996). In support of these *in vitro* effects, in healthy individuals we have shown that carriers of either mutation have modestly higher Tg levels than non carriers, with carriers of the more severe LPL-291S mutation having higher Tg and lower HDL than carriers of the milder LPL-9N mutation (Fisher *et al.*, 1995; Mailly *et al.*, 1995), as well as a slower clearance of lipids from the blood after a fatty meal challenge (Gerdes *et al.*, 1997).

We have now examined the effect of these mutations on CAD risk (Talmud *et al.*, 2000). Over 2700 healthy men from the second Northwick Park Heart Study (NPHSII) have been followed for CAD events for more than 6 years (Miller *et al.*, 1996). The carrier frequencies of the S291 and N9 alleles were 3.9% (95% CI: 3.2–4.7%) and 2.6% (95% CI: 2.0–3.3%), respectively. For both variants, carriers and non carriers did not differ significantly with respect to age, current smoking or plasma cholesterol and apolipoproteins. However, as expected from previous studies, carriers of the S291 allele had mean baseline Tg that was 14.1% higher than that of non carriers ($p=0.014$). A similar trend was seen in N9 carriers who had on average 10.5% higher Tg level than non carriers, although this difference was not statistically significant ($p=0.15$), confirming the more modest effect of this mutation seen previously.

No evidence was found to suggest that the possession of S291 allele had any effect on the risk of CAD ($p=0.73$), either in non smokers or current smokers. By contrast, there was significant ($p=0.05$) evidence for an increased risk of CAD in

LPL-N9 carriers (Hazard Ratio (HR) adjusted for Tg 2.33 (95% CI: 1.08–5.03)). However, as shown in *Figure 2*, there was very strong evidence for an interaction between smoking and LPL genotype in determining risk (p <0.001), that could not be explained by small differences in BMI, cholesterol, or any other measured risk factor *including* Tg levels. Carrying the N9 allele appears to modify the effect of smoking such that carriers who smoke were at significantly higher risk of having a CAD event compared to non carriers who smoke. For the men who were non carriers of N9, smoking increased the hazard of an CAD event by 1.68 (95% CI: 1.1–2.4), similar to that reported earlier (Doll and Hill, 1966). In the group of N9 carriers who were non smokers there were no events. Consequently, the estimate of risk in this group was zero and any attempt to quantify the effect of smoking in carriers of N9 resulted in an infinite estimate. The joint effect of smoking and carrying N9 gave a HR of 10.4 (95%CI: 4.7–22.8) compared to non carriers who did not smoke. Excluding those subjects with sudden death had minimal effect on the overall results or the parameter estimates. To test the strength of the effect, calculations were made to examine if increasing the number of N9 non smokers by 1 (from N=0–1) and reducing the number of N9 smokers by 1 (from N=7–6) would still give a significant likelihood ratio. In this theoretical analysis, the test for interaction between smoking status and D9N genotype gave a statistically significant p-value (p=0.01).

The results of this study suggest that one or more of the injurious effects of smoking are particularly poorly tolerated by subjects with the LPL-N9 variant. Studies *in vivo* have shed light on the molecular basis of this risk (Fisher *et al.*, 2001). Using stably-transfected cell lines we have shown that compared to the wild-type construct, LPL-N9 cells bind and internalize 2–4 fold more LDL and ox-LDL, and that this difference is abolished by prior treatment of the cells with heparin or heparinase. As the enzymatic activity of the two LPLs is similar these

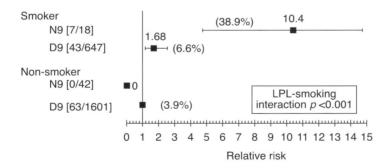

Figure 2. Relative risk of CAD events by smoking and LPL-D9N status. Graph of the estimated HR from the Cox's proportional hazard model, stratified by smoking and D9N genotype. Adjustment has been made for baseline Tg, age, clinic, BMI, Systolic blood pressure, cholesterol, and fibrinogen level (number of events/subjects in each group). The percentage of each group having an event is shown in brackets. Among non carriers of N9, 6.6% (95% CI: 4.9–8.8%) of smokers had an CAD event compared to 3.9% (95% CI: 3.0–5.0%) of non smokers, while in N9 carriers 38.9% (95% CI: 17.3–64.3%) of smokers compared to 0.0% (95% CI: 0.0–8.4%) of non smokers had had an event. (Data from Talmud *et al.*, 2000).

data strongly suggest that the risk-mechanism of the mutation is via its bridging function, and as shown in *Figure 3*, we can speculate on mechanisms by which smoking interacts with LPL-N9 to cause the apparent high risk of CAD. Under normal conditions, LPL is present on the endothelial luminal wall attached to heparan sulfate proteoglycans (HSPG) where Tg hydrolysis takes place. The local production of LPL is now well recognized as an important factor in the developing atherosclerotic lesion. Damage to the endothelium will promote the recruitment of monocyte/macrophages and smooth muscle cells into the lesion area, both of which synthesize LPL (Saxena and Goldberg, 1994). As well as its role in lipoprotein hydrolysis, LPL acts as a bridge between HSPG and lipoproteins and initiates receptor-mediated catabolism of lipoproteins into cells (Beisiegel *et al.*, 1991). The bridging function is also involved in endothelial cell–monocyte interactions, due to the ability of the LPL dimer to promote proteoglycan/proteoglycan interaction (Obunike *et al.*, 1997). In addition, LPL has recently been shown to have a high affinity for oxidized LDL and to promote scavenger receptor uptake of oxidized LDL (Hendriks *et al.*, 1996). As discussed above, *in vitro* studies show LPL-N9 is poorly secreted into the medium, but these studies have not distinguished between secretion deficiency and retention on the cell surface. Based on these observations, we propose that the asparagine for aspartic acid amino acid substitution at residue 9 will increase the 'stickiness' of LPL-N9. This effect, combined with the increased synthesis of LPL-N9 at the site of smoking-related endothelial damage, would lead to greater accumulation, retention or modification of lipoproteins on the endothelium and subendothelium. The enhanced recruitment of monocytes and/or oxidized LDL to the developing lesion, thus stimulates foam cell formation leading to increased atherogenesis. Verification of this hypothesis requires additional *in vitro* and *in vivo* studies.

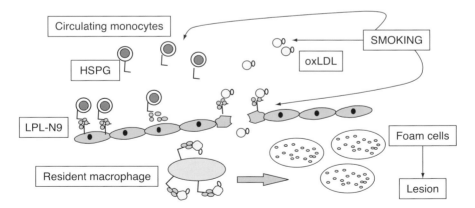

Figure 3. Mechanisms for LPL-D9N: smoking interaction on risk of CAD. LPL-N9 has greater bridging function, which enhances uptake of oxLDL as well as monocyte adhesion to arterial wall. Smoking leads to altered diet and disturbances in lipid metabolism leading to greater plasma levels of small dense LDL which is prone to oxidation. It also increases inflammation leading to more monocytes, endothelial wall damage. Taken together this leads to greater uptake of oxLDL in higher number of macrophages in lesions thus promoting atherosclerosis.

The high risk estimate we have observed in N9 smokers in NPHSII men requires confirmation by further prospective studies; because of the size of this group, the estimate lacks precision. However, as there are a predicted 550 000 men over the age of 16 in the UK who are carriers for this LPL-N9 variant and roughly 25% of these are likely to be smokers, this represents an estimated 150 000 men who, if the data are confirmed, may be at considerable risk of CAD, and who would experience considerable benefit from cessation of smoking. The findings suggest that studies of ways and settings to implement the use of genetic risk information and targeted smoking cessation strategies would be useful.

4.2 Gene–environment interaction: 2 Exercise, LVH and the angiotensin converting enzyme (ACE) gene

As shown in *Figure 4*, The aspartyl protease renin is released from the cells of the juxta-glomerular apparatus under conditions of salt or volume loss or sympathetic activation. Renin cleaves the alpha-2 globulin angiotensinogen (synthesized in the liver) to generate the non pressor decapeptide angiotensin I. The octapeptide angiotensin II is then derived primarily by the action of the dipeptidyl-carboxypeptidase, ACE, which is responsible for the hydrolytic cleavage of dipeptides from the carboxyl terminus his-leu dipeptide. ACE also catalyses inactivation of the nonapeptide bradykinin by two sequential dipeptide hydrolytic steps and in this context, is also known as kininase II. ACE, a zinc

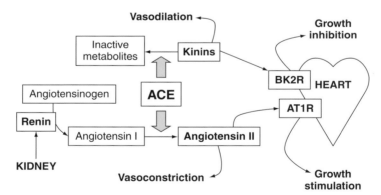

Figure 4. Biochemical pathways by with ACE affects blood pressure and LVH – the RA/kallikrein-kinin system. Because it is the rate-limiting step in the pathway, higher plasma ACE, as is found in individuals lacking the ALU insertion (i.e. the DD genotype), leads to a higher conversion of vasoinactive precursor to the vasoconstrictor product (Angiotensin II), and higher breakdown of the vaso-dilating kinins to inactive metabolites. Both of these contribute to a rapid fall of blood pressure by action on vascular motor tone. In addition, angiotensin II stimulates angiotensinogen type 1 receptors (AT1R) on cardiac tissue which leads to growth stimulation, while the lower levels of kinins leads to less stimulation of the the bradykinin 2 receptor (BK2R) and less growth inhibition. Overall this leads to greater cardiac muscle growth in the presence of high ACE (i.e. the DD genotype) compared to those with low ACE (i.e. II genotype).

metallo-protease, is released from the cell membrane by a carboxypeptidase that cleaves the protein between Arg-663 and Ser-664 to generate circulating ACE (Zisman, 1998). Large interindividual differences in plasma ACE levels exist but are similar within families, suggesting a strong genetic influence. The human *ACE* gene is found on chromosome 17 and contains a restriction fragment length polymorphism consisting of the presence (insertion, I) or absence (deletion, D) of a 287 base pair 'ALU' repeat sequence in intron 16 (Rigat *et al.*, 1990). The association of the I allele with lower ACE activity in both serum and tissues (Danser *et al.*, 1995) has ramifications throughout the renin-angiotensin system (RAS) and kallikrein–kinin system and has stimulated much fascinating work in regard to various pathological and physiological states.

In 1992, the I/D polymorphism was reported to be associated with risk of MI (Cambien *et al.*, 1992), and this effect, though of a more modest effect than originally reported has been confirmed in recent large studies (Keavney *et al.*, 2000) and a meta-analysis (Samani, 1996). Since then, we have carried out several studies to examine the relationship between this polymorphism and cardiovascular health. These have been based on the data that the D allele is associated with higher ACE levels than the I allele in plasma and in tissues, and if RAS regulate LV growth, individuals of DD genotype might show a greater hypertrophic response than those of II genotype. In 1997 we reported results supporting this hypothesis in a study of changes in left ventricular (LV) mass upon exercise training (Montgomery *et al.*,1997). Echocardiographically-determined LV dimensions and mass, and frequency of LVH, were compared at the start and end of a 10 week physical training period in 156 male military recruits. Overall, LV mass increased by 18% ($p < 0.0001$), but response magnitude was strongly associated with *ACE* genotype: mean LV mass altered by $+2.0$ g, $+38.5$ g and $+42.3$ g in II, ID and DD, respectively ($p < 0.0001$). The prevalence of LVH rose significantly only amongst those of DD genotype (from 6/24 before training to 11/24 afterwards: $p < 0.01$). These results point strongly to the mechanism of the *ACE* gene effect seen in previous studies, but indicate that risk of MI may be seen only in those when LVH may have been induced (e.g. by hypertension).

Confirmation of this has been obtained in a second independent study of army recruits (Myerson *et al.*, 2001), which used the strategy of screening over 1200 recruits and selecting only subjects homozygous for the I or D allele for the expensive and time-consuming (but considerably more accurate) method of LV mass determination by magnetic resonance imaging (MRI). As the LVH responses may be mediated through *either* increased activity of the cellular growth factor angiotensin II on the Angiotensin Type 1 receptor (AT1), *or* increased degradation of growth-inhibiting kinins (see *Figure 4*), the study was also designed to clarify the role of the AT1 receptor in this association. The recruits were randomized to receive placebo or a well-known and well tolerated antihypertensive drug losartan (25 mg/day), throughout their 10 week physical training program. This is a subhypotensive dose, but will inhibit tissue AT1 receptors, and thus if LV growth occurs in the presence of the drug it would implicate the other pathways in the *ACE* gene effect. LV mass was not different by genotype at baseline in either group, confirming the previous results (Montgomery *et al.*,

1997) that in healthy non trained subjects ACE genotype is not 'rate-limiting for cardiac size. LV mass increased with training by 8.4 g overall ($p<0.0001$), but with a highly significant difference in men with different ACE genotypes. As shown in *Figure 5* the increase in the placebo limb was 12.1 g vs. 4.8 g for DD vs. II genotype ($p=0.022$). Interestingly, LV growth was similar in the losartan arm, being 11.0 g vs. 3.7 g for DD vs. II genotypes ($p=0.034$). When indexed to lean body mass, LV growth in II subjects was abolished whilst remaining in DD subjects (–0.022 vs. +0.131 g/kg, respectively; $p=0.0009$).

Thus, the ACE genotype-dependence of exercise-induced LVH seen in the earlier study was strongly confirmed. Additionally, LV growth in DD (unlike II) subjects was in excess of the increase in lean body mass. These effects were not influenced by AT1-receptor antagonism using Losartan, suggesting that the 2.4 fold greater LV growth in DD men may be due either to angiotensin II effects on other receptors (e.g. AT4), or lower degradation of growth-inhibitory kinins. It is of relevance that with regard both to LV mass and skeletal muscle efficiency, there was no significant effect associated with *ACE* genotype in 'unstressed' subjects at recruitment but only after the 10 week 'stress' of the training period. This is clearly an example of gene–environment interaction, suggesting *ACE* genotype (and ACE levels) will not be a predictor of LVH in unstressed subjects.

These data suggest that the I allele may be associated with some aspects of endurance performance or muscle efficiency, as the army recruits with the II genotype are able to generate enough cardiac output to complete the training period without a great increase in LV mass, while DD men can only generate this amount of output by increasing heart size. This concept of greater efficiency in II subjects has been supported in several recent studies. Genotype distribution and allele frequency differed significantly between elite climbers with a history of ascents beyond 7000 m (without the use of supplemental inspired oxygen) and controls, with a relative excess of II genotype and deficiency of DD genotype

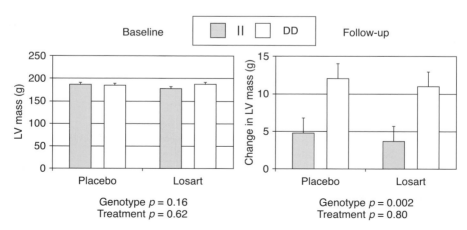

Figure 5. Baseline and change in LV mass by ACE genotype and treatment group in Army Recruits. Data adjusted for height, age, systolic blood pressure (and for change also for pretraining LV mass). Number of subjects in each group were DD placebo 41, Losartan 38, II Placebo 34, Losartan 28 (data from Myerson *et al.*, 2001).

(Montgomery *et al.*, 1998). Amongst the 15 climbers who had ascended beyond 8000 m without oxygen, none was of DD genotype, and ranked by the number of such ascents the top performer was of II genotype (five ascents, compared to a mean of 2.4±0.3). This is a small study, but nevertheless, it suggests a performance advantage conferred by the I allele amongst elite endurance athletes exercising at very high altitude where calorie intake is low, calorie expenditure high, and oxygen supply low.

One might anticipate a similar genetic advantage at sea level amongst endurance athletes. Analysis of 91 British Olympic-standard runners (Myerson *et al.*, 1999) revealed an excess of both the I allele and II genotype compared to controls due to a significant linear trend of increasing I allele frequency with distance run. Finally, duration of loaded repetitive biceps flexion also increased with training 11-fold more amongst those of II genotype compared to those homozygous for the D allele (Montgomery *et al.*, 1999). In the study of the same group of British Army recruits the maximum duration (in seconds) for which they could perform repetitive elbow flexion whilst holding a 15 kg barbell was assessed both before and after their 10 week basic physical training programme. Pre-training performance was independent of genotype, but surprisingly while duration of exercise improved significantly for those of both II and ID genotype (79.4±25.2 and 24.7±8.8 sec, mean±SEM, respectively) this did not occur for the DD subjects (7.1±14.9 sec, mean±SEM). Improvement was thus 11-fold greater for II subjects compared to the DD group.

The precise mechanism by which the ACE genotype may be having its effect is under investigation and may be through both or either the pathway although which ACE influences circulatory homeostasis through the degradation of vasodilator kinins or the formation of vasopressor angiotensin II (Ang II). However, local tissue-based RAS also exists in human myocardium (Dzau, 1988) and it is likely that ACE levels expressed in myocardium would be high in DD and low in II subjects. In support of both pathways, increased ACE gene expression and ACE activity in the myocardium (Ohmichi *et al.*, 1997) significantly increases the rate of local conversion of Ang I to Ang II, and the ACE DD genotype itself tends to show increased conversion of infused Ang I to Ang II in humans (Ohmichi *et al.*, 1997). This study also revealed a significant inverse relationship between the half-life of bradykinin and both serum ACE activity and the conversion of Ang I to Ang II, confirming that the ACE genotype influences bradykinin degradation (Brown *et al.*, 1998). The role of bradykinin in tissue metabolism and vasodilatation seems increasingly recognized as being endothelium-dependent and if one looks at studies on isolated perfused rat hearts (that still have an intact endothelium) there is strong evidence of an improvement in myocardial metabolic efficiency mediated by bradykinin (e.g. Linz *et al.*, 1990).

Despite this evidence it remains possible that the *ACE* gene mediates its observed effects on endurance independent of the RAS or indeed that the ACE *gene* is not directly responsible but other gene or genes in linkage disequilibrium with the ACE locus. As there are many widely used and well tolerated drugs which alter ACE activity, this raises the exciting possibility of the use of these drugs in order to maintain human health in hitherto unexplored ways.

5. Analytical problems for gene–environment interaction studies

A key factor in the identification and study of gene–environment interaction is that an individual carrying such a mutation will develop the phenotype, only if and when they enter the high risk environment. Thus, the mutation will only 'cause' high plasma cholesterol or high fibrinogen in the presence of this environmental challenge. This classical 'lack of penetrance' of a mutation will cause analytical problems and mis-phenotyping which will be particularly problematic with some sampling analytical designs. The problem that this 'context-dependency' of a mutation (i.e. gene × environment effect) creates, from an analytical point of view, has largely been overlooked in the field. The problem occurs when a second study fails to confirm a reported association between a candidate gene polymorphism and levels of intermediate traits or with risk of CAD. Although the failure of a second study to reproduce an association found may cast doubt on the validity of the first report, it may also reflect the presence of a potentially interesting gene–environment interaction that requires further exploration. If, for example, smoking is the key environmental stress that amplifies or diminishes the genotype effect, then the different proportion of smokers in the samples being studied (and possibly the number of cigarettes being smoked) will have a major effect on the power of the sample to detect the association. However, simple power calculations can be carried out knowing the prevalence of the particular environmental factor in the sample under study, and the size of the sample to be studied can be increased accordingly. From a practical point of view this means that large samples are required for interactions to be statistically significant and that the characteristics of the sample being studied must be carefully recorded and compared between samples. As a minimum, it would be reasonable to propose a statistically significant finding should be reproduced in three independent samples (preferably from different laboratories).

One area where environmental manipulation may be available to confirm gene–environment interaction observed in an association study would be to remove the environment (or the individual from the environment) and observe the decay of the induced phenotype effect over a period of time. Although this could not easily be done for the phenotype of atherosclerosis, requiring invasive serial coronary angiography for example, it would be possible for some environments such as smoking or dietary effects on plasma traits such as fibrinogen or cholesterol, or on the consequence of hypertension on LVH after anti-hypertensive therapy. The Tg-raising effect associated with being a carrier of the LPL-N9 mutations, is exacerbated by obesity (Fisher *et al.*, 1997). This predicts that weight loss in carriers should be followed by a greater fall in Tg levels than following a similar weight loss in non carriers, and experiments to confirm this prediction could be an extremely powerful argument in support of causality. This is an area of research which has yet to be examined in detail.

6. Conclusions

It should, in the near future, be possible to include genetic markers in risk algorithms to advise healthy subjects how best to avoid CAD (Khoury and Wagener,

1995). Many associations have proved to be robust, and for example the fibrinogen-raising effect associated with the A-455 allele and the Tg-raising effect associated with the LPL-N9 variant has now been demonstrated beyond any reasonable doubt, although the precise molecular mechanism of both of these effects remains to be elucidated. If gene–environment interactions can also be confirmed, both by repeat studies but also by 'stress-experiments' it should be possible to agree on the necessary criteria for acceptability for a candidate gene common mutation to be deemed proven and useful for inclusion into a CAD-risk algorithms, such has been prepared by Framingham (Wilson *et al.*, 1998) using essentially non genetic factors. This will give us the ability to be able to give genotype-specific life-style advice, or to tailor clinical and therapeutic decisions to an individual's genotype. The identification of the relationship between genetic variants in *LPL* and CAD and *ACE* and LVH points to the rate-limiting and thus key role in the pathological processes of these proteins, with LPL being expressed in the atherosclerotic plaque in foam-cell macrophages or ACE in cardiac tissue. It also reveals potential novel therapeutic possibilities, for example to block LPL bridging functions on these cells, or the novel use of available drugs such as ACE inhibitors, to prevent disease in a molecularly rational manner.

Acknowledgements

The Authors are supported by grants from the British Heart Foundation (RG95007 and SP198003)

References

Allen, D.R., Browse, N.L., Rutt, D.L., Butler, L. and Fletcher, C. (1988) The effect of cigarette smoke, nicotine, and carbon monoxide on the permeability of the arterial wall. *J. Vasc. Surg.* 7: 139–152.

Beisiegel, U., Weber, W. and Bengtsson-Olivecrona, G. (1991) Lipoprotein lipase enhances the binding of chylomicrons to low density lipoprotein receptor-related protein. *Proc.Natl Acad. Sci. USA.* 88: 8342–8346.

Blann, A.D. (1992) The acute influence of smoking on the endothelium. *Atherosclerosis* 96: 249–250.

Brown, N.J., Blais, C., Gandhi, S.K. and Adam, A. (1998) ACE insertion/deletion genotype affects bradykinin metabolism. *J. Cardiovasc. Pharmacol.* 32: 373–377.

Brull, D., Montgomery, H.E., Sanders, J., Dhamrait, S., Luong, L-A., Rumley, A., Lowe, G.D.O. and Humphries, S.E. (2001) Interleukin-6 gene −174G>C and −152G>C polymorphisms are strong predictors of plasma IL6 levels after cardiopulmonary bypass. *Atheroscler. Thromb. Vasc. Biol.* (In press).

Cambien, F., Poirier, O. and Lecerf, L. *et al.* (1992) Deletion polymorphism in the gene for angiotensin-converting enzyme is a potent risk factor for myocardial infarction. *Nature* 359: 641.

Castelli, S.P., Garrison, R.J., Wilson, P.W., Abbott, R.D., Kalousdian, S. and Kannel, W.B. (1986) Incidence of coronary heart disease and lipoprotein cholesterol levels. The Framingham Study. *JAMA* 256: 2835–2838.

Cotton, J.M., Webb, K.E., Mathur, A., Martin, J.F. and Humphries, S.E. (2000) Impact of the −455G>A promoter polymorphism in the B fibrinogen gene on stimulated fibrinogen production following bypass surgery. *Thromb. Haemost.* 84: 926–927.

Dalmon, J., Laurent, M. and Courtois, G. (1993) The human β fibrinogen promoter contains a hepatocyte nuclear factor 1-dependent interleukin-6 responsive element. *Mol.CellBiol.* 13: 1183–1193.

Danser, A.H., Schalekamp, M.A. and Bax, W.A. *et al.* (1995) Angiotensin-converting enzyme in the human heart. Effect of the deletion/insertion polymorphism. *Circulation* **92**: 1387–1388.

Dekou, V., Whincup, P., Papacosta, O., Lennon, L., Ebrahim, S., Humphries, S.E. and Gudnason, V. (2001) The effect of C677T and C1298A polymorphisms in methylenetetrahydrofolate reductase on plasma homocysteine levels in elderly men and women from the British Regional Heart Study. *Atherosclerosis* (in press).

Doll, R. and Hill, A.B. (1966) Mortality of British doctors in relation to smoking; observation on coronary thrombosis. *Natl. Cancer. Inst. Monogr.* **99**: 205–268.

Dzau, V.J. (1988) Circulating vs local renin-angiotensin system in cardiovascular homeostasis. *Circulation* **77(1)**: 4–13.

Eliasson, B., Mero, N., Taskinen, M.R. and Smith, U. (1997) The insulin resistance syndrome and postprandial lipid intolerance in smokers. *Atherosclerosis* **129**: 79–88.

Fickle, H., Van Antwerpen, V.L. and Richards, G.A. *et al.* (1996) Increased levels of autoantibodies to cardiolipin and oxidised low density lipoprotein are inversely associated with plasma vitamin C status in cigarette smokers. *Atherosclerosis* **124**: 75–81.

Fisher, R.M., Mailly, F. and Peacock, R.E. *et al.* (1995) Interaction of the lipoprotein lipase asparagine 291Öserine mutation with body mass index determines elevated plasma triacylglycerol concentrations: a study in hyperlipidemic subjects, myocardial infarction survivors, and healthy adults. *J. Lipid Res.* **36**: 2104–2112.

Fisher, R.M., Humphries, S.E. and Talmud, J. (1997) Common variation in the lipoprotein lipase gene: effects on plasma lipids and risk of atherosclerosis. *Atherosclerosis* **13**: 145–159.

Fisher, R.M., Benhizia, F. and Schreiber, R. *et al.* (2001) Lipoprotein lipase N9 stable cell lines show enhanced binding and internalisation of LDL: insights into increased risk of coronary artery disease in N9 carriers. *J. Biol. Chem.* (in press).

Fishman, D., Faulds, G., Jeffery, R., Mohamed-Ali, V., Yudkin, J.S., Humphries, S. and Woo, P. (1998). The effect of novel polymorphisms in the interleukin-6 (IL-6) gene on IL-6 transcription and plasma IL-6 levels, and an association with systemic onset juvenile chronic arthritis. *J. Clin. Invest.* **102**: 1369–1376.

Fumeron, F., Betoulle, D. and Luc, G. *et al.* (1995) Alcohol intake modulates the effect of a polymorphism of the cholesteryl ester transfer gene on plasma high density lipoprotein and the risk of myocardial infarction. *J.Clin. Invest.* **96**: 1664–1671.

Gardemann, A., Schwartz, O. and Haberbosch, W. *et al.* (1997) Positive association of the β fibrinogen H1/H2 gene variation to basal fibrinogen levels and to the increase in fibrinogen concentration during acute phase reaction but not to coronary artery disease and myocardial infarction. *Thromb. Haemost.* **77**: 1120–1126.

Gerdes, C., Fisher, R.M., Nicaud, V., Boer, J., Humphries, S.E., Talmud, P.J. and Faergeman, O. *On behalf of the EARS Group.* (1997) Lipoprotein lipase variants D9N and N291S are associated with increased plasma triglyceride and lower high-density lipoprotein cholesterol concentrations: studies in the fasting and post-prandial states; the European Atherosclerosis Research Studies. *Circulation* **96**: 733–740.

Gudnason, V., Thormar, K. and Humphries, S.E. (1997) Interaction of the cholesterol ester transfer protein 1405V polymorphism with alcohol consumption in smoking and non-smoking healthy men, and the effect on plasma HDL cholesterol and apoAI concentration. *Clin. Genet.* **51**: 15–21.

Hokanson, J.E. and Austin, M.A. (1996) Plasma triglyceride level is a risk factor for cardiovascular disease independent of high-density lipoprotein cholesterol level: a meta-analysis of population-based prospective studies. *J. Cardiovasc. Risk* **3**: 213–219.

Hendriks, W.L., van der Boom, H. and van Vark, L.C. *et al.* (1996) Lipoprotein lipase stimulates the binding and uptake of moderately oxidized low-density lipoprotein by J774 macrophages. *Biochem. J.* **314**: 563–568.

Humphries, S.E., Thomas, A., Montgomery, H.E., Green, F., Winder, A. and Miller, G. (1997) Gene-environment interaction in the determination of plasma levels of fibrinongen. *Fibrinolysis Proteolysis* **11**: 3–7.

Keavney, B., McKenzie, C. and Delépine, M. *et al.* (2000) Large-scale test of hypothesised associations between the angiotensin-converting-enzyme insertion/deletion polymorphism and myocardial infarction in about 5000 cases and 6000 controls. *Lancet* **355**: 434–442.

Khoury, M.J. and Wagener, D.K. (1995) Epidemiological evaluation of the use of genetics to improve the predictive value of disease risk factors. *Am. J. Hum. Genet.* **56**: 835–844.

Levy, D., Garrison, R., Savage, D., Kannel, W. and Castelli, W. (1990) Prognostic implications of echocardiographically determined left ventricular mass in the Framingham Heart Study. *N. Engl. J. Med.* **322**: 1561–1566.

Linz, W., Martorana, P.A. and Scholkens, B.A. (1990) Local inhibition of bradykinin degradation in ischemic hearts. *J. Cardiovasc. Pharmacol.* **15**: S99–S109.

Lowe, G.D. (1993) Blood viscosity and cardiovascular risk. *Curr. Opin. Lipidol.* **4**: 283–287.

Ma, J., Stampfer, M.J. and Hennekens, C.H. *et al.* (1996) Methylenetetrahydrofolate reductase polymorphism, plasma folate, homocysteine, and risk of myocardial infarction in US physicians. *Circulation* **94**: 2410–2416.

Mailly, F., Tugrul, Y. and Reymer, P.W. *et al.* (1995) A common variant in the gene for lipoprotein lipase (Asp9ŌAsn). Functional implications and prevalence in normal and hyperlipidemic subjects. *Arterioscler. Thromb. Vasc. Biol.* **15**: 468–478.

Meade, T.W., Mellows, S. and Brozovic, M. *et al.* (1986) Haemostatic function and ischaemic heart disease: principal results of the Northwick Park Heart Study. *Lancet* **ii**:

Miller, G.J., Bauer, K.A. and Barzegar, S. *et al.* (1996) Increased activation of the haemostaic system in men at high risk of fatal coronary heart disease. *Thromb. Haemost.* **75**: 767–771.

Montgomery, H.E., Clarkson, P. and Nwose, O.M. *et al.* (1996) The acute rise in serum fibrinogen concentration with exercise is influenced by the G-453-A polymorphism of the beta-fibrinogen gene. *Arterioscler. Thromb. Vasc. Biol.* **16**: 386–391.

Montgomery, H., Clarkson, P. and Dollery, C.M. *et al.* (1997) Association of angiotensin-converting enzyme gene I/D polymorphism with change in left ventricular mass in response to physical training. *Circulation* **96**: 741–747.

Montgomery, H., Barnard, M.L., Bell, J.J.D. *et al.* (1998) Human gene for physical performance. *Nature* **393**: 221–222.

Montgomery, H., Clarkson, P. and Barnard, M. *et al.* (1999) The ACE gene insertion/deletion polymorphism and the response to physical training. *Lancet* **353**: 541–545.

Mulder, M., Lombardi, P. and Jansen, H. *et al.* (1993) Low density lipoprotein receptor internalizes low density and very low density lipoproteins that are bound to heparan sulfate proteoglycans via lipoprotein lipase. *J. Biol. Chem.* **268**: 9369–9375.

Myerson, S., Hemingway, H., Budget, R., Martin, J., Humphries, S. and Montgomery, H. (1999) Human angiotensin I-converting enzyme gene and endurance performance. *J. Appl. Physiol.* **87**: 1313–1316.

Myerson, S.G., Montgomery, H.E., Whittingham, M., Jubb, M., World, M.J., Humphries,S.E. and Pennell, D.J. (2001) Left ventricular hypertrophy with exercise and the angiotensin converting enzyme gene I/D polymorphism: a randomised controlled trial with losartan. *Circulation* **103**: 226–230.

Obunike, J.C., Paka, S. and Pillarisetti, S. *et al.* (1997) Lipoprotein lipase can function as a monocyte adhesion protein. *Arterioscler. Thromb. Vasc. Biol.* **17**: 1414–1420.

Ohmichi, N., Iwai, N. and Kinoshita, M. (1997) Expression of angiotensin converting enzyme and chymase in human atria. *J. Hypertens.* **15**: 935–943.

Olivecrona, G. and Olivecrona, T. (1995) Triglyceride lipases and atherosclerosis. *Curr. Opin. Lipidol.* **6**: 291–305.

Ordovas, J.M. and Schaefer, E.J. (2000) Genetic determinants of plasma lipid response to dietary intervention: the role of the APOA1/C3/A4 gene cluster and the APOE gene. *Br. J. Nutr.* **83**: 127–136.

Ridker, P.M., Hennekens, C.H., Buring, J.E., Sc, D. and Rifai, N. (2000a) C-reactive protein and other markers of inflammation in the prediction of cardiovascular disease in women. *New Eng. J. Med.* **342**: 836–843.

Ridker, P.M., Rifai, N., Stampfer, M.J. and Hennekens, C.H. (2000b) Plasma concentration of interleukin-6 and the risk of future myocardial infarction among apparently healthy men. *Circulation* **101**: 1767–1772.

Rigat, B., Hubert, C., Alhenc-Gelas, F., Cambien, F., Corvol, P. and Soubrier, F. (1990) An insertion/deletion polymorphism in the angiotensin I-converting enzyme gene accounting for half the variance of serum enzyme levels. *J. Clin. Invest.* **86**: 1343–1346.

Ross, R. (1999) Atherosclerosis – an inflammatory disease. *N. Engl. J. Med.* **340**: 115–126.

Saxena, U. and Goldberg, I.J. (1994) Endothelial cells and atherosclerosis: lipoprotein metabolism, matrix interactions, and monocyte recruitment. *Curr. Opin. Lipidol.* **5**: 316–322.

Samani, N.J., Thompson, J.R., O'Toole, L., Channer, K. and Woods, K.L. (1996) A meta-analysis of the association of the deletion allele of the angiotensin-converting enzyme gene with myocardial infarction. *Circulation* **94**: 708–712.

Sigurdsson, G. Jr., Gudnason, V., Sigurdsson, G. and Humphries, S.E. (1992) Interaction between a polymorphisms of the apo A-I promoter region and smoking determines plasma levels of HDL and apo A-I. *Atheroscler. Thromb. Vasc. Biol.* **12**: 1017–1022.

Talmud, P., Bujac, S.R., Hall, S., Miller, G.J. and Humphries, S.E. (2000) Substitution of asparagine for aspartic acid at residue 9 (D9N) of lipoprotein lipase markedly augments risk of CAD in male smokers. *Atherosclerosis* **149**: 75–81.

van't Hooft, F.M., von Bahr, S.J., Silveira, A., Iliadou, A., Eriksson, P. and Hamsten, A. (1999) Two common, functional polymorphisms in the promoter region of the beta-fibrinogen gene contribute to regulation of plasma fibrinogen concentration. *Arterioscler. Thromb. Vasc. Biol.* **19**: 3063–3070.

Wallace, A.J., Mann, J.I., Sutherland, W.H.F., Williams, S., Chisholm, A., Skeaff, M., Gudnason, V., Talmud, P.J. and Humphries, S.E. (2000) Variants in the cholesterol ester transfer protein and lipoprotein lipase genes are predictors of plasma cholesterol response to dietary change. *Atherosclerosis* **15**: 327–336.

Waterworth, D.M., Talmud, P.J., Bujac, S.R., Fisher, R.M., Miller, G.J. and Humphries, S.E. (2000) The contribution of apoCIII variants to the determination of triglyceride levels and interaction with smoking in middle-aged men. *Arterioscler. Thromb. Vasc. Biol.* **20**: 2663–2669.

Wilson, P.W., Abbott, R.D. and Castelli, W.P. *et al.* (1996) High density lipoprotein cholesterol and mortality. *Arteriosclerosis* **276**: 544–548.

Wilson, P.W., D'Agostino, R.B., Levy, D., Belanger, A.M., Silbershatz, H. and Kannel, W.B. (1998) Prediction of coronary heart disease using risk factor categories. *Circulation* **97**: 1837–1847.

Zhang, H., Henderson, H., Gagne, S.E., Clee, S.M., Miao, L., Liu, G. and Hayden, M.R. (1996) Common sequence variance of lipoprotein lipase: standardized studies of in vitro expression and catalytic function. *Biochim. Biophys. Acta.* **1302**: 159–166.

Zisman, L.S. (1998). Inhibiting tissue Angiotensin-converting enzyme. A pound of flesh without the blood? *Circulation* **98**: 2788–2790.

5

Pharmacogenomics

Ann K. Daly

1. Introduction

Pharmacogenomics can be defined as the study of the genetic factors that determine drug efficacy and toxicity and is closely allied to the subject of pharmacogenetics, which up to the present has been mainly concerned with the study of the genetics of drug metabolism, but also encompasses genetic factors affecting drug targets such as receptors. Using DNA sequence data, it is increasingly possible to consider the consequences of inter-individual variation in sequences of genes encoding drug targets as well as drug metabolizing enzymes and use this information both to design new drugs and to individualize drug therapy with existing drugs. However, for this approach to be successful, it is necessary to understand the exact genetic defects that give rise to specific diseases and also to understand the basis of inter-individual variation in both pharmacokinetics (the way the body deals with a drug) and pharmacodynamics (the effect the drug has on the body) for specific drugs. Though the application of pharmacogenomics to clinical treatment is still at a very early stage, there are already some examples of the use of knowledge of a patient's genotype for specific genes encoding proteins involved in either drug disposition, or directly in disease pathogenesis to determine whether use of a particular drug is indicated, or the most appropriate drug dose to be administered (see sections 2 and 3). Many pharmacogenetic polymorphisms show considerable ethnic variation with differences in frequency and/or the types of mutation present and this may need to be considered in the context of individualized prescribing of drugs.

2. Genetic polymorphisms relating to drug metabolism and disposition

The majority of prescription drugs together with a range of other foreign compounds from the diet or the environment need to undergo metabolism to more hydrophilic forms to facilitate their elimination from the body. Failure to metabolize a foreign compound normally may result in toxicity. Polymorphisms affecting drug metabolism have been detected in a number of different animal species but this chapter will cover only those in humans. The earliest example of a drug where variability in response due to a genetic polymorphism affecting

Genotype to Phenotype second edition, edited by S. Malcolm and J. Goodship.

metabolism occurred, was a rare defect in the enzyme serum cholinesterase which gave rise to prolonged muscle relaxation after administration of suxamethonium due to an inability to hydrolyse succinylcholine in the plasma at the normal rate (Kalow *et al.*, 1956). Evidence for the existence of several other common polymorphisms affecting drug metabolism later emerged. Much of the earlier work in this area was concerned with analysis of phenotype either by measuring enzyme activity in a tissue sample, or analyzing patterns of metabolites seen following administration of a probe drug. The genotypic basis for most of these phenotypic polymorphisms has now been determined. In addition, cloning and sequencing of other genes relevant to drug metabolism and disposition has resulted in the detection of functionally significant polymorphisms in additional genes. *Table 1* lists a number of genes relevant to drug disposition in which relatively common polymorphisms have been shown to exist and result in functionally significant effects on enzyme activity or protein levels. There is also evidence for the existence of considerable variation in levels of other xenobiotic metabolizing enzymes in populations, but only the genes listed in *Table 1* show confirmed functional polymorphisms on the basis of *in vivo* phenotypic measurements. The high level of polymorphism and variability in xenobiotic-metabolizing enzymes is likely to be due to their nonessential role in normal metabolism.

Drug metabolism often proceeds in two phases termed phase I and phase II and typically involves an initial phase I oxidation reaction which facilitates a subsequent phase II conjugation reaction. Phase I enzymes include the cytochromes P450, products of a multigene family with a capacity to oxidize a variety of xenobiotics, and phase II enzymes include conjugating enzymes such as the acetyltransferases, glutathione *S*-transferases and the UDP-glucuronosyltransferases. Genetic polymorphism in the cytochrome P450 genes and in the genes encoding the phase II enzymes *N*-acetyltransferase 2 and thiopurine methyltransferase is discussed here in detail as there is a reasonable understanding of genetic polymorphism in these genes and its consequences.

Table 1. Functionally significant polymorphisms in genes relevant to drug metabolism and disposition

Gene	Drug substrate	Reference
CYP2A6	Nicotine	Oscarson *et al.* (1999)
CYP2C9	*S*-warfarin	Aithal *et al.* (1999)
CYP2C19	Omeprazole	Furuta *et al.* (1998)
CYP2D6	Amitryptyline	Wolf *et al.* (2000)
CYP3A4	Cyclosporin	Sato *et al.* (2000)
CYP3A5	Cyclosporin	Paulussen *et al.* (2000)
Serum cholinesterase	Suxamethonium	Kalow *et al.* (1956)
NAT2	Isoniazid	Grant *et al.* (1997)
UGT1A1	Irinotecan	Iyer *et al.* (1998)
TPMT	Azathioprine	Krynetski *et al.* (1996)
DPD	5-fluorouracil	Gonzalez and Fernandez-Salguero (1995)
MDR-1	Digoxin	Hoffmeyer *et al.* (2000)

CYP, cytochrome P450; UGT, UDP-glucuronosyltransferase; TPMT, thiopurine *S*-methyltransferase; DPD, dihydropyrimidine dehydrogenase; NAT2, *N*-acetyltransferase 2. Drug substrates are examples only.

2.1 Cytochrome P450s polymorphisms

Cytochrome P450s are the most important type of phase I metabolizing enzyme and also represent the best studied group of enzymes from the pharmacogenetic standpoint. This supergene family consists of both genes encoding enzymes that metabolize endogenous compounds such as steroids, and genes encoding enzymes that metabolize xenobiotics. There is overlap between these activities in the case of some of the enzymes. At least 18 different cytochrome P450s (all encoded by separate genes) that metabolize xenobiotics have been identified in humans and together these enzymes have the ability to oxidize a large variety of chemical structures. Functionally significant polymorphisms have been detected in a number of different cytochrome P450 genes (*Table 1*). The two best studied polymorphisms are those in *CYP2D6* and *CYP2C19* where significant percentages of individuals are unable to metabolize certain drugs due to complete absence of active enzyme, but there is also increasing evidence for the existence of variation in the activity of certain other cytochrome P450s owing to polymorphisms in both coding regions and in upstream sequences.

In 1977, it was demonstrated that 5–10% of Caucasians were unable to hydroxylate the drugs debrisoquine and sparteine (Eichelbaum *et al.*, 1979; Mahgoub *et al.*, 1977). The trait appeared to be inherited in an autosomal recessive manner and individuals with the deficiency were termed 'poor metabolizers'. Subsequently, the hydroxylation reaction was demonstrated to be catalyzed by a cytochrome P450 enzyme which is now termed CYP2D6 and metabolism of at least 30 commonly prescribed drugs, mainly cardiovascular and psychotropic agents, has also been shown to be impaired in poor metabolizers (Wolf *et al.*, 2000). There is ethnic variation in the frequency of the poor metabolizer phenotype with Oriental populations showing a frequency of 0–1% for this phenotype compared with approximately 6% of Europeans. The gene encoding CYP2D6 has been cloned, sequenced and localized to chromosome 22q13.1. Analysis of genomic DNA from poor metabolizers has demonstrated that in Europeans there are four common CYP2D6 alleles associated with absence of activity (see *Table 2*). Up to 98% of poor metabolizers can be identified by screening for these alleles. Individuals heterozygous for these alleles have been demonstrated to show slower metabolism of CYP2D6 substrates compared with those with two wild-type alleles and in some cases, this may be clinically relevant. In addition, alleles associated with impaired metabolism rather than absence of activity exist and one of these, *CYP2D6*10*, is common in Orientals. A subgroup of individuals termed ultrarapid metabolizers who show faster than normal metabolism of CYP2D6 substrates have also been identified (Ingelman-Sundberg *et al.*, 1999). The basis of the rapid metabolism is the amplification of *CYP2D6* in germline DNA. Ultrarapid metabolizers have most commonly one and more rarely up to 12 additional copies of *CYP2D6* present in tandem with the normal gene and this results in enzyme levels being higher than normal.

There is increasing evidence that CYP2D6 genotyping may be helpful in individualizing drug prescription, either in deciding whether a CYP2D6 substrate should be prescribed at all and, if so, the most appropriate dosage. For example, the analgesic codeine is a prodrug which requires activation to morphine by CYP2D6 for effective analgesia. It has been demonstated that CYP2D6 poor

Table 2. Variant alleles in *CYP2D6*, *CYP2C19* and *CYP2C9*

Allele	Main nucleotide change(s)	Effect
CYP2D6		
Alleles associated with normal or near normal activity		
CYP2D6*1	None	
*CYP2D6*2*	G1661C; C2850T; G4180C	R296C; S486T
Alleles associated with increased activity		
Duplications and other amplifications (*n* = 2, 3, 4, 5 or 13) of *CYP2D6*1, 2* and *4*		
Alleles associated with absence of activity		
*CYP2D6*3*	A2549 deletion	Frameshift
*CYP2D6*4*	G1846A	Splicing defect
*CYP2D6*5*	*CYP2D6* deleted	*CYP2D6* deleted
*CYP2D6*6*	T1707 deleted	Frameshift
*CYP2D6*7*	A2935C	H324P
*CYP2D6*8*	G1758T	Stop codon
*CYP2D6*11*	G883C	Splicing defect
*CYP2D6*12*	G124A	G42R
*CYP2D6*13*	*CYP2D7P/CYP2D6* hybrid	Frameshift
*CYP2D6*14*	G1758A	P34S
*CYP2D6*15*	T183 insertion	Frameshift
*CYP2D6*16*	*CYP2D7P/CYP2D6* hybrid	Frameshift
CYP2D6*18	9 bp insertion in exon 9	Insertion
*CYP2D6*19*	A2539-T2542 deleted	Frameshift
*CYP2D6*20*	G1973 insertion	Frameshift
*CYP2D6*38*	G2587-T2590 deleted	Frameshift
Alleles associated with decreased activity		
*CYP2D6*9*	A2613-A2615	K281 deleted
*CYP2D6*10*	C100T	P34S
*CYP2D6*17*	C1023T;C2850T	T107I; R296S
CYP2C19		
Alleles associated with normal activity		
*CYP2C19*1*	None	
Alleles associated with absence of activity		
*CYP2C19*2*	G681A	Splicing defect
*CYP2C19*3*	G636A	Stop codon
*CYP2C19*4*	G1A	GTG initiation codon
*CYP2C19*5*	C1297T	R433W
*CYP2C19*6*	G395A	G395A
*CYP2C19*7*	Intron 5 T to A inversion	Splicing defect
Alleles associated with decreased activity		
*CYP2C19*8*	T358C	W120R
CYP2C9		
Alleles associated with normal activity		
*CYP2C9*1*	None	
Alleles associated with decreased activity		
*CYP2C9*2*	C430T	R144C
*CYP2C9*3*	A1075C	I359L

Numbering is based on the cDNA sequence from the start of the open reading frame. Effect on activity may be substrate dependent. Only the main functionally significant polymorphisms are indicated. More detailed information can be obtained from http://www.imm.ki.se/CYPalleles/.

metabolizers show a poor response to codeine analgesia and therefore use of an alternative analgesic may be appropriate (Sindrup *et al.*, 1993). Poor metabolizers undergoing psychiatric treatment appear to be particularly at risk from adverse drug reactions because of the large number of anti-depressant and anti-psychotic drugs that are CYP2D6 substrates. It has been suggested that CYP2D6 genotyping to set dosage prior to initiation of therapy might avoid these problems and be cost effective (Chou *et al.*, 2000).

In the early 1980s, it was demonstrated by analysis of plasma drug levels that hydroxylation of the *S*-enantiomer of mephenytoin, an anti-convulsant drug, was polymorphic with 2–6% of Europeans and 15–20% of Orientals unable to metabolize this compound (Kupfer and Preisig, 1984). The pattern of inheritance of the deficiency was autosomal recessive. Identification of the precise enzyme responsible for *S*-mephenytoin hydroxylation initially proved difficult, but this has now been demonstrated to be CYP2C19 and a number of variant alleles associated with absence of activity have been identified (see *Table 2*). CYP2C19 has a smaller range of drug substrates than CYP2D6, but examples of clinically relevant substrates include the proton pump inhibitor omeprazole and other benzimidazoles and the anti-malarial proguanil. Several recent studies have demonstrated that individuals lacking CYP2C19 show a better therapeutic response to treatment of peptic ulcer with omeprazole than individuals with normal levels of activity (Furuta *et al.*, 1998). CYP2C19 genotyping might therefore be beneficial in determining optimal clinical dose for this drug.

The CYP2D6 and CYP2C19 polymorphisms were the main cytochrome P450 polymorphisms whose existence was well established from phenotypic analysis prior to the widespread application of gene cloning and other molecular biology techniques to this area. Sequence comparisons and the detailed analysis of DNA from individuals who appear to metabolize some compounds in an ususual manner, has resulted in the identification of polymorphisms which lead to functionally significant effects in several additional P450s including CYP2C9, CYP2A6, CYP3A4 and CYP3A5. The relationship between genotype and phenotype has been best studied in the case of CYP2C9. CYP2C9 is responsible for the oxidative metabolism of a range of drugs including the anti-coagulant warfarin, the anti-convulsant phenytoin and several nonsteroidal anti-inflammatory drugs. There are two common CYP2C9 variant alleles termed *CYP2C9*2* and *CYP2C9*3* (see *Table 2*). Individuals heterozygous or homozygous for these alleles have a lower than average dose requirement for the anti-coagulant warfarin owing to having lower levels of CYP2C9 activity which results in slower metabolism of the active form of the drug (Aithal *et al.*, 1999). Warfarin is a drug with a narrow therapeutic index where dosage is already individualized on the basis of response and, in the future, determination of CYP2C9 genotype at initiation of therapy may assist in optimizing dosage more rapidly and effectively.

As the cytochromes P450 have the ability to oxidize a variety of xenobiotics including potential carcinogens, it has been proposed that the genotype for certain P450 polymorphisms might predict individual susceptibility to cancer and other diseases associated with xenobiotic exposure. However, with a few exceptions most studies on the relationship between particular P450 genotypes and cancer susceptibility have not detected any altered susceptibility associated with the presence of a particular variant allele (d'Errico *et al.*, 1999). In spite of largely negative results up to the present, it remains possible that additional polymorphisms in *P450* genes occur and that some of these might be risk factors for cancer and other diseases.

2.3 N-acetyltransferase 2 polymorphism

The existence of a polymorphism in the metabolism of the anti-tuberculosis drug isoniazid was confirmed in a family study approximately 40 years ago (Evans *et al.*, 1960). This drug was subsequently demonstrated to undergo acetylation with an acetyl group transferred from acetyl coenzyme A by an enzyme known as *N*-acetyltransferase. The approximately 50% of Europeans who were unable to carry out this reaction were termed slow acetylators. Slow acetylators also acetylate certain other drugs including sulphamethoxazole and caffeine (termed polymorphic substrates) poorly, but carry out acetylation of other compounds such as 4-aminobenzoic acid (termed monomorphic substrates) normally. It has now been demonstrated that there are two forms of *N*-acetyltransferase termed *N*-acetyltransferase 1 and 2 (NAT1 and NAT2) and that the *NAT2* gene, which is intronless, codes for the polymorphic enzyme. A number of different polymorphisms in NAT2 give rise to amino acid substitutions and these have been demonstrated to result in absence of catalytic activity *in vitro* (Grant *et al.*, 1997). Like the polymorphic cytochrome P450 genes, a relatively small number of polymorphisms are associated with the defective phenotype, and screening for three variant alleles in Caucasians results in the detection of the vast majority, with the slow acetylator phenotype, which is inherited as an autosomal recessive trait.

Most drug substrates for the polymorphic NAT2 are not widely used in modern medicine, though isoniazid remains an important drug in the treatment of tuberculosis and sulphamethoxazole is used in the treatment of secondary infections in AIDS patients. Knowledge of the patient's NAT2 genotype might be helpful therefore in certain circumstances. It is well established that slow acetylators are more likely to suffer side effects when prescribed isoniazid, though there is also evidence that these individuals' overall response to therapy may be better due to being exposed to higher drug levels for longer. The NAT2 polymorphism may also modulate risk of lung, bladder, breast and colon cancer owing to NAT2 acetylating aromatic amines found in tobacco smoke and cooked food. In general studies on the possible relationship between NAT2 genotype and cancer susceptibility have been more positive and convincing than similar studies in relation to P450 polymorphisms (d'Errico *et al.*, 1999). Where positive associations have been detected, the increased risks seen for particular genotypes are relatively small, but it is possible that NAT2 genotyping combined with advice for susceptible individuals on avoiding, for example, smoking or consumption of high levels of barbecued meat, might be of benefit in preventing cancer.

2.4 Thiopurine S-methyltransferase polymorphism

Thiopurine *S*-methyltransferase (TPMT) is another phase II drug metabolizing enzyme which methylates thiol groups. Its substrate specificity is narrow, with only three widely used drug substrates, 6-mercaptopurine and 6-thioguanine which are used in the treatment of childhood leukemia and a related drug, azathioprine, which is used as an immunosuppressant drug particularly in organ transplant patients. Approximately 1 in 300 Caucasians lack TPMT activity and are at risk of serious toxicity if given the normal recommended doses of 6-mercaptopurine, 6-thioguanine or azathioprine (Krynetski *et al.*, 1996). Several variant alleles which encode defective enzyme forms have been identified and the deficiency has been demonstrated to be inherited in an autosomal recessive manner. Defective individuals can now be identified either by genotyping or by measuring TPMT levels in erythrocytes. It appears likely that analysis of TPMT genotype or phenotype prior to prescription of 6-mercaptopurine, 6-thioguanine or azathioprine will be widely adopted in the future. Adjusting the dose on the basis of genotype may also be beneficial to individuals with activities in the normal range as there is evidence that individuals heterozygous for variant TPMT alleles and therefore with relatively low enzyme activity show a better response to 6-mercaptopurine treatment than homozygous wild-type individuals (Krynetski *et al.*, 1996).

2.5 Polymorphisms in other genes affecting drug metabolism and disposition

There is much less information at present on polymorphism in other genes encoding enzymes of drug metabolism and other proteins important in drug disposition such as drug transporters which have roles in drug absorption, transport to certain sites such as the brain and drug excretion. Many of the relevant genes have now been identified and, as more is understood about their individual functions, information about the existence of genetic polymorphisms and their functional effects is also becoming available. The drug transporter P-glycoprotein (Pgp), which is a product of the *MDR-1* gene, exports a variety of compounds including drugs and endogenous compounds such as steroids from cells in an energy dependent manner. This protein has a wider substrate specificity than most drug metabolizing enzymes with typical substrates including most of the commonly prescribed anti-cancer drugs, and appears to play an important role in determining the rate of absorption of these substrates in the intestine and also in determining whether they will cross the blood–brain barrier. Detailed analysis of the *MDR-1* gene for polymorphism has recently been performed and 15 different single nucleotide polymorphisms have been detected (Hoffmeyer *et al.*, 2000). Individuals homozygous for a particular variant (approximately 24% of a German population) had a higher level of expression of Pgp and this was associated with lower plasma levels of the drug digoxin when administered orally. It is now feasible to carry out similar analysis on other drug transporters and it is likely that other polymorphisms that may affect plasma levels of drugs will emerge. Whether screening for such polymorphisms will be useful in individualizing drug prescribing is still unclear, but is likely to depend both on the overall effect of the polymorphism on plasma drug levels (a 10 fold difference between two individuals may be more important than a 2 fold difference) and also on the therapeutic index and other properties of the individual drug.

3. Polymorphisms affecting drug targets

Targets for drug action include both intracellular and cell surface receptors for hormones, neurotransmitters and other signalling molecules, enzymes, ion channels and carrier molecules. Genetic polymorphisms in a drug target may modify the response of that target to the drug with either toxicity due to an exaggerated response or lack of response at the normal therapeutic dose. At present, much less is known about polymorphisms in genes encoding drug targets compared with those in genes encoding proteins that affect drug disposition. There are a number of reasons for this. Firstly, because of the essential metabolic role of most drug targets, polymorphisms with biologically significant effects are likely to be less common. Secondly, where polymorphisms exist studies on the relationship between genotype and phenotype are generally more complex to perform than those for polymorphisms affecting drug metabolizing enzymes. However, there is now a reasonable amount of information available on the existence of polymorphism in some of the common drug targets and in some cases these polymorphisms have been shown to affect either drug response or drug toxicity. *Table 3* lists a range of examples and a few of these are discussed in detail below.

3.1 Polymorphisms affecting drug response

An ability to predict likely response to particular drugs on the basis of patient genotype would be helpful in determining the most appropriate drug to prescribe. As yet there are few examples of this approach. The β2-adrenergic receptor is an important target for a number of commonly prescribed drugs including the β-receptor agonists such as salbutamol used to treat asthma. Polymorphism in the gene encoding the β2-adrenergic receptor has been widely studied and a number

Table 3. Polymorphisms affecting drug targets and responses

Gene	Product	Drug	Reference
DRD3	Dopamine D3 receptor	Antipsychotics	Steen *et al.* (1996)
5-HT2A	5-HT receptor 2A	Clozapine	Arranz *et al.* (2000)
β2-AR	β2-adrenergic receptor	β2 agonists	Liggett *et al.* (2000)
ALOX5	Lipoxygenase	ABT-761	Drazen *et al.* (1999)
ACE	Angiotensin converting enzyme	β antagonist/ ACE inhibitor	van Essen *et al.* (1996)
KCNE2	MiRP1 potassium channel	Various	Sesti *et al.* (2000)
ApoE	Apolipoprotein E Tacrine	Poirier *et al.* (1995)	
CETP	Cholesteryl ester transfer protein	Pravastatin	Kuivenhoven *et al.* (1998)
Mitochondrial 16S rRNA 16S rRNA gene		Aminoglycoside antibiotics	Guan *et al.* (2000)

of both upstream and coding region polymorphisms have been identified. Some of these appear to be of functional significance (Liggett, 2000). A number of studies suggest that genotype for certain polymorphisms may determine both the severity of disease and the response to β-agonists with individuals positive for certain allelic variants reported to be more likely to show good initial and long-term responses to these drugs. However, there are still unanswered questions especially as most studies have not determined overall haplotype but concentrated on single polymorphisms. It is not yet clear whether the response of a particular genotype will be the same for all β-agonists or whether certain compounds might be more beneficial for certain genotypes. It is also possible that the phenotype for all β-agonists may be similar for a particular genotype and that for individuals of certain genotypes, treatment of asthma by alternative therapeutic agents may be most beneficial.

Another possible asthma treatment involves the use of inhibitors of 5-lipoxygenase, an enzyme which produces leukotrienes. Within the 5-lipoxygenase (ALOX5) gene promoter region, there are between three and six tandem repeats of a six base-pair Sp1-binding motif, with the most common genotype of five repeats being regarded as the wild-type genotype. The five repeat wild-type sequence shows highest activity with reporter gene constructs. When response to treatment with a novel 5-lipoxygenase inhibitor was examined with respect to ALOX5 genotype, individuals without at least one copy of the wild-type sequence showed no response to treatment, whereas a good response was generally seen in homozygous wild-type or heterozygous individuals (Drazen et al., 1999).

Optimizing the drug treatment of schizophrenia remains difficult with both disease resistance to treatment with anti-psychotic drugs and toxicity problems with these agents common. The atyptical anti-psychotic drug clozapine is beneficial in the treatment of patients with resistant disease but not all patients show a good response to this drug. It has been suggested that genotyping for polymorphisms in the 5-hydroxytryptamine (serotonin) 2A receptor gene (5-HT_{2A}) can identify patients who are likely to show a good response. Patients positive for an upstream polymorphism and homozygous for a coding region polymorphism comprised 50% of the total group but accounted for 80% of the patients who showed a substantial improvement in treatment (Arranz et al., 2000). At present, these findings need to be treated with caution as they could not be reproduced in another Caucasian group (Schumacher et al., 2000).

Drugs such as tacrine which affect the supply of acetylcholine in the brain can alleviate some symptoms of Alzheimer disease. However, response to tacrine appears to be linked to the genotype for the gene encoding apolipoprotein E4 (ApoE4), which mediates the binding of lipoproteins to the low density lipoprotein receptor, and appears to have a key role in lipid transport in the brain. It is well established that possession of one or two variant ApoE alleles of the ApoE4 type is associated with an increased risk of Alzheimer disease development. Individuals with these genotypes also show a poorer response to tacrine treatment compared with Alzheimer disease patients of other ApoE genotypes (Poirier et al., 1995). These findings could be helpful in determining the most effective strategy for treatment of Alzheimer disease, especially if drugs with targets other than the cholinergic transmission system become available though there are potential ethical problems with approaches of this type, particularly since some ApoE4 carriers did show a response to treatment.

3.2 Polymorphisms affecting drug toxicity

A wide range of drugs have been associated with drug toxicities which are not predictable on the basis of their plasma levels. These toxicities vary in frequency from very common to rare, but are currently difficult to predict and often have serious consequences. A common example of such a toxicity is the side effect of tardive dyskinesia, a movement disorder which can occur in schizophrenics undergoing treatment with anti-psychotic drugs. It appears that homozygosity for a variant allele of the gene encoding the dopamine D3 receptor (DRD3) which results in an amino acid substitution is associated with a significantly increased risk of this toxicity (Steen *et al.*, 1996).

Drug-induced long QT syndrome is a life-threatening cardiac arrhythmia which can be induced in susceptible individuals by a number of different drugs. It now appears that these individuals carry mutations in genes encoding potassium channels which result in small, but not normally problematic, prolongation of QT interval unless a drug which induces some blockage of K^+ channels is taken. Up to the present, population screening for the indentification of at risk individuals for drug-induced long QT syndrome has been difficult owing to most cases being found to be positive for isolated mutations in one of several K^+ channel genes (Priori *et al.*, 1999). However, a polymorphism in the *KCNE2* gene which encodes a subunit of the cardiac K^+ channel I_{kr} has now been identified. The variant allele is seen in approximately 1.6% of Caucasians and results in an amino acid substitution which is apparently associated with altered K^+ flux in the presence of sulphamethoxazole, though in the absence of the drug no abnormality is seen (Sesti *et al.*, 2000). Screening for this polymorphism prior to prescription of drugs associated with cardiac arrythmia might decrease the frequency of the problem.

Use of aminoglycoside antibiotics such as gentamycin can give rise to irreversible hearing loss in a small number of individuals. It has been established that in the inherited form of this disorder, the defect is maternally transmitted due to the occurrence of a mitochrondrial gene mutation. This mutation has now been found to be a A–G transition in the 12S rRNA gene which appears to create a new binding site for the aminoglycoside, resulting in effects by the drug on protein synthesis (Guan *et al.*, 2000). The mutation is also associated with congenital deafness.

4. The future for pharmacogenomics

The application of pharmacogenomics to patient treatment is still at a very early stage, but it is likely that the development of simple high throughput genotyping methods may facilitate its transfer from the research laboratory to the clinic. In the future, genotyping to determine which drug will benefit the patient best, and what dose should be used may become widespread. The genes which contribute to common polygenic disorders such as diabetes mellitus, asthma and hypertension are in the process of being identified. The current search for single nucleotide polymorphisms (SNPs) throughout the genome is likely to facilitate our understanding of polygenic diseases and enable strategies for individualized prescribing to be developed (Roses, 2000). It may in the future be possible to develop new drugs which will target particular gene defects and use these to treat patients on an individualized basis. Already high throughput sequencing has enabled the identification of novel drug targets and the design of new compounds (Debouck and Metcalf, 2000). The possibility of developing drugs suitable for a particular disease genotype and the need for large scale genetic testing in this context raise certain ethical problems, however, which require further consideration and discussion (Issa, 2000).

References

Aithal, G.P., Day, C.P., Kesteven, P.J.L. and Daly, A.K. (1999) Association of polymorphisms in the cytochrome P450 CYP2C9 with warfarin dose requirement and risk of bleeding complications. *Lancet* **353**: 717–719.

Arranz, M.J., Munro, J., Birkett, J. *et al.* (2000) Pharmacogenetic prediction of clozapine response. *Lancet* **355**: 1615–1616.

Chou, W.H., Fan, F.-X., de Leon, J. *et al.* (2000) An extension of a pilot study: impact from the cytochrome P450–2D6 (CYP2D6) polymorphism on outcome and costs in severe mental illness. *J. Clin. Psychopharmacol.* **20**: 246–251.

Debouck, C. and Metcalf, B. (2000) The impact of genomics on drug discovery. *Ann. Rev. Pharmacol. Toxicol.* **40**: 193–208.

d'Errico. A., Malats, N., Vineis, P. and Boffetta, P. (1999) Review of studies of selected metabolic polymorphisms and cancer. In: Vineis, P., Malats, N., Lang, M., d'Errico, A., Caporaso, N., Cuzick, J. and Boffetta, P. (eds), *Metabolic Polymorphisms and Susceptibility to Cancer.* International Agency for Research on Cancer, Lyon, France, pp. 323–394.

Drazen, J.M., Yandava, C.N., Dube, L. *et al.* (1999) Pharmacogenetic association between *ALOX5* promoter genotype and the response to anti-asthma treatment. *Nat. Genet.* **22**: 168–170.

Eichelbaum, M., Spannbrucker, N., Steincke, B. and Dengler, H.J. (1979) Defective N-oxidation of sparteine in man: a new pharmacogenetic defect. *Eur. J. Clin. Pharmacol.* **16**: 183–187.

Evans, D.A.P., Manley, K.A. and McKusick, V.A. (1960) Genetic control of isoniazid metabolism in man. *Br. Med. J.* **2**: 485–491.

Furuta, T., Ohashi, K., Kamata, T. *et al.* (1998) Effect of genetic differences in omeprazole metabolism on cure rates for Helicobacter pylori infection and peptic ulcer. *Ann. Intern. Med.* **129**: 1027–1030.

Gonzalez, F.J. and Fernandez-Salguero, P. (1995) Diagnostic analysis, clinical importance and molecular-basis of dihydropyrimidine dehydrogenase-deficiency. *Trends Pharmacol. Sci.* **16**: 325–327.

Grant, D.M., Hughes, N.C., Janezic, S.A., Goodfellow, G.H., Chen, H.J., Gaedigk, A., Yu, V.L. and Grewal, R. (1997) Human acetyltransferase polymorphisms. *Mutation Res.* **376**: 61–70.

Guan, M.-X., Fischel-Ghodsian, N. and Attardi, G. (2000) A biochemical basis for the inherited susceptibility to aminoglycoside ototoxicity. *Hum. Mol. Genet.* **9**: 1787–1793.

Hoffmeyer, S., Burk, O., von Richter, O. *et al.* (2000) Functional polymorphisms of the human multidrug-resistance gene: multiple sequence variations and correlation of one allele with P-glycoprotein expression and activity *in vivo. Proc Natl Acad. Sci. USA* **97**: 3473–3478.

Ingelman-Sundberg, M., Oscarson, M. and McLellan R.A. (1999) Polymorphic human cytochrome P450 enzymes: an opportunity for individualized drug treatment. *Trends Pharmacol. Sci.* **20**: 342–349.

Issa, A.M. (2000) Ethical considerations in clinical pharmacogenomics research. *Trends Pharmacol. Sci.* **21**: 247–249.

Iyer, L., King, C.D., Whitington, P.F., Green, M.D., Roy, S.K., Tephly, T.R., Coffman, B.L. and Ratain, M.J. (1998) Genetic predisposition to the metabolism of irinotecan (CPT-11): role of uridine diphosphate glucuronosyltransferase isoform 1A1 in the glucuronidation of its active metabolite (SN-38) in human liver microsomes. *J. Clin. Invest.* **101**: 847–854.

Kalow, W., Genest, K. and Staron, N. (1956) Kinetic studies on the hydrolysis of benzoylcholine by human serum cholinesterase. *Can. J. Biochem. Physiol.* **34**: 637–653.

Krynetski, E.Y., Tai, H.L., Yates, C.R., Fessing, M.Y., Loennechen, T., Schuetz, J.D., Relling, M.V. and Evans, W.E. (1996) Genetic polymorphism of thiopurine S-methyltransferase: clinical importance and molecular mechanisms. *Pharmacogenetics* **6**: 279–290.

Kuivenhoven, J.A., Jukema, J.W., Zwinderman, A.H., de Knijff, P., McPherson, R., Bruschke, V.G., Lie, K.I. and Kastelein, J.J.P. (1998) The role of a common variant of the cholesterol ester transfer protein gene in the progression of coronary atherosclerosis. *New Engl. J. Med.* **338**: 86–93.

Kupfer, A. and Preisig, R. (1984) Pharmacogenetics of mephenytoin – a new drug hydroxylation polymorphism in man. *Eur. J. Clin. Pharmacol.* **26**: 753–759.

Liggett, S.B. (2000) β_2-adrenergic receptor pharmacogenetics. *Am. J. Respir. Crit. Care Med.* **161**: S197–S201.

Mahgoub, A., Idle, J.R., Dring, L.G., Lancaster, R.L. and Smith, R.L. (1977) Polymorphic hydroxylation of debrisoquine in man. *Lancet* **2**: 584–586.

Oscarson, M., McLellan, R.A., Gullsten, H., Agundez, J.A.G., Benitez, J., Rautio, A., Raunio, H., Pelkonen, O. and Ingelman-Sundberg, M. (1999) Identification and characterisation of novel polymorphisms in the CYP2A locus: implications for nicotine metabolism. *FEBS Lett.* **460**: 321–327.

Paulussen, A., Lavrijsen, K., Bohets, H., Hendrickx, J., Verhasselt, P., Luyten, W., Konings, F. and Armstrong, M. (2000) Two linked mutations in transcriptional regulatory elements of the *CYP3A5* gene constitute the major genetic determinant of polymorphic activity in humans. *Pharmacogenetics* **10**: 415–424.

Poirier, J., Delile, M.-C., Quirion, R. *et al.* (1995) Apolipoprotein E4 allele as a predictor of cholinergic deficits and treatment outocme in Alzheimer disease. *Proc. Natl Acad. Sci. USA* **92**: 12260–12264.

Priori, S.G., Barhanin, J., Hauer, R.N.W. *et al.* (1999) Genetic and molecular basis of cardiac arrhythmias: Impact on clinical management parts I and II. *Circulation* **99**: 518–528.

Roses, A.D. (2000) Pharmacogenetics and future drug development and delivery. *Lancet* **355**: 1358–1361.

Sata, F., Sapone, A., Elizondo, G., Stocker, P., Miller, V.P., Zheng, W., Raunio, H., Crespi, C.L. and Gonzalez, F.J. (2000) CYP3A4 allelic variants with amino acid substitutions in exons 7 and 12: Evidence for an allelic variant with altered catalytic activity. *Clin. Pharmacol. Ther.* **67**: 48–56.

Schumacher, J., Schulze, T.G., Wienker, T.F., Rietschel, M. and Nothen, M.M. (2000) Pharmacogenetics of clozapine response. *Lancet* **356**: 506–507.

Sesti, F., Abbott, G.W., Wei, J. *et al.* (2000) A common polymorphism associated with antibiotic-induced cardiac arrhythmia. *Proc. Natl Acad. Sci. USA* **97**: 10613–10618.

Sindrup, S.H., Poulsen, L., Brosen, K., Arendt-Nielsen, L. and Gram, L.F. (1993) Are poor metabolizers of sparteine debrisoquine less pain tolerant than extensive metabolizers? *Pain* **53**: 335–339.

Steen, V.M., Lovlie, R., MacEwan, T. and McCreadie, R.G. (1996) Dopamine D3-receptor gene variant and susceptibility to tardive dyskinesia in schizophrenic patients. *Mol. Psychiatry* **2**: 139–146.

van Essen, G.G., Rensma, P.L., de Zeeuw, D., Sluiter, W.J., Scheffer, H., Apperloo, A.J. and deJong, P.E. (1996) Association between angiotensin-converting-enzyme gene polymorphism and failure of renoprotective therapy. *Lancet* **347**: 94–95.

Wolf, C.R., Smith, G. and Smith, R.L. (2000) Pharmacogenetics. *Br. Med. J.* **320**: 987–990.

6

Mitochondrial genetics

P.F. Chinnery and D.M. Turnbull

1. Introduction

Mitochondria are ubiquitous intracellular organelles that are essential for aerobic metabolism. The mitochondrial respiratory chain is a group of over 70 polypeptide subunits that form five enzyme complexes situated on the inner mito-chondrial membrane. Most of these subunits are synthesized within the cytosol and targeted to the inner mitochondrial matrix by a short peptide pre-sequence, however 13 polypeptide subunits are synthesized within the mitochondrial matrix from the mitochondrial DNA (mtDNA, *Figure 1*). A complex system of nuclear-encoded proteins is involved in the assembly of both nuclear and mtDNA subunits to form a functional respiratory chain (Tiranti *et al.*, 1998; Zhu *et al.*, 1998). As a result, mitochondrial respiratory chain disease may be due to genetic defects within the nuclear and mitochondrial genomes (*Table 1*). Respiratory chain dysfunction is important in the pathogenesis of a number of other nuclear genetic disorders such as Friedreich's ataxia (Rotig *et al.*, 1997), Wilson's disease (Lutsenko and Cooper, 1998) and autosomal recessive hereditary spastic para-plegia (SPG7; Casari *et al.*, 1998). The genetics of these nuclear-encoded mito-chondrial disorders is beyond the scope of this chapter. Low levels of somatic mtDNA mutations accumulate throughout life and in association with certain diseases (Nagley and Wei, 1998). The role of these secondary mutations in aging is highly contentious (Lightowlers *et al.*, 1999), and will also not be discussed here. In this chapter we will describe the recent advances in our understanding of mito-chondrial genetics that are directly relevant to our understanding of disease that is directly due to defects of the mitochondrial genome.

2. Basic mitochondrial genetics and biochemistry

The human mtDNA sequence was published in 1981 (Anderson *et al.*, 1981), and is referred to as the Cambridge Reference Sequence (CRS). MtDNA mutations are generally described with reference to the 16569 base pair CRS light (L)-strand. The original placental DNA sample used to deduce the CRS was recently re-sequenced (Andrews *et al.*, 1999), and a number of errors were identified, including a mis-read CC doublet at position 3106. As a result, the revised CRS

Genotype to Phenotype second edition, edited by S. Malcolm and J. Goodship.
© 2001 BIOS Scientific Publishers Ltd, Oxford.

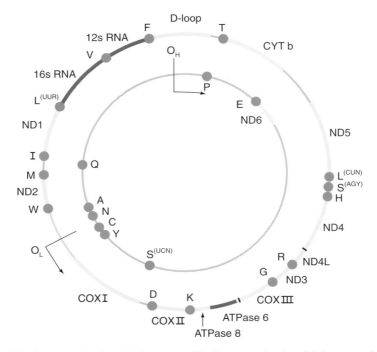

Figure 1. The human mitochondrial genome. The human mitochondrial genome (mtDNA) is a small 16.5 kb molecule of double stranded DNA. The 1.1 kb D-loop is involved in the regulation of transcription and replication of the molecule, and is the only region not directly involved in the synthesis of respiratory chain polypeptides. MtDNA encodes for 13 essential components of the respiratory chain. ND1-ND6, and ND4L encode 7 subunits of complex I. Cyt b is the only mtDNA encoded complex III subunit. COX I to III encode for three of the complex IV (cytochrome *c* oxidase, or COX) subunits, and the ATPase 6 and ATPase 8 genes encode for two subunits of complex V. Two ribosomal RNA genes (12S and 16S rRNA), and 22 transfer RNA genes are interspaced between the protein-encoding genes. These provide the necessary RNA components for intra-mitochondrial protein synthesis. O_H and O_L are the origins of heavy and light strand mtDNA replication.

(rCRS) has only 16568 base pairs, but the original CRS numbering system has been retained to avoid confusion.

Each diploid human cell contains between 1000 and 100 000 copies of mtDNA, depending upon the cell type (Larsson and Clayton, 1995; Lightowlers *et al.*, 1997; Robin and Wong, 1988; Veltri *et al.*, 1990; *Figure 1*). Each mitochondrial genome is a circular 16.5 kb molecule of double-stranded DNA that encodes for seven complex I subunits (NADH-ubiquinone oxidoreductase), one of the complex III subunits (ubiquinol-cytochrome *c* oxidoreductase), three of the complex IV subunits (cytochrome *c* oxidase, or COX) subunits, and the ATPase 6 and ATPase 8 subunits of complex V. Inter-spaced between the protein-encoding genes are two ribosomal RNA genes (12S and 16S rRNA), and 22 transfer RNA genes that provide the necessary RNA components for the mitochondrial translation machinery. The remaining polypeptides, including *all* of the complex II subunits, are synthesized from nuclear gene transcripts. There are also numerous other

Table 1. Genetic basis of mitochondrial disease

Nuclear DNA defects
Nuclear genetic disorders of the mitochondrial respiratory chain
Leigh syndrome (complex I deficiency – mutations in AQDQ subunit on Chr 5)
Optic atrophy and ataxia (complex II deficiency – mutations in Fp subunit of SDH on Chr 3)
Leigh syndrome (complex IV deficiency – mutations in *SURF I* gene on Chr 9q1)

Nuclear genetic disorders associated with multiple mtDNA deletions
Autosomal dominant external ophthalmoplegia (point mutations in *Ant1, POLG* and Twinkle)
Mitochondrial neuro-gastrointestinal encephalomyopathy (thymidine phosphorylase
deficiency – mutations in thymidine phosphorylase gene on Chr 22q13.32–qter)

Mitochondrial DNA defects
Rearrangements (deletions and duplications)
Chronic progressive external ophthalmoplegia (CPEO)
Kearns Sayre syndrome
Diabetes and deafness

Point mutations[*]
Protein-encoding genes
  LHON (G11778A, T14484C, G3460A)
  NARP/Leigh syndrome (T8993G/C)
  Exercise intolerance and myoglobinuria (cyt *b* mutations)
tRNA genes
  MELAS (A3243G, T3271C, A3251G)
  MERRF (A8344G, T8356C)
  CPEO (A3243G, A4269G)
  Myopathy (T14709C, A12320G)
  Cardiomyopathy (A3243G, A4269G)
  Diabetes and deafness (A3243G, C12258A)
  Encephalomyopathy (G1606A, T10010C)
  Non-syndromic sensorineural deafness (A7445G)
rRNA genes
  Aminoglycoside induced non-syndromic deafness (A1555G)

---

[*]mtDNA nucleotide positions refer to the L-chain, cyt *b* = cytochrome *b*.

enzymes and co-factors that are essential for mtDNA replication and transcription which are also encoded by nuclear genes (Schadel and Clayton, 1997).

The intermediary metabolism of carbohydrates, amino acids, and fatty acids generates reduced co-factors that transfer electrons to complexes I and II of the respiratory chain. As the electrons are passed through complexes I to IV, protons are pumped out of the mitochondrial matrix into the inter-membrane space. This creates an electrochemical gradient that is harnessed by mitochondrial ATP synthase (Complex V) to generate ATP from ADP and inorganic phosphate. The newly generated ATP is then exported into the cytosol in exchange for ADP by the adenine nucleotide translocator (ANT).

3. Polymorphic variability of the mitochondrial genome

The mutation rate of mtDNA is thought to be 10–16 fold higher than in the nuclear genome. This is probably because mtDNA is tethered to the inner mitochondrial membrane adjacent to the respiratory chain which is a potent source of free radicals (Bandy and Davison, 1990; Wallace, 1992). In addition, mtDNA is not protected by histones, and the DNA repair mechanisms within mitochondria are probably not as efficient as those found within the nucleus (Wallace, 1992). MtDNA is exclusively transmitted down the maternal line, and the traditional view is that inter-molecular mtDNA recombination does not occur *in vivo* (Howell, 1997c; although see section 6.4). As a result, mtDNA lineages are clonal and gradually accumulate new mutations over time. This is responsible for the variability in mitochondrial genomes within the population – unrelated individuals differ by an average total of 25 base pair substitutions. These polymorphisms are not uniformly distributed throughout the genome. Specific regions of the protein encoding (http://www3.ebi.ac.uk/Research/Mitbase/mitbase.pl) and tRNA (Sprintzl *et al.*, 1989) genes are highly conserved between and within species, but some regions are hypervariable (HVR; Howell *et al.*, 1996), particularly in the non coding displacement loop (D-loop). This degree of polymorphism is useful in forensic medicine (Ivanov *et al.*, 1996; Parsons *et al.*, 1997), and the genomic sequence of the HVRI and HVR2 can be used to identify the remains of trauma victims with defined levels of confidence. MtDNA sequence changes can also be used to analyze population history (Helgason *et al.*, 2000; van Holst Pellakaan *et al.*, 1998; Wallace, 1995; Wallace and Torroni, 1992; Watson *et al.*, 1997), but the accuracy of this approach depends upon the rate of mutation (or 'evolution') of the mitochondrial genome (the mitochondrial genetic 'clock'). There have been numerous attempts to measure the rate of mutation of the human mitochondrial genome, but no one method is ideal (Lynch and Jarrell, 1993; Wise *et al.*, 1997). Differences in the mutation rate at different sites has a profound effect on phylogenetic estimates of divergence rates (Yang, 1996), and as a result different rates of mtDNA evolution have been reported by different laboratories (Howell *et al.*, 1996; Jazin *et al.*, 1998; Parsons *et al.*, 1997; Siguroardottir *et al.*, 2000), and no one rate is accepted by all. The high rate of sequence diversity does create difficulties when investigating patients with suspected mtDNA point mutation disease: it can be difficult to establish that a novel point mutation is pathogenic and not a simple private polymorphism.

4. Pathogenic mitochondrial DNA defects and disease

Adults who present with mitochondrial disease are often found to have a defect of mtDNA. Children often present with different clinical features and are more likely to have a nuclear genetic defect. Pathogenic mtDNA defects affect at least 1 in 15 000 of the adult population (Chinnery *et al.*, 2000). These disorders are clinically and genetically heterogeneous (Chinnery and Turnbull, 1999; Leonard and Schapira, 2000a,b). An identical clinical syndrome can be due to a range of different mtDNA defects, and a specific mtDNA defect may present in a variety of different ways (*Figure 2*). It may be possible to recognize a well-defined clinical syndrome (*Table 2*), but many patients present with a collection of clinical features that are highly suggestive of mtDNA disease, but do not fit into a discrete clinical category (*Figure 2*). Pathogenic mtDNA defects fall into two groups: deletions and point mutations (*Table 1*). In the North East of England, approximately half of the

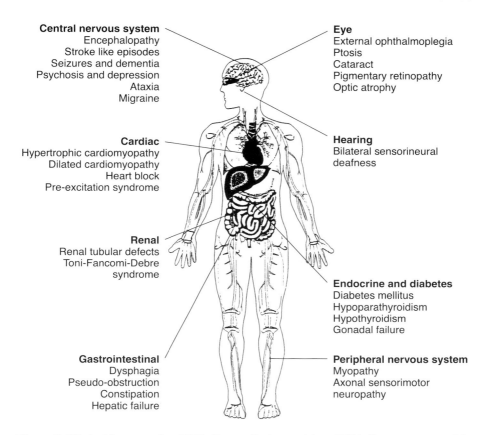

Central nervous system
Encephalopathy
Stroke like episodes
Seizures and dementia
Psychosis and depression
Ataxia
Migraine

Eye
External ophthalmoplegia
Ptosis
Cataract
Pigmentary retinopathy
Optic atrophy

Cardiac
Hypertrophic cardiomyopathy
Dilated cardiomyopathy
Heart block
Pre-excitation syndrome

Hearing
Bilateral sensorineural
deafness

Renal
Renal tubular defects
Toni-Fancomi-Debre
syndrome

Endocrine and diabetes
Diabetes mellitus
Hypoparathyroidism
Hypothyroidism
Gonadal failure

Gastrointestinal
Dysphagia
Pseudo-obstruction
Constipation
Hepatic failure

Peripheral nervous system
Myopathy
Axonal sensorimotor
neuropathy

Figure 2. Clinical features of mtDNA disease. Patients with mtDNA disease present with a range of clinical features. These may be clustered in to form recognizable clinical syndrome (*Table 1*). Certain features are highly suspicious of mtDNA disease, even in isolation (e.g. chronic progressive external ophthalmoplegia). Many features are not specific for mtDNA disease, and a diagnosis of mtDNA disease may only come to light when there are multiple organs involved, of there is a relevant family history. Ocular features are particularly common in patients with mtDNA disease.

adults with mtDNA disease have a point mutation causing Leber's hereditary optic neuropathy (LHON), 30% have another point mutation causing a variety of different phenotypes, and 20% have a mtDNA rearrangement causing Kearns–Sayre syndrome (KSS) or chronic progressive external ophthalmoplegia (CPEO, *Figure 3*; Chinnery *et al.*, 2000).

4.1 MtDNA rearrangements

Large deletions of mtDNA (from a few hundred base pairs to >10 kb) generally cause sporadic disorders such as CPEO and KSS (*Tables 1* and *2*; Holt *et al.*, 1988; Moraes *et al.*, 1989; Zeviani *et al.*, 1988). CPEO is a mild disease, causing minimal disability over many decades. The signs may be limited to the restriction in eye movements, and occasionally patients develop a mild proximal myopathy. By

Table 2. Clinical syndromes due to mtDNA defects

Disorder	Primary features	Additional features
CPEO	External ophthalmoplegia and bilateral ptosis	Mild proximal myopathy
KSS	PEO onset before age 20 with pigmentary retinopathy Plus one of the following: CSF protein greater than 1g/l, cerebellar ataxia, heart block	Bilateral deafness Myopathy Dysphagia Diabetes mellitus Hypoparathyroidism Dementia
Pearson's syndrome	Sideroblastic anemia of childhood Pancytopenia Exocrine pancreatic failure	Renal tubular defects
MELAS	Stroke-like episodes before age 40 years Seizures and/or dementia Ragged-red fibers and/or lactic acidosis	Diabetes mellitus Cardiomyopathy (hypertrophic leading to dilated) Bilateral deafness Pigmentary retinopathy Cerebellar ataxia
MERRF	Myoclonus Seizures Cerebellar ataxia Myopathy	Dementia Optic atrophy Bilateral deafness Peripheral neuropathy Spasticity Multiple lipomata
LHON	Subacute bilateral visual failure Males:females approx. 4:1 Median age of onset 24 years	Dystonia Cardiac pre-excitation syndromes
Leigh syndrome	Subacute relapsing encephalopathy with cerebellar and brain-stem signs	Basal ganglia lucencies

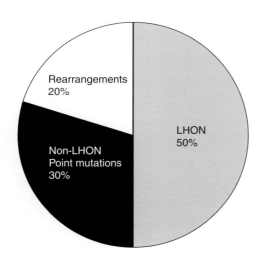

Figure 3. Relative prevalence of mtDNA disease in adults. Adapted from (Chinnery *et al.*, 2000). LHON = Leber's hereditary optic neuropathy.

contrast, KSS is a severe progressive neurodegenerative disorder that presents below the age of 20 years with ataxia, pigmentary retinopathy, deafness, and cardiac conduction defects (heart block). Endocrine disorders, such as diabetes mellitus and hypoparathyroidism are reported in KSS, and late in the disease patients may develop respiratory muscle weakness and dysphagia. The general view is that there is no correlation between the size of the mtDNA deletion and the severity of the clinical phenotype, but careful scrutiny of a large series of patients suggests that this is not the case (Hanna *et al.*, 1998a). At one extreme, small deletions (<300 bp) are usually associated with pure CPEO, whereas large deletions often cause additional features such as a proximal myopathy and ataxia. Large deletions of mtDNA are also found in children presenting with sideroblastic anemia, pancytopenia, exocrine pancreatic failure and renal tubular defects (Pearson's syndrome; Rotig *et al.*, 1990). Many of these children die, but better supportive care means that some survive into adulthood, when they may develop a KSS-like phenotype.

Other mtDNA rearrangements may cause disease, and recently a mtDNA inversion was described in a patient with a myopathy and myoglobinuria (Musumeci *et al.*, 2000). In contrast, not all mtDNA rearrangements are pathogenic. A mtDNA triplication has also been found in a healthy patient (Tengan and Moraes, 1998). MtDNA duplications are also found in some patients with mtDNA disease (Poulton *et al.*, 1989). Duplications are not primarily pathogenic, but they may lead to the formation of deletions that can cause disease (Poulton *et al.*, 1993).

4.2 MtDNA point mutations

Over 70 different mtDNA point mutations have been cataloged over the last decade, affecting both protein encoding genes and RNA genes (*Table 1*; Servidei, 2000).

Mis-sense mutations. The most common mis-sense mutations are found in families with LHON, which usually presents in young adult life with painless subacute bilateral visual failure with a predilection for males (Riordan-Eva *et al.*, 1995). Individuals with LHON are usually part of large multi-generation maternal pedigree, but approximately 30% of new cases of LHON presenting with visual loss do not report a family history (Harding *et al.*, 1995). In one series, over 95% of individuals with a clinical diagnosis of LHON were found to have one of three mtDNA complex I subunit 1 (ND1) gene point mutations (Howell *et al.*, 1991a; Johns *et al.*, 1992; Mackey *et al.*, 1996; Wallace *et al.*, 1988), but numerous additional LHON mutations have been identified (Servidei, 2000). Some of these mutations have only been found in pedigrees with LHON, but some (for example T4216C, G13708A, G15257A), are present in high frequency within the general population (Howell *et al.*, 1995). The etiological relevance of these so-called 'secondary' LHON mutations is currently uncertain.

Mis-sense mutations in the ATPase genes can cause a subacute necrotizing encephalopathy with ataxia and brain stem signs in children (Leigh syndrome) (Ciafaloni *et al.*, 1993). The same mutation may also present later in childhood or adult life with a milder phenotype (neurogenic weakness with ataxia and retinitis pigmentosa, NARP; Holt *et al.*, 1990). In a recent multi-center study, five patients with exercise intolerance and myoglobinuria were found to have point mutations in the cytochrome *b* (cyt *b*) gene (Andreu *et al.*, 1999). Although the mutations were scattered throughout the protein, the cyt *b* mutations cause a strikingly similar phenotype.

RNA gene mutations. The most common tRNA gene point mutation, A3243G, involves the Leu (UUR) tRNA gene (Goto *et al.*, 1990). It was first described in a patient with mitochondrial encephalomyopathy with lactic acidosis and stroke-like episodes (MELAS; Ciafaloni *et al.*, 1992), but it may also cause diabetes and deafness (Reardon *et al.*, 1992), CPEO (Moraes *et al.*, 1993) and cardiomyopathy (Santorelli *et al.*, 1996). Numerous other point mutations involving tRNA genes have been described in association with a range of different clinical phenotypes, including myoclonic epilepsy with ragged-red fibers (MERRF; Shoffner *et al.*, 1990). Mutations in the ribosomal RNA genes cause non syndromic sensorineural deafness, and an enhanced cochlear sensitivity to aminoglycoside antibiotics (Prezant *et al.*, 1993).

5. From genotype to biochemical phenotype

Patients with mtDNA disease often harbor a mixture of mutated and wild-type mtDNA (heteroplasmy). In the early 1990s, heteroplasmic trans-mitochondrial cybrids were generated by fusing enucleated cells from patients with mtDNA defects with a host cell line. Cybrid cell lines were used to study the relationship between the proportion of mutant mtDNA and mitochondrial function *in vitro* (reviewed in Attardi *et al.*, 1995). In general, the cell line only expressed a biochemical defect when the proportion of mutant mtDNA was high. For point mutations this 'threshold' was typically over 85% mutant (Chomyn *et al.*, 1991). Further evidence for a high genetic threshold level came from single muscle fiber studies. Patients with heteroplasmic mtDNA defects often show a mosaic pattern for cytochrome *c* oxidase (COX) activity in skeletal muscle. Adjacent fibers can have normal COX activity or no COX activity. The percentage level of mutant mtDNA in the COX deficient fibers is usually very high (>85%), and the level in COX positive cells is considerably less (Boulet *et al.*, 1992; Moraes *et al.*, 1992a,b). As a consequence of these studies, it is generally accepted that mtDNA mutations are highly recessive. This may be due to functional complementation of mutant genomes by small numbers of wild-type mtDNA molecules (Schon *et al.*, 1997). When the proportion of mutant mtDNA exceeds the threshold level, functional complementation cannot be sustained, and the genetic defect results in a biochemical deficiency of the mitochondrial respiratory chain.

5.1 MtDNA deletions

MtDNA deletions can affect the synthesis of mtDNA encoded respiratory chain units either directly, through deletion of a particular gene, or through a generalized defect of intra-mitochondrial protein synthesis (Shoubridge *et al.*, 1990). This occurs because of the topology of the mtDNA molecule. Large deletions usually involve both protein encoding and RNA genes, possibly explaining why there is not a tight correlation between the size of the deletion and the severity of disease. Very small deletions may cause a milder phenotype such as CPEO or recurrent myoglobinuria (Keightley *et al.*, 1996). In cybrid cells, the genetic threshold for mtDNA deletions appears to be lower than for most point mutations, at around 60% (Hayashi *et al.*, 1991), but there have been conflicting reports as to whether the proportion of mutant mtDNA (Moraes *et al.*, 1992b) or the amount of mutant mtDNA (Shoubridge *et al.*, 1990) is behind the respiratory chain deficiency.

5.2 MtDNA point mutations

Mis-sense (Wallace *et al.*, 1988), stop (Hanna *et al.*, 1998b) and initiation (Clark *et al.*, 1999) codon mutations have been identified in the structural genes in patients with a range of different disorders associated with the deficiency of a single respiratory chain complex (*Table 1* and Servidei, 2000). Point mutations of the mtDNA ATPase genes may also cause disease (Holt *et al.*, 1990). TRNA gene mutations appear to be particularly common, causing a generalized defect of intra-mitochondrial protein synthesis. Pathogenic mutations have been described at sites throughout the theoretical clover-leaf structure of the tRNA molecule, affecting tRNA charging and also the anti-codon region. tRNA mutations may have other effects, because the mtDNA molecule is so compact and different regions may have more than one function. For example, the A3243G mutation may influence the termination of transcription, tRNA function and possibly cause ribosomal stalling during translation (reviewed in Schon *et al.*, 1997). Mutations in the ribosomal RNA genes affect the aminoglycoside binding site, and may cause non syndromic deafness or enhanced sensitivity to these ototoxic drugs (Prezant *et al.*, 1993).

The pathogenic mechanism of the primary LHON mis-sense mutations is perplexing. Over 70% of individuals with LHON are homoplasmic for mutant mtDNA. The G11778A mutation results in an arginine to histidine substitution at amino acid position 340 (R340H) of the complex I subunit 4 (ND4) subunit (Wallace *et al.*, 1988), the G3460A mutation causes an A52T substitution in the ND1 subunit (Howell *et al.*, 1991a), and the T14484C mutation causes a M64V substitution in the complex I subunit 6 (ND6) subunit (Johns *et al.*, 1992), but these mutations are not always associated with a biochemical defect in the laboratory (Brown, 1999). In addition, the results of the biochemical assays performed on isolated mitochondria do not correspond to the *in vivo* bioenergetic defects which have been detected in skeletal muscle with ^{31}P-magnetic resonance spectroscopy (Lodi *et al.*, 1997). The relationship between the primary LHON mutation and the biochemical defect is complex, and nuclear (Cock *et al.*, 1998) and additional mitochondrial genetic factors may be important (El Meziane *et al.*, 1998; Howell *et al.*, 1991b).

6. From genotype to clinical phenotype

MtDNA mutations cause disease through an effect on respiratory chain function. It is easy to understand how different mtDNA defects can cause a similar disorder, but it is somewhat harder to explain the striking *difference* in phenotype seen amongst different patients with an *identical* genetic defect.

6.1 Variation in the level of heteroplasmy and tissue-specific thresholds

For heteroplasmic mtDNA mutations, the percentage level of mutant mtDNA may vary both between and within individuals with mtDNA disease (Macmillan *et al.*, 1993). This variability, coupled with tissue-specific differences in the threshold and the varied dependence of different organs on oxidative metabolism, is part of the explanation for the varied clinical phenotype seen in patients with mtDNA disease (Wallace, 1994). Two mechanisms may lead to differences in the level of mutant mtDNA in adjacent cells. First, during cell division, different amounts of mutant mtDNA may be delivered to the daughter cells (vegetative segregation; Shoubridge, 1995). This may result in different levels of mutant mtDNA in adjacent cells during embryonic development and between cells that continue to proliferate throughout life. Second, unlike nuclear DNA, mtDNA is continuously replicating, even in non dividing cells (relaxed replication; Clayton, 1992). This means that after cell division has finished, mutant and wild-type mtDNA may replicate at a different rate, which may also alter the level of heteroplasmy within the cell (intracellular drift; Birky, 1994). The effect of these processes can be greatly magnified if there is a selection pressure operating for or against mutant mtDNA. In general, mutant genomes are lost from rapidly dividing tissues such as the bone marrow. By comparison, post-mitotic (non dividing) tissues such as neurons, skeletal and cardiac muscle, and endocrine organs accumulate mutant genomes throughout life. This may contribute to the late presentation and slow progression seen in some patients with mtDNA disease.

6.2 Nuclear and environmental factors

Differences in heteroplasmy and tissue vulnerability can only be part of the explanation, and additional genetic and environmental factors are probably also important. Cybrid studies have shown that differences in the nuclear background can affect the genetic threshold and the biochemical defect (Cock *et al.*, 1998; Dunbar *et al.*, 1995, 1996), and additional mutations within the mitochondrial genome may suppress the effects of an established pathogenic mtDNA mutation (El Meziane *et al.*, 1998). There is also circumstantial evidence that nuclear genetic factors also influence the clinical expression of mtDNA mutations (Howell *et al.*, 1991b). The A3243G mutation may cause distinct phenotypes in different pedigrees. In some families, the predominant features are diabetes and deafness (Reardon *et al.*, 1992), in other families it may be recurrent strokes and cardiomyopathy (Ciafaloni *et al.*, 1992), and in other families, it may be CPEO (Moraes *et al.*, 1993). LHON provides a more complex example. LHON is primarily due to one of three homoplasmic mtDNA mis-sense mutations affecting the complex I (ND) genes (*Table 1*; Howell *et al.*, 1991a; Johns *et al.*, 1992; Wallace *et al.*, 1988), but only ~ 40% of males and ~10% of females with a primary LHON mutation develop the disease (Harding *et al.*, 1995). The sex bias may be due to an as yet unidentified X-chromosomal modifier locus, or an additional gender-associated autosomal modifier, possibly exerting its effect through gender-specific anatomical or physiological mechanisms (reviewed in Howell, 1997b). In UK pedigrees, the penetrance of LHON has not changed over the last 20 years, but in some Australian pedigrees, there has been a marked reduction in penetrance (Howell and Mackey,

1998). Some individuals develop the optic neuropathy after stress, trauma or in association with an excess alcohol intake (Riordan-Eva *et al.*, 1995). Although a systematic study of these environmental factors is difficult, it is likely that they do affect the expression of the mtDNA mutation.

6.3 Unanswered questions

Although we have some understanding of the relationship between mtDNA heteroplasmy and dysfunction at the level of the cell and organ, and we are beginning to understand the role of nuclear genes and environmental factors in the pathophysiology of mtDNA disease, there are still fundamental gaps in our knowledge. It is easy to understand how a mtDNA deletion in skeletal muscle might lead to muscle fatigue and weakness through a defect of oxidative metabolism, but many mtDNA disorders do not present in this way. This is illustrated by a simple consideration of the most common mtDNA point mutations. How do specific complex I gene mis-sense mutations cause a disease that only affects the optic nerve, and suddenly presents in young adults with normal vision up to that time (Howell, 1997a)? Similarly, what is the basis of the stroke-like episodes in patients with the A3243G Leu (UUR) tRNA gene mutation, and why do patients with this mutation have a relapsing/remitting course, unlike many individuals with tRNA gene mutations? These basic questions concerning the pathogenesis of the disease remain unanswered.

6.4 Correlation between the level of heteroplasmy and clinical features of mtDNA disease

Despite these complexities, when it has been possible to accumulate sufficient data to study the relationship between mutation load and clinical phenotype, it appears that, at least for some mutations, the proportion of mutant mtDNA is the overriding factor in determining clinical phenotype. In certain tissues, mutation load correlates with the severity of the clinical phenotype (Chinnery *et al.*, 1997a; White *et al.*, 1999a). For the two most common heteroplasmic mtDNA point mutations: A3243G and A8344G, high levels of mutant mtDNA in muscle are associated with a more severe clinical phenotype (Chinnery *et al.*, 1997a). For example, stroke-like episodes are unusual in patients with <50% A3243G mutation in their muscle, but the vast majority of individuals with >90% A3243G in their muscle suffer from recurrent stroke-like episodes (*Figure 2a* of Chinnery *et al.*, 1997a). This suggests that measuring the level of mutant mtDNA in skeletal muscle may provide a useful guide to prognosis. The relationship between the level of A3243G mutation in blood and clinical phenotype is less clear. This is probably because the percentage level in skeletal muscle corresponds to the level in the clinically relevant tissues such as the central nervous system. The level of the A3243G mutation decreases in blood over time ('t Hart *et al.*, 1996), and in some individuals it may fall to levels undetectable by some techniques (Chinnery *et al.*, 1997b). By contrast, levels of the T8993G/C mutations do not vary significantly from tissue to tissue (White *et al.*, 1999c), and a measurement of heteroplasmy in blood does correlate with the clinical phenotype (White *et al.*, 1999b).

7. The inheritance of pathogenic mtDNA defects

7.1 Maternal inheritance and the genetic bottleneck

After fertilization of the oocyte, sperm mtDNA is actively degraded (Shitara *et al.*, 1998), possibly mediated through ubiquitin tagging of sperm mitochondria (Sutovsky *et al.*, 1999). As a consequence, mtDNA is transmitted exclusively down the maternal line. This means that affected males with mtDNA disease cannot transmit the genetic defect. For reasons that are not well understood, mtDNA deletions are rarely, if ever, transmitted from clinically affected females to their offspring. In contrast, a female harboring a heteroplasmic mtDNA point mutation, or mtDNA duplications, may transmit a variable amount of mutated mtDNA to her children (Poulton and Turnbull, 2000).

It is thought that early during development of the female germ line, the number of mtDNA molecules within each oocyte is dramatically reduced before being subsequently amplified to reach final number of around 100 000 in each mature oocyte (Chen *et al.*, 1995; *Figure 4*). This restriction and amplification (also called the mitochondrial 'genetic bottleneck') leads to major differences in the level of heteroplasmy between individual oocytes, and the different levels of mutant mtDNA seen in the offspring of a single female (Poulton *et al.*, 1998). As a result, a single female harboring a pathogenic mtDNA defect may transmit a low level of mutant mtDNA to some offspring and high levels of the next offspring. Some of these offspring may be severely affected and die at a young age, whilst some may remain asymptomatic throughout their life.

The precise mechanisms of transmission are only just being elucidated. A number of groups have developed mice transmitting heteroplasmic mtDNA polymorphisms (Jenuth *et al.*, 1996; Meirelles and Smith, 1997, 1998), and it has been possible to study the level of heteroplasmy in cells within the developing

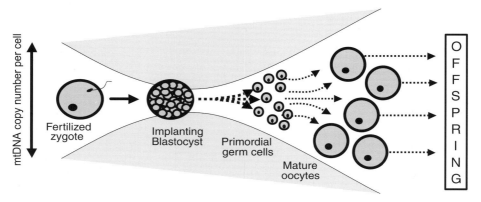

Figure 4. The mitochondrial genetic bottleneck. The mitochondrial genetic bottleneck provides an explanation for the different percentage levels of mutant mtDNA found between siblings. It is thought that a restriction in the number of mtDNA molecules within the cell occurs early in the development of the female germ line. This leads to marked differences in the level of heteroplasmy between primary oocytes within the same female, and accounts for the variation amongst offspring.

female germ line (Jenuth *et al.*, 1996). Most of the variation between the offspring of a single mother is present within her primary oocytes, which were formed when she was developing within her mother. These findings indicate that events during early female embryogenesis are critical in determining the level of heteroplasmy transmitted to the next generation (Jenuth *et al.*, 1996; *Figure 4*).

7.2 Genetic counseling for mtDNA diseases

Despite this variability, retrospective studies have revealed a relationship between the level of mutant mtDNA in a woman harboring a pathogenic mtDNA mutation and the risk of having affected offspring. For the most common mtDNA point mutation, A3243G, all women are at high risk of having clinically affected children. In contrast, for the A8344G mutation, women with less than 40% mutant mtDNA in their blood are considerably less likely to have a clinically affected child than a woman with ≥40% mutant mtDNA in her blood. For women with >40% A8344G mutation in their blood, the risk of having an affected child rises in proportion to the maternal blood mutation load (Chinnery *et al.*, 1998). A retrospective analysis for the T8993G/C mutations also demonstrates the potential benefit of measuring maternal mutation load for genetic counseling (White *et al.*, 1999a), although this mutation is somewhat different because the rate of *de novo* mutations causing isolated severely affected offspring is high (White *et al.*, 1999a).

Although it is likely that similar studies for other point mutations may reveal mutation-specific risks associated with a particular maternal mutation load, there have been insufficient numbers to calculate reliable statistics. The available evidence indicates that all pathogenic mtDNA mutations are transmitted by the same basic mechanism, but differences in the segregation and expression of the mutations will lead to characteristic inheritance patterns for each mutation (Chinnery *et al.*, 1998).

There are a significant proportion of females who harbor mtDNA point mutations that do not appear to be transmitted to their offspring. These females usually do not have a family history of mtDNA disease. They have a novel mtDNA mutation that is present in skeletal muscle, but *not* present in all tissues including blood, and presumably the ovaries (for examples see Taylor *et al.*, 1998 and Weber *et al.*, 1997). These individuals probably acquired a somatic mtDNA mutation in a purely somatic cell lineage early during embryogenesis. If these women come for counseling, the absence of a family history and the absence of the mutation in blood provide some reassurance that they are unlikely to have affected offspring. One approach to this problem is to stimulate ovulation, collect the oocytes, and determine whether any of them contain mutant mtDNA (discussed in Poulton and Turnbull, 2000).

Prenatal diagnosis is possible for some mtDNA diseases, but variation in the tissue level of mutant mtDNA before and after chorionic villus sampling may complicate the situation (Poulton *et al.*, 1998). There is an urgent need for prospective studies of the inheritance of heteroplasmic mtDNA mutations and prenatal diagnosis for mtDNA disorders.

7.3 Autosomally inherited mtDNA disorders

Finally, in a number of families, the propensity to form mtDNA deletions in muscle may be inherited as an autosomal trait. Both autosomal dominant and autosomal recessive CPEO has been linked to nuclear genetic loci (Chromosome 10q23.3-q24.3, and 3p14.1–21.2, *Table 1*; Kaukonen *et al.*, 1996; Suomalainen *et al.*, 1995). Recently, point mutations in three nuclear genes (*Ant1*, POLG and Twinkle) have been identified in families with autosomal dominant CPEO (Kaukonen *et al.*, 2000; Spelbrink *et al.*, 2001; van Goethem et al., 2001). Mitochondrial neuro-gastrointestinal encephalomyopathy (MNGIE) is a recessive disorder due to mutations in the thymidine phosphorylase gene on chromosome 22q13.32-qter (Nishino *et al.*, 1999).

7.4 mtDNA recombination and the paternal transmission of mtDNA: is it important?

The traditional view is that mammalian mtDNA is inherited strictly down the maternal line, but recent evidence has brought this into question, raising the possibility that the might be paternal mtDNA 'leakage'. The careful analysis of a large collection of mitochondrial genomes revealed an interesting finding (Awadalla *et al.*, 1999). First, linkage disequilibrium appeared to decline with distance (common polymorphic sites were less likely to be associated if they were widely separated); and second, based upon phylogenetic analysis, there was an excess of homoplasies (parallel mutations) in the analyzed sequences. Intermolecular mtDNA recombination is one possible explanation for these findings – but they have been hotly contested (Jorde and Banmshad, 2000; Parsons and Irwin, 2000). There are other explanations for the observed decline in linkage disequilibrium, and the study must be repeated before we accept that mtDNA recombination occurs. If there is mtDNA recombination it is extremely rare and it is unlikely to be of any practical significance for families transmitting pathogenic mtDNA mutations (Morris and Lightowlers, 2000).

8. The investigation of patients with suspected mtDNA disease

Some patients with mtDNA disease have an instantly recognizable clinical phenotype, such as LHON. In these individuals it is possible to make a molecular diagnosis from mtDNA analysis of a blood sample. For many other patients this is not possible. A huge number of different genetic defects may be responsible, some of which will not be detectable in blood. The investigation of patients with a suspected mitochondrial encephalomyopathy therefore involves the careful assimilation of clinical and laboratory data, before proceeding to molecular genetic testing.

Investigations fall into two main groups: clinical investigations used to characterize the pattern and nature of the different organs involved, and specific investigations looking for the underlying biochemical or genetic abnormality.

8.1 General clinical investigations

It is essential to look for common clinical features associated with mtDNA disease. Assessment should include an electrocardiogram and echocardiography, and an endocrine assessment (oral glucose tolerance test, thyroid function tests, alkaline phosphotase, fasting calcium and parathyroid hormone levels). Urine organic and amino acids may be abnormal even in the absence of overt tubular dysfunction. Measuring blood and CSF lactate levels is more helpful in the investigation of children than adults. These measurements must be interpreted with caution because there are many causes of blood and CSF lactic acidosis, including fever, sepsis, dehydration, seizures and stroke. The CSF protein may be elevated. The serum creatine kinase level may be raised, but is often normal. Neurophysiological studies may identify a myopathy or neuropathy. Electroencephalography may reveal diffuse slow-wave activity consistent with a subacute encephalopathy, or evidence of seizure activity. Cerebral imaging may be abnormal, showing lesions of the basal ganglia, high signal in the white matter on MRI, or generalized cerebral atrophy.

8.2 Specific investigations

A skeletal muscle biopsy is invaluable in the investigation of a suspected mtDNA defect. Histochemical and biochemical investigations, in conjunction with the clinical assessment, often indicate where the underlying genetic abnormality must lie.

Histochemistry and biochemistry. Histochemical analysis may reveal subsarcolemmal accumulation of mitochondria (so-called 'ragged-red' fibers), or COX deficiency. A mosaic of COX positive and COX negative muscle fibers suggests an underlying mtDNA defect. Patients who have COX deficiency due to a nuclear genetic defect usually have a global deficiency of COX affecting all muscle fibers. Electron microscopy may identify paracrystalline inclusions in the inter-membrane space, but these are non-specific and may be seen in other non mito-chondrial disorders. Respiratory chain complex assays can be carried out on various tissues. Skeletal muscle is preferable, but cultured fibroblasts are useful in the investigation of childhood mitochondrial disease. Measurement of the individual respiratory chain complexes determines whether an individual has multiple complex defects that would suggest an underlying mtDNA defect, involving either a tRNA gene or a large deletion. Isolated complex defects may be due to mutations in either mitochondrial or nuclear genes.

Molecular genetic investigations. For some mtDNA defects (particularly mtDNA deletions) the abnormality is not detectable in a DNA sample extracted from blood, and the analysis of DNA extracted from muscle is essential to establish the diagnosis. The first stage is to look for mtDNA rearrangements or mtDNA depletion by Southern blot analysis and long-range polymerase chain reaction (PCR). This is followed by PCR/RFLP analysis looking for common point mutations. Many patients with mitochondrial disease have a previously unrecognized mtDNA defect and it is necessary to directly sequence the mito-chondrial genome. Interpretation of the sequence data can be extremely difficult because mtDNA is highly polymorphic (Section 2). A mutation can only be considered to be pathogenic if it has arisen multiple times in the population, it is not seen in controls, and it is associated with a potential disease mechanism.

These stringent criteria depend upon a good knowledge of polymorphic sites in the background population. If a novel base change is heteroplasmic, this suggests that it is of relatively recent onset. Family, tissue segregation, and single cell studies may show that higher levels of the mutation are associated with mitochondrial dysfunction and disease, which strongly suggests that the mutation is causing the disease.

9. Therapeutic manipulation of the mitochondrial genome

Although there is currently no effective disease modifying therapy for patients with mtDNA disease, careful clinical management can minimise the burden of disease in families transmitting these disorders. Pharmacological treatment, using drugs such as ubiquinone (co-enzyme Q10), antioxidant vitamin supplements, thiamine and dichloracetate have had benefits in isolated cases, but there is a lack of randomized control trial data to show a convincing benefit (Taylor *et al.*, 1997a). The efficacy of dichloracetate is currently under investigation.

9.1 Exercise training

An exercise training programme can lead to a subjective improvement in muscle-related symptoms, enhanced aerobic exercise capacity, and greater muscle strength (Taivassalo *et al.*, 1997, 1999a). This may be of great practical benefit for the minority of patients with a predominantly myopathic phenotype.

9.2 Gene shifting

Concentric (shortening contractions) exercise training may also lead to a reduction in the mutation load within specific muscles, and a corresponding reduction in the proportion of COX negative fibers (Taivassalo *et al.*, 1999b). This strategy has been called 'gene shifting', and it is based upon the remarkable observation that in some patients, the amount of mutant mtDNA in muscle cell precursors (satellite cells) is undetectable. Concentric exercise induces the proliferation of the satellite cell population, which fuse with mature muscle fibers, 'diluting' the mutation load to sub-threshold levels. Satellite cell proliferation can also be induced by muscle toxins including bupivacaine and other local anesthetics. Fusion of these satellite cells with the mature muscle shifts normal mtDNA into COX deficient fibers, leading to a restoration of COX activity (Clark *et al.*, 1997; Fu *et al.*, 1996).

9.3 Gene therapy

A number of groups are targeting the mitochondrial genome. One approach is to specifically inhibit the replication of mutant mtDNA. Because mtDNA is continuously turning-over, this approach may eventually lead to a reduction in the amount of mutation within cells and a reversal of the biochemical defect (Taylor *et al.*, 1997b). Other laboratories have focused on the direct pharmacological methods of reducing mutant mtDNA, or delivering a self-replicating plasmid into mitochondria (Seibel *et al.*, 1995). Although these strategies show great potential *in vitro*, they are all plagued by the same problem – how can the therapeutic agent be systemically delivered into mitochondria within the relevant cells? Although our understanding of normal mitochondrial protein import has assisted the development of techniques to specifically target molecules into the mitochondrial compartment (Chinnery *et al.*, 1999), as with nuclear gene therapy, the problems associated with systemic delivery still remain.

10. Conclusion

In the decade following the identification of the first human pathogenic mtDNA mutations, a huge number of different mtDNA mutations were associated with disease. This led to major advances in our knowledge of cellular pathophysiology, but we are only just beginning to understand the complex relationship between the genetic defect and the clinical features of mtDNA disease. Some of these developments are of practical use for the clinician and the patient, leading to improved genetic and prognostic counseling, and better disease monitoring and surveillance. The scientific developments are also leading to novel treatment strategies, and a definitive therapy is only just beyond the horizon.

Acknowledgments

PFC is funded by Wellcome Trust. DMT is funded by the Wellcome Trust, The MRC (UK), and the Muscular Dystrophy Group of Great Britain.

References

Anderson, S., Bankier, A.T., Barrell, BG. *et al.* (1981) Sequence and organization of the human mitochondrial genome. *Nature* **290**: 457–465.
Andreu, A.L., Hanna, M.G., Reichmann, H. *et al.* (1999) Exercise intolerance due to mutations in the cytochrome b gene of mitochondrial DNA. *N. Engl. J. Med.* **341**: 1037–1044.
Andrews, R.M., Kubacka, I., Chinnery, P.F., Turnbull, D.M., Lightowlers, R.N. and Howell, N. (1999) Reanalysis and revision of the Cambridge Reference Sequence. *Nat. Genet.* **23**: 147.
Attardi, G., Yoneda, M. and Chomyn, A. (1995) Complementation and segregation behavior of disease causing mitochondrial DNA mutations in cellular model systems. *Biochim. Biophys. Acta* **1271**: 241–248.
Awadalla, P., Eyre-Walker, A. and Maynard Smith, J. (1999) Linkage disequilibrium and recombination in hominid mitochondrial DNA. *Science* **286**: 2524–2525.
Bandy, B. and Davison, A.J. (1990) Mitochondrial mutations may increase oxidative stress: implications for carcinogenesis and aging? *Free Radic. Biol. Med.* **8**: 523–539.
Birky, C.W. (1994) Relaxed and stringent genomes: why cytoplasmic genes don't obey Mendel's laws. *J. Heredity* **85**: 355–365.
Boulet, L., Karpati, G. and Shoubridge, E.A. (1992) Distribution and threshold expression of the tRNA(Lys) mutation in skeletal muscle of patients with myoclonic epilepsy and ragged-red fibers (MERRF). *Am. J. Hum. Genet.* **51**: 1187–1200.
Brown, M.D. (1999) The enigmatic relationship between mitochondrial dysfunction and Leber's hereditary optic neuropathy. *J. Neurol. Sci.* **165**: 1–5.

Casari, G., De Fusco, M., Ciarmatori, S. *et al.* (1998) Spastic paraplegia and OXPHOS impairment caused by mutations in paraplegin, a nuclear-encoded mitochondrial metalloprotease. *Cell* **93**: 973–983.

Chen, X., Prosser, R., Simonetti, S., Sadlock, J., Jagiello, G. and Schon, E.A. (1995) Rearranged mitochondrial genomes are present in human oocytes. *Am. J. Hum. Genet.* **57**: 239–247.

Chinnery, P.F. and Turnbull, D.M. (1999) Mitochondrial DNA and disease. *Lancet* **354** (suppl I): 17–21.

Chinnery, P., Howell, N., Lightowlers, R. and Turnbull, D. (1997a) Molecular pathology of MELAS and MERRF: the relationship between mutation load and clinical phenotype. *Brain* **120**: 1713–1721.

Chinnery, P.F., Reading, P.J., Walls, T.J. and Turnbull, D.M. (1997b) Recurrent strokes in a 34 year old man. *Lancet* **350**: 560.

Chinnery, P.F., Howell, N., Lightowlers, R.N. and Turnbull, D.M. (1998) MELAS and MERRF: the relationship between maternal mutation load and the frequency of clinically affected offspring. *Brain* **121**: 1889–1994.

Chinnery, P.F., Johnson, M.A., Wardell, T.M., Singh-Kler, R., Hayes, C., Taylor, R.W., Bindoff, L.A. and Turnbull, D.M. (2000) Epidemiology of pathogenic mitochondrial DNA mutations. *Ann. Neurol.* **48**: 188–193.

Chinnery, P.F., Taylor, R.W., Diekert, K., Lill, R., Turnbull, D.M. and Lightowlers, R.N. (1999) Peptide nucleic acid delivery into human mitochondria. *Gene Ther.* **6**: 1919–1928.

Chomyn, A., Meola, G., Bresolin, N., Lai, S.T., Scarlato, G. and Attardi, G. (1991) In vitro genetic transfer of protein synthesis and respiration defects to mitochondrial DNA-less cells with myopathy-patient mitochondria. *Mol Cell. Biol.* **11**: 2236–2244.

Ciafaloni, E., Ricci, E., Shanske, S. *et al.* (1992) MELAS: clinical features, biochemistry, and molecular genetics. *Ann. Neurol.* **31**: 391–398.

Ciafaloni, E., Santorelli, F.M., Shanske, S., Deonna, T., Roulet, E., Janzer, C., Pescia, G., DiMauro, S. (1993) Maternally inherited Leigh syndrome. *J. Pediatr.* **122**: 419–422.

Clark, K., Bindoff, L.A., Lightowlers, R.N., Andrews, R.M., Griffiths, P.G., Johnson, M.A., Brierley, E.J. and Turnbull, D.M. (1997) Correction of a mitochondrial DNA defect in human skeletal muscle. *Nat. Genet.* **16**: 222–224.

Clark, K.M., Taylor, R.W., Johnson, M.A. *et al.* (1999) An mtDNA mutations in the initiation codon of cytochondrome *C* oxidase subunit II results in lower levels of the protein and a mitochondrial encephalomyopathy. *Am. J. Hum. Genet.* **64**: 1330–1339.

Clayton, D.A. (1992) Replication and transcription of vertebrate mitochondrial DNA. *Ann. Rev. Cell Biol.* **7**: 453–478.

Cock, H.R., Tabrizi, S.J., Cooper, J.M. and Schapira, A.H. (1998) The influence of nuclear background on the biochemical expression of 3460 Leber's hereditary optic neuropathy. *Ann. Neurol.* **44**: 187–193.

Dunbar, D.R., Moonie, P.A., Jacobs, H.T. and Holt, I.J. (1995) Different cellular backgrounds confer a marked advantage to either mutant or wild-type mitochondrial genomes. *Proc. Natl Acad. Sci. USA* **92**: 6562–6566.

Dunbar, D.R., Moonie, P.A., Zeviani, M. and Holt, I.J. (1996) Complex I deficiency is associated with 3243G: C mitochondrial DNA in osteosarcome cell cybrids. *Hum. Mol. Genet.* **5**: 123–129.

El Meziane, A., Lehtinen, S.K., Hance, N., Nijtmans, L.G.J., Dunbar, D., Holt, I.J. and Jacobs, H.T. (1998) A tRNA suppressor mutation in human mitochondria. *Nat. Genet.* **18**: 350–353.

Fu, K., Hartlen, R., Johns, T., Genge, A., Karpati, G. and Shoubridge, E.A. (1996) A novel heteroplasmic tRNAleu(CUN) mtDNA point mutation in a sporadic patient with mitochondrial encephalomyopathy segregates rapidly in skeletal muscle and suggests an approach to therapy. *Hum. Mol. Genet.* **5**: 1835–1840.

Goto, Y., Nonaka, I. and Horai, S. (1990) A mutation in the tRNA(Leu)(UUR) gene associated with the MELAS subgroup of mitochondrial encephalomyopathies. *Nature* **348**: 651–653

Hanna, M.G., Brockington, M., Sweeney, M.G., Lamont, P.J., Morgan-Hughes, J.A. and Wood, N.W. (1998a) Mitochondrial DNA rearrangements in human disease: clinical and genetic correlations. *J. Neurol. Neurosurg. Psychiatry* **64**: 696–697.

Hanna, M.G., Nelson, I.P., Rahman, S. *et al.* (1998b) Cytochrome c oxidase deficiency associated with the first stop-codon point mutation in human mtDNA. *Am. J. Hum. Genet.* **63**: 29–36.

Harding, A.E., Sweeney, M.G., Govan, G.G. and Riordan-Eva, P. (1995) Pedigree analysis in Leber hereditary optic neuropathy families with a pathogenic mtDNA mutation. *Am. J. Hum. Genet.* **57**: 77–86.

Hayashi, J., Ohta, S., Kikuchi, A., Takemitsu, M., Goto, Y. and Nonaka, I. (1991) Introduction of disease-related mitochondrial DNA deletions into HeLa cells lacking mitochondrial DNA results in mitochondrial dysfunction. *Proc. Natl Acad. Sci. USA* **88**: 10614–10618.

Helgason, A., Siguroardottir, S., Gulcher, J.R., Ward, R. and Stefanson, K. (2000) mtDNA and the origin of the Icelanders: deciphering signals of recent population history. *Am. J. Hum. Genet.* **66**: 999–1016.

Holt, I., Harding, A.E. and Morgan-Hughes, J.A. (1988) Deletion of muscle mitochondrial DNA in patients with mitochondrial myopathies. *Nature* **331**: 717–719.

Holt, I.J., Harding, A.E., Petty, R.K. and Morgan-Hughes, J.A. (1990) A new mitochondrial disease associated with mitochondrial DNA heteroplasmy. *Am. J. Hum. Genet.* **46**: 428–433.

Howell, N. (1997a) Leber hereditary optic neuropathy: how do mitochondrial DNA mutations cause degeneration of the optic nerve? *J. Bioenerget. Biomembr.* **29**: 165–173.

Howell, N. (1997b) Leber hereditary optic neuropathy: mitochondrial mutations and degeneration of the optic nerve. *Vision Res.* **37**: 3495–3507.

Howell, N. (1997c) mtDNA recombination: what do the in vitro data mean? *Am. J. Hum. Genet.* **61**: 18–22.

Howell, N., Bindoff, L.A., McCullough, D.A., Kubacka, I., Poulton, J., Mackey, D., Taylor, L. and Turnbull, D.M. (1991a) Leber hereditary optic neuropathy: identification of the same mitochondrial ND1 mutation in six pedigrees. *Am. J. Hum. Genet.* **49**: 939–950.

Howell, N., Kubacka, I., Halvorson, S., Howell, B., McCullough, D.A. and Mackey, D. (1995) Phylogenetic analysis of the mitochondrial genomes from Leber hereditary optic neuropathy pedigrees. *Genetics* **140**: 285–302.

Howell, N., Kubacka, I. and Mackey, D.A. (1996) How rapidly does the human mitochondrial genome evolve? *Am. J. Hum. Genet.* **59**: 501–509.

Howell, N., Kubacka, I., Xu, M. and McCullough, D.A. (1991b) Leber hereditary optic neuropathy: involvement of the mitochondrial ND1 gene and evidence for an intragenic suppressor mutation. *Am. J. Hum. Genet.* **48**: 935–942.

Howell, N. and Mackey, D.A. (1998) Low-penetrance branches in matrilineal pedigrees with Leber Hereditary Optic Neuropathy. *Am. J. Hum. Genet.* **63**: 1220–1224.

Ivanov, P.L., Wadhams, M.J., Roby, R.K., Holland, M.M., Weedn, V.W. and Parsons, T.J. (1996) Mitochondrial DNA sequence heteroplasmy in the Grand Duke of Russia Georgij Romanov establishes the authenticity of the remains of Tsar Nicholas II. *Nat. Genet.* **12**: 417–420.

Jazin, E., Soodyall, H., Jalonen, P., Lindholm, E., Stoneking, M. and Gyllensten, U. (1998) Mitochondrial mutation rate revisited: hot spots and polymorphism. *Nat. Genet.* **18**: 109–110.

Jenuth, J., Peterson, A.C., Fu, K. and Shoubridge, E.A. (1996) Random genetic drift in the female germ line explains the rapid segregation of mammalian mitochondrial DNA. *Nat. Genet.* **14**: 146–151.

Johns, D.R., Neufeld, M.J. and Park, R.D. (1992) An ND-6 mitochondrial DNA mutation associated with Leber hereditary optic neuropathy. *Biochem. Biophys. Res. Commun.* **187**: 1551–1557.

Jorde, L.B. and Banmshad, M. (2000) Questioning evidence for recombination in mitochondrial DNA. *Science* **288**: 1931a.

Kaukonen, J.A., Amati, P., Suomalainen, A. et al. (1996) An autosomal locus predisposing to multiple deletions of mtDNA on chromosome 3p. *Am. J. Hum. Genet.* **58**: 763–769.

Kaukonen, J., Juselius, J.K., Tiranti, V., Kyttala, A., Zeviani, M., Comi, G.P., Keranen, S., Peltonen, L. and Suomalainen, A. (2000) Role of adenine nucleotide translocator 1 in mtDNA maintainance. *Science* **289**: 782–785.

Keightley, J.A., Hoffbuhr, K.C., Burton, M.D., Salas, V.M., Johnston, W.S., Penn, A.M., Buist, N.R. and Kennaway, N.G. (1996) A microdeletion in cytochrome c oxidase (COX) subunit III associated with COX deficiency and recurrent myoglobinuria. *Nat. Genet.* **12**: 410–416.

Larsson, N.-G. and Clayton, D.A. (1995) Molecular genetic aspects of human mitochondrial disorders. *Ann. Rev. Genet.* **29**: 151–178.

Leonard, J.V. and Schapira, A.V.H. (2000a) Mitochondrial respiratory chain disorders I: mitochondrial DNA defects. *Lancet* **355**: 299–304.

Leonard, J.V. and Schapira, A.V.H. (2000b) Mitochondrial respiratory chain disorders II: neurodegenerative disorders and nuclear gene defects. *Lancet* **355**: 389–394.

Lightowlers, R.N., Chinnery, P.F., Turnbull, D.M. and Howell, N. (1997) Mammalian mitochondrial genetics: heredity, heteroplasmy and disease. *Trends Genet.* **13**: 450–455.

Lightowlers, R.N., Jacobs, H.T. and Kajander, O.A. (1999) Mitochondrial DNA – all things bad? *Trends Genet.* **15**: 91–93.

Lodi, R., Taylor, D.J., Tabrizi, S.J., Kumar, S., Sweeney, M., Wood, N., Styles, P., Radda, G.K. and Schapira, A.V.H. (1997) In vivo skeletal muscle mitochondrial function in Leber's hereditary optic neuropathy assessed by ^{31}P magnetic resonance spectroscopy. *Ann. Neurol.* **42**: 573–579.

Lutsenko, S. and Cooper, M.J. (1998) Localization of the Wilson's disease protein product to mitochondria. *Proc. Natl Acad. Sci. USA* **95**: 6004–6009.

Lynch, M. and Jarrell, P.E. (1993) A method for calibrating molecular clocks and its application to animal mitochondrial DNA. *Genetics* **135**: 1197–1208.

Mackey, D.A., Oostra, R.-J., Rosenberg, T. *et al.* (1996) Primary pathogenic mtDNA mutations in multigeneration pedigrees with Leber hereditary optic neuropathy. *Am. J. Hum. Genet.* **59**: 481–485.

Macmillan, C., Lach, B. and Shoubridge, E.A. (1993) Variable distribution of mutant mitochondrial DNAs (tRNA(Leu[3243])) in tissues of symptomatic relatives with MELAS: the role of mitotic segregation. *Neurology* **43**: 1586–1590.

Meirelles, F. and Smith, L.C. (1997) Mitochondrial genotype segregation in a mouse heteroplasmic lineage produced by embryonic karyoplast transplantation. *Genetics* **145**: 445–451.

Meirelles, F. and Smith, L.C. (1998) Mitochondrial genotype segregation during preimplantation development in mouse heteroplasmic embryos. *Genetics* **148**: 877–883.

Moraes, C.T., Ciacci, F., Silvestri, G., Shanske, S., Sciacco, M., Hirano, M., Schon, E.A., Bonilla, E. and DiMauro, S. (1993) Atypical clinical presentations associated with the MELAS mutation at position 3243 of human mitochondrial DNA. *Neuromusc. Disord.* **3**: 43–50.

Moraes, C.T., DiMauro, S., Zeviani, M. *et al.* (1989) Mitochondrial DNA deletions in progressive external ophthalmoplegia and Kearns-Sayre syndrome. *N. Engl. J. Med.* **320**: 1293–1299.

Moraes, C.T., Ricci, E., Bonilla, E., DiMauro, S. and Schon, E.A. (1992a) The mitochondrial tRNA(Leu(UUR)) mutation in mitochondrial encephalomyopathy, lactic acidosis, and strokelike episodes (MELAS): genetic, biochemical, and morphological correlations in skeletal muscle. *Am. J. Hum. Genet.* **50**: 934–949.

Moraes, C.T., Ricci, E., Petruzzella, V., Shanske, S., DiMauro, S., Schon, E.A. and Bonilla, E. (1992b) Molecular analysis of the muscle pathology associated with mitochondrial DNA deletions. *Nat. Genet.* **1**: 359–367.

Morris, A.A. and Lightowlers, R.N. (2000) Can paternal mtDNA be inherited? *Lancet* **355**: 1290–1291.

Musumeci, O., Andreu, A.L., Shanske, S., Bresolin, N., Comi, G.P., Rothstein, R., Schon, E.A. and DiMauro, S. (2000) Intragenic Inversion of mtDNA: A New Type of Pathogenic Mutation in a Patient with Mitochondrial Myopathy. *Am. J. Hum. Genet.* **66**: 1900–1904.

Nagley, P. and Wei, Y.H. (1998) Ageing and mammalian mitochondrial genetics. *Trends Genet.* **14**: 513–517.

Nishino, I., Spinazzola, A. and Hirano, M. (1999) Thymidine phosphorylase gene mutations in MNGIE, a human mitochondrial disorder. *Science* **283**: 689–692.

Parsons, T.J. and Irwin, J.A. (2000) Questioning evidence for recombination in mitochondrial DNA. *Science* **288**: 1931a.

Parsons, T.J., Muniec, D.S., Sullivan, K. *et al.* (1997) A high observed substitution rate in the human mitochondrial DNA control region. *Nat. Genet.* **15**: 363–368

Poulton, J., Deadman, M.E., Bindoff, L., Morten, K., Land, J. and Brown, G. (1993) Families of mtDNA re-arrangements can be detected in patients with mtDNA deletions: duplications may be a transient intermediate form. *Hum. Mol. Genet.* **2**: 23–30.

Poulton, J., Deadman, M.E. and Gardiner, R.M. (1989) Tandem direct duplications of mitochondrial DNA in mitochondrial myopathy: analysis of nucleotide sequence and tissue distribution. *Nucleic Acids Res.* **17**: 10223–10229.

Poulton, J., Macaulay, V. and Marchington, D.R. (1998) Mitochondrial genetics '98: Is the bottleneck cracked? *Am. J. Hum. Genet.* **62**: 752–757.

Poulton, J. and Turnbull, D.M. (2000) 74th ENMC International workshop: Mitochondrial diseases. *Neuromusc. Disord.* **10**: 460–462.

Prezant, T.R., Agapian, J.V., Bohlman, M.C. *et al.* (1993) Mitochondrial ribosomal RNA mutations associated with both antibiotic-induced and non-sydromic deafness. *Nat. Genet.* **4**: 289–294.

Reardon, W., Ross, R.J., Sweeney, M.G., Luxon, L.M., Pembrey, M.E., Harding, A.E. and Trembath, R.C. (1992) Diabetes mellitus associated with a pathogenic point mutation in mitochondrial DNA. *Lancet* **340**: 1376–1379.

Riordan-Eva, P., Sanders, M.D., Govan, G.G., Sweeney, M.G., Da Costa, J. and Harding, A.E. (1995) The clinical features of Leber's hereditary optic neuropathy defined by the presence of a pathogenic mitochondrial DNA mutation. *Brain* **118**: 319–337.

Robin, E.D. and Wong, R. (1988) Mitochondrial DNA molecules and virtual number of mitochondria per cell in mammalian cells. *J. Cell Physiol.* **136**: 507–513.

Rotig, A., Cormier, V., Blanche, S., Bonnefont, J.-P. and Ledeist, F. (1990) Pearson's marrow pancreas syndrome. A multisystem mitochondrial disorder of infancy. *J. Clin. Invest.* **86**: 1601–1608.

Rotig, A., de Lonlay, P., Chretien, D., Foury, F., Koenig, M., Sidi, D., Munnich, A. and Rustin, P. (1997) Aconitase and mitochondrial iron-sulphur protein deficiency in Friedreich ataxia. *Nat. Genet.* **17**: 215–217.

Santorelli, F.M., Mak, S.C., El-Schahawi, M., Casali, C., Shanske, S., Baram, T.Z., Madrid, R.E. and DiMauro, S. (1996) Maternally inherited cardiomyopathy and hearing loss associated with a novel mutation in the mitochondrial tRNA(Lys) gene (G8363A). *Am. J. Hum. Genet.* **58**: 933–939.

Schadel, G.S. and Clayton, D.A. (1997) Mitochondrial DNA maintenance in vertebrates. *Ann. Rev. Biochem.* **66**: 409–435.

Schon, E.A., Bonilla, E. and DiMauro, S. (1997) Mitochondrial DNA mutations and pathogenesis. *J. Bioenerget. Biomembr.* **29**: 131–149.

Seibel, P., Trappe, J., Villani, G., Klopstock, T., Papa, S. and Reichmann, H. (1995) Transfection of mitochondria: strategy towards a gene therapy of mitochondrial DNA diseases. *Nucleic Acids Res.* **23**: 10–17.

Servidei, S. (2000) Mitochondrial encephalomyopathies: gene mutation. *Neuromusc. Disord.* **10**: X–XVI.

Shitara, H., Hayashi, J.I., Takahama, S., Kaneda, H. and Yonekawa, H. (1998) Maternal inheritance of mouse mtDNA in interspecific hybrids: segregation of the leaked paternal mtDNA followed by the prevention of subsequent paternal leakage. *Genetics* **148**: 851–857.

Shoffner, J.M, Lott, M.T., Lezza, A.M., Seibel, P., Ballinger, S.W. and Wallace, D.C. (1990) Myoclonic epilepsy and ragged-red fiber disease (MERRF) is associated with a mitochondrial DNA tRNA(Lys) mutation. *Cell* **61**: 931–937.

Shoubridge, E.A. (1995) Segregation of mitochondrial DNAs carrying a pathogenic point mutation (tRNA(leu3243)) in cybrid cells. *Biochem. Biophys. Res. Commun.* **213**: 189–195.

Shoubridge, E.A., Karpati, G. and Hastings, K.E. (1990) Deletion mutants are functionally dominant over wild-type mitochondrial genomes in skeletal muscle fiber segments in mitochondrial disease. *Cell* **62**: 43–49.

Siguroardottir, S., Helgason, A., Gulcher, J.R., Stefanson, K. and Donnelly, P. (2000) The mutation rate of the human mtDNA control region. *Am. J. Hum. Genet.* **66**: 1599–1609.

Spelbrink, J.N., Li, F.Y., Tiranti, V. *et al.* (2001) Human mitochondrial DNA deletions associated with mutations in the gene encoding Twinkle, a phage TY gene 4-like protein localised in the mitochondria. *Nat. Genet.* **28**: 223–231.

Sprintzl, M., Hartman, T., Weber, J., Blank, J. and Zeidler, R. (1989) Compilation of tRNA sequences and sequences of tRNA genes. *Nucl. Acid Res.* **17** (suppl): r1–r172.

Suomalainen, A., Kaukonen, J., Amati, P., Timonen, R., Haltia, M., Weissenbach, J., Zeviani, M., Somer, H. and Peltonen, L. (1995) An autosomal locus predisposing to deletions of mitochondrial DNA. *Nat. Genet.* **9**: 146–151.

Sutovsky, P., Moreno, R.D., Ramalho-Santos, J., Domiko, T., Simerly, C. and Schatten, G. (1999) Ubiquitin tag for sperm mitochondria. *Nature* **403**: 371–372.

't Hart, L.M., Jansen, J.J., Lemkes, H.H.P.J., de Knijff, P. and Maassen, J.A. (1996) Heteroplasmy levels of a mitochondrial gene mutation associated with diabetes mellitus decrease in leucocyte DNA upon aging. *Hum. Mut.* **7**: 193–197.

Taivassalo, T., De Stefano, N., Matthews, P.M., Argov, Z., Genge, A., Karpati, G. and Arnold, D.L. (1997) Aerobic training benefits patients with mitochondrial myopathies more than other chronic myopathies. *Neurology* **48**: A214.

Taivassalo, T., De Stefano, N., Chen, J., Karpati, G., Arnold, D.L. and Argov, Z. (1999a) Short-term aerobic training response in chronic myopathies. *Muscle Nerve* **22**: 1239–1243.

Taivassalo, T., Fu, K., Johns, T., Arnold, D., Karpati, G. and Shoubridge, E.A. (1999b) Gene shifting: a novel therapy for mitochondrial myopathy. *Hum. Mol. Genet.* **8**: 1047–1052.

Taylor, R.W., Chinnery, P.F., Bates, M.J.D., Jackson, M.J., Johnson, M.A., Andrews, R.M. and Turnbull, D.M. (1998) A novel mitochondrial DNA point mutation in the tRNA[Ile] gene: studies in a patient presenting with chronic progressive external ophthalmoplegia and multiple sclerosis. *Biochem. Biophys. Res. Com.* **243**: 47–51.

Taylor, R.W., Chinnery, P.F., Clark, K.M., Lightowlers, R.N. and Turnbull, D.M. (1997a) Treatment of mitochondrial disease. *J. Bioenerget. Biomembr.* **29**: 195–205.

Taylor, R.W., Chinnery, P.F., Turnbull, D.M. and Lightowlers, R.N. (1997b) Selective inhibition of mutant human mitochondrial DNA replication in vitro by peptide nucleic acids. *Nat. Genet.* **15**: 212–215.

Tengan, C.H. and Moraes, C.T. (1998) Duplication and triplication with staggered breakpoints in human mitochondrial DNA. *Biochim. Biophys. Acta* **1406**: 73–80.

Tiranti, V., Hoertnagel, K., Carrozzo, R. *et al.* (1998) Mutations of SURF-1 in Leigh disease associated with cytochrome c oxidase deficiency. *Am. J. Hum. Genet.* **63**: 1609–1621.

van Goethem, G., Dermaut, B., Loggren, A., Martin, J-J. and van Broeckhoven, C. (2001) Mutation of *POLG* is associated with progressive external ophthalmoplegia characterized by mtDNA deletions. *Nat. Genet.* **28**: 211–212.

van Holst Pellakaan, S.M., Frommer, M., Sved, J. and Boettcher, B. (1998) Mitochondrial control-sequence variation in aboriginal australians. *Am. J. Hum. Genet.* **62**: 435–449.

Veltri, K.L., Espiritu, M. and Singh, G. (1990) Distinct genomic copy number in mitochondria of different mammalian organs. *J. Cell Physiol.* **143**: 160–164.

Wallace, D.C. (1992) Mitochondrial genetics: a paradigm for aging and degenerative diseases? *Science* **256**: 628–632.

Wallace, D.C. (1994) Mitochondrial DNA mutations in diseases of energy metabolism. *J. Bioenerg. Biomembr.* **26**: 241–250.

Wallace, D.C. (1995) 1994 William Allan Award Address. Mitochondrial DNA variation in human evolution, degenerative disease, and aging. *Am. J. Hum. Genet.* **57**: 201–223.

Wallace, D.C. , Singh, G., Lott, M.T., Hodge, J.A., Schurr, T.G., Lezza, A.M., Elsas, L.J.D. and Nikoskelainen, E.K. (1988) Mitochondrial DNA mutation associated with Leber's hereditary optic neuropathy. *Science* **242**: 1427–1430.

Wallace, D.C. and Torroni, A. (1992) American Indian prehistory as written in the mitochondrial DNA: a review. *Hum. Biol.* **64**: 403–416.

Watson, E., Forster, P., Richards, M. and Bandelt, H.-J. (1997) Mitochondrial footprints of human expansion in africa. *Am. J. Hum. Genet.* **61**: 691–704.

Weber, K., Wilson, J.N., Taylor, L., Brierley, E., Johnson, M.A., Turnbull, D.M. and Bindoff, L.A. (1997) A new mtDNA mutation showing accumulation with time and restriction to skeletal muscle. *Am. J. Hum. Genet.* **60**: 373–380.

White, S.L., Collins, V.A., Woolfe, R., Cleary, M.A., Shanske, S., DiMauro, S., Dahl, H.M. and Thorburn, D.R. (1999a) Genetic counseling and prenatal diagnosis for the mitochondrial DNA mutations at nucleotide 8993. *Am. J. Hum. Genet.* **65**: 474–482.

White, S.L., Collins, V.R., Wolfe, R., Cleary, M.A., Shanske, S., DiMauro, S., Dahl, H.H. and Thorburn, D.R. (1999b) Genetic counseling and prenatal diagnosis for the mitochondrial DNA mutations at nucleotide 8993. *Am. J. Hum. Genet.* **65**: 474–482.

White, S.L., Shanske, S., McGill, J.J., Mountain, H., Geraghty, M.T., DiMauro, S., Dahl, H.-H.M. and Thorburn, D.R. (1999c) Mitochondrial DNA mutations at nucleotide 8993 show a lack of tissue of age-related variation. *J. Inher. Metab. Dis.* **22**: 899–914.

Wise, C.A., Sraml, M., Rubinsztein, D.C. and Easteal, S. (1997) Comparative nuclear and mitochondrial genome diversity in humans and chimpanzees. *Mol. Biol. Evol.* **14**: 707–716.

Yang, Z. (1996) Among-site rate variation and its impact on phylogenetic analyses. *Trends Ecol. Evol.* **11**: 367–372.

Zeviani, M., Moraes, C.T., DiMauro, S., Nakase, H., Bonilla, E., Schon, E.A. and Rowland, L.P. (1988) Deletions of mitochondrial DNA in Kearns-Sayre syndrome. *Neurology* **38**: 1339–1346.

Zhu, Z., Yao, J., Johns, T. *et al.* (1998) *SURF*1, encoding a factor involved in the biogenesis of cytochrome c oxidase, is mutated in Leigh syndrome. *Nat. Genet.* **20**: 337–343.

7

Identification of disease susceptibility genes (modifiers) in mouse models: cancer and infectious diseases

Tom van Wezel, Marie Lipoldovà, Peter Demant

1. Introduction

The common, non Mendelian, diseases are responsible for the largest part of morbidity and mortality, health care costs and the economic loss due to disease (Risch and Merikangas, 1996). The identification of quantitative trait loci (QTLs) controlling their development will soon be greatly facilitated by the possibility of combining the information about their chromosomal position with the corresponding genome sequence. Moreover, the single nucleotide polymorphisms (SNPs; Wang *et al.*, 1998) will permit genotyping of the multiple QTLs affecting disease susceptibility. As the mapping of QTLs in humans is difficult even if they have sizable effects (Ponder, 1990), strategies aiming to map and clone the relevant QTLs, first in the mouse and subsequently to define their human homologs, are being widely pursued (Bedell *et al.*, 1997). The disease susceptibility genes, called also 'modifier genes' because they modify effects of major Mendelian disease genes, have been the subject of intensive investigation in the past decade, as the availability of microsatellite markers (Dietrich *et al.*, 1996) made detailed linkage studies possible. In this chapter we will elucidate the main features of modifier genes and approaches to their study using the examples of cancer modifier genes and genes that modify T-cell activation and response to infection, to a considerable extent based on our own work.

Cancer is one of the major causes of morbidity and mortality in Western societies. Cancer development is a multi-stage process driven by somatic alterations in the genetic make-up of the prospective cancer cells. The accumulation of mutations

Genotype to Phenotype second edition, edited by S. Malcolm and J. Goodship.

which increase the selective advantage of the cancer cells, or of their precursors, over normal cells is the basis of cancer progression and leads to the gradual escape of cancer cells from the regulatory factors responsible for the orderly maintenance of the cell populations in organs and tissues. A number of genes whose somatic mutations contribute to cancer phenotype have been described. These include oncogenes, which are activated in tumor cells, and suppressor genes, which are inactivated in tumor cells. In addition, defective function of genes involved in repair of DNA damage, such as point mutations or chromosome breaks, contributes to genetic instability of cells and hence can enhance tumorigenesis. It is not surprising that if mutations in such genes occur in the germ line, their carriers may be more prone to develop cancer. This is indeed the case with tumor suppressor genes such as Rb (retinoblastoma) APC (adenomatous polyposis coli) or MSH2 (mut S homolog 2) and others (Ponder, 2001).

The elucidation of the genetic basis of cancer may contribute to improved prevention and treatment strategies, as well as to a better insight into the mechanisms of carcinogenesis. The genetic basis of cancer is evident in numerous familial cancer syndromes with Mendelian inheritance of cancer predisposition. The genetic mutations responsible for these syndromes have very high penetrance; virtually all carriers of such gene defects are likely to develop the syndrome. However, statistical considerations (Peto, 1980; Ponder, 1990; Lichtenstein et al., 2000) as well as large differences in incidence of tumors between inbred strains of animals (Murphy, 1966; van der Valk, 1981) indicate that a large proportion of apparently sporadic cancers occur in genetically predisposed individuals. Mutations in the responsible susceptibility genes have a lower penetrance than those causing the familial cancer. Detection of low-penetrance genetic component is probably virtually impossible in humans, but it is feasible in the mice and rats, by analyzing tumor susceptibility as a quantitative trait. Subsequently one can search for homologous loci in humans.

1.1 Recombinant congenic strains

The multiplicity of genes affecting common diseases such as cancer, infectious and autoimmune diseases, renders the mapping of the QTLs difficult, as each of the multiple genes accounts, by definition, for only a small part of genetic variance. Therefore, a genetic tool has been developed, the recombinant congenic strains (RCS), which permits analysis of small subsets of the relevant genes in the mouse. The RCS are a uniquely effective tool for analysis of tumor susceptibility genes in the mouse, which aids their detection and offers the possibility to study their function in tumorigenesis (Demant, 1992; Demant and Hart, 1986). The RCS were produced from two standard inbred strains, one of which serves as a background strain, the other as a donor strain. Two generations of back-crossing to the background strain, followed by brother–sister mating, produce a series of new homozygous strains (the RC strains), each of which carries a random fraction of only about 12.5% of the genome from the donor strain and 87.5% of the genome from the common background strain. Three series of RC strains of mice were produced: BALB/c-c-STS/Dem (CcS), C3H-c-C57BL/10/Dem (HcB) and O20-c-B10.O20/Dem (OcB). A series of RCS comprises approximately 20 homozygous strains, all produced by back-crossing and inbreeding from two parental inbred strains: a 'background' strain and a 'donor' strain. Each RC strain of a series contains about 87.5% genes of the common background strain and about 12.5% genes of the common donor strain. As each RC strain carries a different, partly overlapping, random subset of genes of the donor strain on the genetic background of the background strain, the number of non linked susceptibility genes of the donor origin in each RC strain will be greatly reduced (*Figure 1*). Thus, the RCS system transforms a multigenic difference into a set of single gene or oligogenic

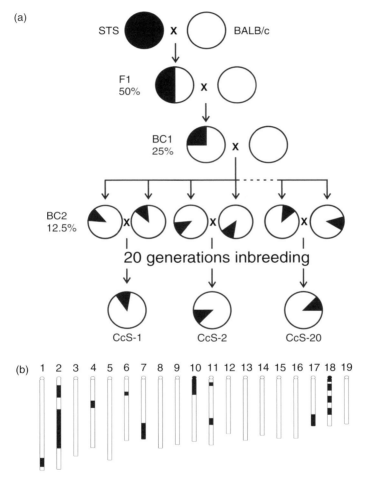

Figure 1. (a) Construction of the CcS/Dem Recombinant Congenic Strains (RCS) from the donorstrain STS and background strain BALB/c. Strain STS is backcrossed twice to strain BALB/c. Subsequently, pairs of backcross 2 mice were inbred to generate a series of 20 different CcS strains. Each CcS strain has inherited approximately 12.5% of the genes from the donor strain and 87.5% from the background strain. The other RCS series; OcB and Hcb, were constructed from strains O20 and B10.O20 and from C3H and B10 respectively. (b) Example of the distribution pattern of the 12.5% of STS-derived segments (black bar) in strain CcS-19. The segments are spread over 9 of the 19 autosomes.

differences. The use of RCS increases the efficiency of the mapping of individual and multiple QTL through two basic mechanisms: (1) a diminished residual genetic variance; (2) a lower threshold for significance of linkage. Thus, the use of RCS can considerably enhance our capacity to map the QTLs, including the tumor susceptibility genes, by achieving higher LOD score (or F-ratio) for the same locus as compared with the total genome cross of the same size. In addition, the LOD score (or F-ratio) obtained with an RCS cross is more significant than the same value obtained in a total genome cross because of less stringent correction for multiple testing due to segregation of $\frac{1}{8}$th of the genome only. In order to map a QTL in an RC strain, it is necessary to establish, using polymorphic markers, which segments of its genome originate from the background strain and which from the donor strain. The three series of RC strains were characterized extensively with more than 600 markers for the CcS- and OcB-series and 130 markers for the HcB series (Stassen *et al.*, 1996).

1.2 Cancer susceptibility

The availability of inbred strains of mice and rats and their more sophisticated derivatives, congenic strains, recombinant inbred strains (RIS), RCS and consomic strains, allowed a more detailed study of the phenomenon of tumor susceptibility than observations of family syndromes or population studies in man. The results of this analysis in different species and with spontaneous or induced tumors in different organs, revealed several common features observed in studies of different tumors (reviewed by Balmain and Nagase, 1998; Demant, 1992; Demant *et al.*, 1989; Drinkwater and Bennett, 1991; Murphy, 1966):

Polymorphism. The genetic differences in tumor susceptibility have been detected in all species studied. Because with most tumor types a test of a limited number of inbred strains is sufficient to detect a susceptible and a resistant strain, there must be considerable polymorphism among the tumor susceptibility genes. Whether this represents a large number of tumor susceptibility genes with a small number of alleles, or a limited number of tumor susceptibility genes with a large number of alleles at each locus remains to be investigated.

Organ specificity or organ limitation. Susceptibility to tumors in different organs usually exhibits a different strain distribution pattern, indicating that different subsets of genes affect the tumor susceptibility in each organ. For example, the BALB/c mice are susceptible to mammary tumors, but resistant to tumors of the small intestine, but the reverse is true for C57BL/6 (B6) mice, while both these strains are resistant to colon tumors, to which STS/A mice are susceptible. To what extent these different sets of genes share certain common loci remains to be seen. In view of numerous familial syndromes with increased cancer incidence in several organs, this is not unlikely.

Tumor susceptibility genes. These operate within the target organ, rather than systemically, as indicated in experiments with tumor induction in transplanted organs or in chimeric mice. However, there is evidence that at least some tumor susceptibility genes have a systemic effect (Dux and Demant, 1987).

Individual tumor susceptibility genes. These affect some but not all aspects of the tumor phenotype. Thus, colon tumor multiplicity appears to be controlled by a different subset of genes than tumor size (Moen *et al.*, 1991; Fijneman *et al.*, unpublished data). Similarly, histological type or propensity for malignant progression may be under control of a different subset of genes than tumor multiplicity or incidence.

2. Cancer genetics in mouse strains, current QTLs

In mouse models, numerous susceptibility genes have been identified over the past few years for the complex traits of cancer susceptibility. These loci are involved in the control of many different aspects of cancer genetics such as tumor multiplicity, tumor size, progression, or histological type. In this chapter, we will give an overview of the status of mapped cancer QTLs, both in the RCS model and in other models, listed in *Table 1* and *Figure 2*. The CcS series has been used for the dissection of colon cancer susceptibility (nine different loci) and the OcB series have been used for the identification of lung cancer susceptibility loci (14 different loci). Using mostly whole genome crosses, additional QTLs have been mapped for lung and colon cancer and for other cancers such as cancer of the skin and liver, lymphomas and teratomas.

2.1 Colon cancer

Colon tumors can be induced by repeated administration of DMH to the mice, these tumors form a model for sporadic colorectal cancer. Mouse inbred-strains differ widely in their susceptibility to DMH induced colon cancer and experiments with RIS have shown that multiple unlinked QTLs influence this susceptibility (Fleiszer *et al.*, 1988).

The CcS series of the RCS is based on strain STS, which is highly susceptible to colon cancer, and the resistant strain BALB/c. Tests for colon tumor susceptibility in the 20 CcS/Dem strains revealed a number of susceptibility loci polymorphic between BALB/c and STS (Moen *et al.*, 1991). In linkage tests using crosses of the susceptible strains CcS-3, -5, -11 and -19, with the resistant background strain BALB/c, nine QTLs have been mapped to date: *Scc1* and *Scc2* (Moen *et al.*, 1992, 1996), *Scc3–Scc5* (van Wezel *et al.*, 1996), and *Scc6–Scc9* (van Wezel *et al.*, 1999). *Scc1* was confirmed and fine-mapped using the strains CcS-16 and CcS-17 (Moen *et al.*, 1996). In addition to *Scc1*, this revealed a closely linked QTL, *Scc2*, centromeric to *Scc1* on chromosome 2.

Interestingly two pairs of loci; *Scc4–Scc5* and *Scc7–Scc8* are involved in a reciprocal genetic interaction (van Wezel *et al.*, 1996, 1999).

In whole genome crosses of the mouse strains, ICR, B6 and CBA, two loci, *Ccs1* and *Ccs2* were identified (Angel *et al.*, 2000; Jacoby *et al.*, 1994). *Ccs2* maps to the same location as *Scc7*; so it is likely that *Ccs2* is identical to *Scc7*.

The Min mouse forms a model for familial adenomatous polyposis (FAP), a familial form of colon cancer. Min mice carry a dominant mutation in the Apc gene and develop multiple intestinal neoplasia (Su *et al.*, 1992). The number of

Table 1. Overview of all cancer QTLs per cancer type. Chromosome and centimorgan position (cM) for each QTL (or the nearest marker) were taken from the Mouse Genome Database (Blake, 2000). Only those loci with a significant linkage were included, e.g. $p < 0.003$ or LOD > 3.1, for a whole genome cross. A graphical overview of all the cancer QTLs per chromosome is shown in *Figure 1*.

	Locus	Chr.	Pos.		Description
Lung cancer	*Pas8*	1	81.6	Urethane	Pulmonary adenoma susceptibility 8
	Sluc5	1	87.9	ENU	Susceptibility to lung cancer 5
	Sluc2 (Pas6)	2	38	ENU	Susceptibility to lung cancer 2
	Pas9	4	42.5	Urethane	Pulmonary adenoma susceptibility 9
	Papg1	4	49.6	Urethane	Pulmonary adenoma progression 1
	Sluc6	4	67	Urethane	Susceptibility to lung cancer 6
	Par4	6	3.3	Urethane	Pulmonary adenoma resistance 4
	Sluc7	6	6	ENU	Susceptibility to lung cancer 7
	Pas1c	6	37	Urethane	Pulmonary adenoma susceptibility 1c
	Sluc3 (Pas1b)	6	61	ENU	Susceptibility to lung cancer 3
	Pas1	6	72	Urethane	Pulmonary adenoma susceptibility 1
	Sluc8	7	72	Urethane	Susceptibility to lung cancer 8
	Sluc9	8	59	Urethane	Susceptibility to lung cancer 9
	Pas4	9	12	Urethane	Pulmonary adenoma susceptibility 4
	Sluc10	9	17	ENU	Susceptibility to lung cancer 10
	Sluc11	9	55	ENU	Susceptibility to lung cancer 11
	Sluc4 (Pas5)	11	40	ENU	Pulmonary adenoma susceptibility 5
	Par1 IPas5b	11	56	Urethane	Pulmonary adenoma resistance 1
	Par3	12	37	Urethane	Pulmonary adenoma resistance 3
	Sluc12	12	59	ENU	Susceptibility to lung cancer 12
	Sluc13	14	12.5	ENU	Susceptibility to lung cancer 13
	Lts	17	19	ENU	Lung tumor susceptibility
	Sluc14	18	20	ENU	Susceptibility to lung cancer 14
	Par2/Pas7	18	44	Urethane	Pulmonary adenoma resistance 2
	Pas3	19	5	Urethane	Pulmonary adenoma susceptibility 3
	Sluc1/Pas3b	19	47	ENU	Susceptibility to lung cancer 1
Intestinal Cancer	*Scc3*	1	101.5	DMH/ENU	Susceptibility to colon cancer 3
Colon	*Scc2*	2	28	DMH	Susceptibility to colon cancer 2
	Scc1	2	49.5	DMH	Susceptibility to colon cancer 1
	Scc7/Ccs2	3	76.2	DMH	Susceptibility to colon cancer 7
	Scc8	8	4.4	DMH	Susceptibility to colon cancer 8
	Scc9	10	63	DMH	Susceptibility to colon cancer 9
	Scc6	11	2.4	DMH	Susceptibility to colon cancer 6
	Ccs1	12	38	DMH	Colon cancer susceptibility 1
	Scc4	17	47.4	DMH/ENU	Susceptibility to colon cancer 4
	Scc5	18	25	DMH/ENU	Susceptibility to colon cancer 5
Intestine	*Mom1*	4	67	Min mutation	Modifier of Min 1
	Ssic1	4	67	ENU	Susceptibility to small intestinal cancer 1

Table 1. *continued*

	Locus	Chr.	Pos.		Description
Liver cancer	*Hcf2*	1	82	DEN	Hepatocarcinogenesis in females 2
	Hcs7	1	84	Urethane	Hepatocarcinogenesis susceptibility 7
	Hcs4	2	99	Urethane	Hepatocarcinogenesis susceptibility 4
	Hcr1	4	51	DEN	Hepatocarcinogen-resistance 1
	Hcs5	5	49	Urethane	Hepatocarcinogenesis susceptibility 5
	Hcs1	7	24	Urethane	Hepatocarcinogenesis susceptibility 1
	Hcs2	8	56	Urethane	Hepatocarcinogenesis susceptibility 2
	Hcr2	10	32	DEN	Hepatocarcinogen-resistance 2
	Hcs3	12	59	Urethane	Hepatocarcinogenesis susceptibility 3
	Hcf1	17	19	DEN	Hepatocarcinogenesis in females 1
	Hsc6	19	32	Urethane	Hepatocarcinogenesis susceptibility 6
Lymphomas	*Lyr2*	4	14.5	Endogenous virus	Lymphoma resistance 2
	Gct1	4	69	DHEA	Granulosa cell tumorigenesis 1
	Tlsm1	7	66	Endogenous virus	Thymic lymphoma susceptible 1
	Gct2	12	22	DHEA	Granulosa cell tumorigenesis 2
	Msmr1	17	22	Endogenous virus	Pre-B lymphoma resistance 1
	Msmr2	18	32	Endogenous virus	Pre-B lymphoma resistance 2
	Gct4	x	38	DHEA	Granulosa cell tumorigenesis 4
	Pctm	1	80	Pristine	Plasmacytoma susceptibility 3
	Pctr1	4	70	Pristine	Plasmacytoma susceptibility 1
	Pctr2	4	79	Pristine	Plasmacytoma susceptibility 2
Skin cancer	*Skts8*	1	79	DMBA/TPA	Skin tumor susceptibility 8
	Skts7	4	56	DMBA/TPA	Skin tumor susceptibility 7
	Skts3	5	24	DMBA/TPA	Skin tumor susceptibility 3
	Skts4	5	64	DMBA/TPA	Skin tumor susceptibility 4
	Skts12	6	74	DMBA/TPA	Skin tumor susceptibility 12
	Skts2	7	27	DMBA/TPA	Skin tumor susceptibility 2
	Skts1	7	64	DMBA/TPA	Skin tumor susceptibility 1
	Skts6	9	49	DMBA/TPA	Skin tumor susceptibility 6
	Ps/1	9	61	MNNG/TPA	Skin tumor promotion
	Skts5	12	17	DMBA/TPA	Skin tumor susceptibility 5
	Skts9	16	14	DMBA/TPA	Skin tumor susceptibility 9
	Skts10	17	32	DMBA/TPA	Skin tumor susceptibility 10
Teratomas	*Ots1*	6	40	Spontaneous	Ovarian teratoma susceptibility 1
	Pgct1	13	38	p53–/–	Primordial germ cell tumors 1
	Ter	18	20	Spontaneous	Testicular germ cell tumors
	Tgct1	19	12	Spontaneous	Testicular germ cell tumors 1

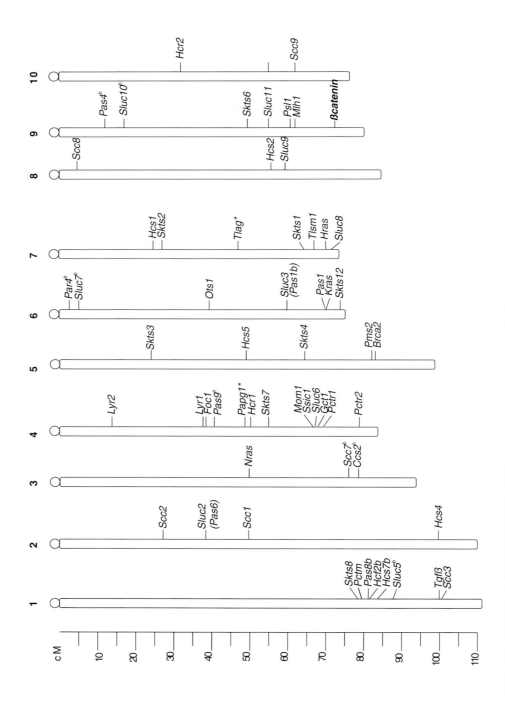

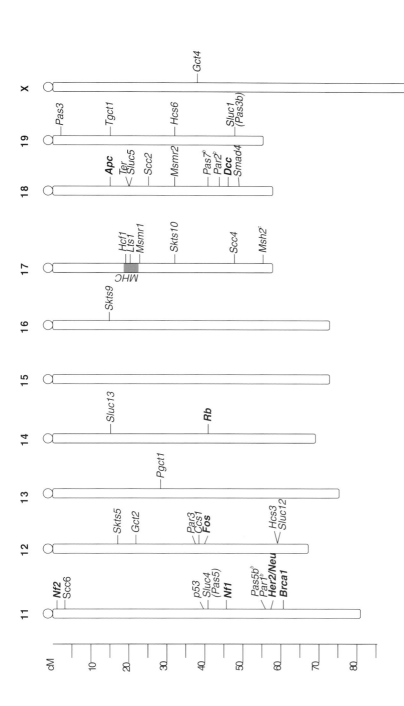

Figure 2. Graphical overview of cancer susceptibility QTLs (see *Table 1*) and cancer related genes (bold) per chromosome. The QTLs and genes are placed at the centimorgan positions reported by the mouse genome database (Blake, 2000) or at the position of the marker with the highest linkage in the mapping interval. For all QTLs the 90% confidence interval typically is between 15 and 25 cM around the indicated position. Several susceptibility QTLs for lung cancer a, liver cancer b and colon cancer c map at the same position, possibly being the same locus. Two suggestive linkages to lymphoma susceptibility d map at the same position, supporting their mapping.

intestinal tumors is strongly influenced by the genetic background. Min mice develop, on average, 4–5 times more tumors on the genetic background of B6 than on the background of AKR. Genetic mapping revealed a dominant acting susceptibility locus *Mom1*, distal on chromosome 4 (Dietrich *et al.*, 1993). A locus influencing ENU induced intestinal tumors, *ssic1*, was mapped to the same position as *Mom1* (Fijneman *et al.*, 1995) and is probably identical to *Mom1*. Pla2s was identified as a strong candidate gene for *Mom1*; pla2s shows full concordance between susceptibility and a non functional allele (MacPhee *et al.*, 1995). It appears, however, that the effect of Mom1 is due to two linked genes (Cormier *et al.*, 2000).

2.2 Lung cancer

The OcB series of the RCS has been used to study lung cancer susceptibility. The two parental strains of the OcB series, O20 and B10.O20 differ significantly in their ENU induced lung tumors (O20 develops larger tumors). The major histocompatibility complex (MHC) is known to be involved in lung tumor susceptibility (Oomen *et al.*, 1983); at least three distinct loci within the MHC influence lung cancer susceptibility. A susceptibility locus for alveolar lung tumors was mapped to a 50 kb interval in the MHC (Fijneman *et al.*, 1995). O20 and B10.O20 share the same MHC region, thus excluding this segment in the analysis of susceptibility in the OcB series. In crosses between strain O20 and each of the RC-strains OcB-4, -6 and -9, 14 different QTLs influencing susceptibility to lung cancer (Sluc) were identified. These QTLs are involved in the control of both lung tumor size (*Sluc1-5, 7, 9-12* and *-14*) and lung tumor number(*Sluc6, 8* and *13*; Fijneman *et al.*, 1996, 1998). Strikingly, all *Sluc* loci, except *Sluc13*, are involved in one or more genetic interactions, demonstrating that the occurrence of interactions in lung tumor susceptibility is as common a phenomenon, as it is in colon cancer.

Several studies have compared lung cancer susceptibility of different inbred strains relative to each other. These strains vary in susceptibility, both in quantitative and qualitative aspects and strain A is always the most susceptible strain (independent of the method of induction), whereas strains B6 and C3H are usually at the other end of the spectrum (reviewed in Demant *et al.*, 1989).

Different studies on the lung cancer susceptibility have used whole genome crosses involving the highly susceptible strain A and resistant strains such as C3H and BALB/c. In these studies 11 different lung cancer QTLs have been identified, these loci were either designated *Pas* (pulmonary adenoma susceptibility; Devereux *et al.*, 1994; Festing *et al.*, 1994, 1998; Gariboldi *et al.*, 1993) or *Par* (pulmonary adenoma resistance; Manenti *et al.*, 1994, 1997; Obata *et al.*, 1996; Pataer *et al.*, 1996). The *Pas/Par* loci are involved in lung tumor multiplicity, another locus, *Papg1*, controls lung tumor size (Manenti *et al.*, 1997). The large number of different susceptibility loci clearly show that the *Pas1* locus is not the major lung tumor susceptibility locus it is often considered to be. The likely candidate for *Pas1*, *Kras2* has been implicated in lung tumor susceptibility (Ryan *et al.*, 1987) and *Pas1* maps close to *Kras* on chromosome 6.

The 26 different lung cancer susceptibility QTLs have been mapped to 14 different autosomes, no QTLs have been identified (yet) on chromosomes 3, 5, 10, 13 and 15. Four pairs of QTLs map to the same chromosomal segments (*Figure 1*):

Sluc5 and *Pas8* distal on chromosome 1, *Papg1* and *Pas9* on chromosome 4, *Sluc7* and *Par4* on chromosome 6 (different from Kras) and *Sluc10* and *Pas4* on chromosome 9. It seems likely that some of these QTLs might be identical. Fine mapping the QTLs and subsequent cloning of the underlying genes will prove if these QTLs are identical.

2.3 Skin cancer

To date, 12 skin cancer susceptibility loci have been identified in different segregating crosses. These QTLs are mainly involved in the control of papilloma and carcinoma multiplicity, but also in skin tumor promotion and survival time. Skin cancer is mostly induced in a two-stage carcinogenesis protocol where mice receive a single dose of an initiating carcinogen, usually 7,12-dimethylbenz(a)anthracene (DMBA) applied to the skin, followed by repeated application of the promoting agent 12-O-tetradecanoylphorbol-13-acetate (TPA). Mouse inbred strains differ widely in their susceptibility to skin cancer, strains such as CD1, DBA/2 and SENCAR are relatively susceptible, strains BALB/c and B6 are resistant to skin cancer. Several aspects of two-stage skin carcinogenesis, such as tumor incidence and multiplicity, were analyzed up to 80 weeks after initiation in a large interspecific back-cross between Mus Spretus and the NIH inbred strain (Nagase *et al.*, 1995, 1999). In this way, 10 different skin tumor susceptibility (Skts) loci were mapped. *Skts1*, *Skts2* and *Skts5* to *10* are involved in papilloma multiplicity, *Skts3* controls papilloma incidence whereas *Skts12* is involved in survival time. *Skts10* was only significant in an interaction with other loci. The susceptibility locus *Skts4*, central on chromosome 5, was identified in the control of papilloma multiplicity with the use of the SENCARA/Pt inbred strain; a sub strain derived from the SENCAR outbred stock (Mock *et al.*, 1998). SENCAR mice are specifically produced for increased susceptibility to skin tumors by selective breeding of outbred mice. When combining different crosses between the resistant B6 strain and the susceptible strain DBA/2, a single skin cancer susceptibility locus on chromosome 4 was mapped (*Psl1*, Angel *et al.*, 1997). The DBA/2 allele of *Psl1* was associated with both skin tumor incidence and multiplicity.

2.4 Teratomas

Spontaneous ovarian teratomas are extremely rare in most inbred mouse strains. In strain LT/Sv however, approximately half the females develop spontaneous ovarian teratomas. Analysis on a B6 × LT/Sv intercross indicated a multigenic control of the susceptibility to ovarian teratomas, linkage was found to a single locus on chromosome 6, *Ots1*. All females with teratomas carry at least one LT derived allele at this locus (Lee *et al.*, 1997).

The development of testicular teratomas is under multigenic control, it is suggested by at least 13 genes (Collin *et al.*, 1996). Testicular teratoma incidence is increased by the *Ter* mutation on the 129/Sv background. The *Ter* mutation has been mapped to chromosome 18 (Asada *et al.*, 1994; Noguchi *et al.*, 1985). An intersubspecific cross with 129/Ter and MOLF yielded suggestive linkage to chromosome 19 (Collin *et al.*, 1996). This linkage was confirmed with a consomic strain where chromosome 19 of 129/Sv was substituted by the MOLF homolog (Matin *et al.*, 1999). However, the complex genetics, with possibly multiple loci on chromosome 19 and genetic interactions involved, show the limitations in the use of CCS for dissecting multigenic traits. Another locus was recently identified in crosses between B6 and 129 mice that lack functional p53 protein. The locus, *Pgct1* resides on chromosome 13 (Muller *et al.*, 2000).

2.5 Liver cancer

Large strain differences exist between mouse inbred strains in their susceptibility to liver tumors induced by ENU or DEN, The ENU/DEN susceptibility of these inbred strains is largely similar for spontaneous liver tumors and other carcinogens (Drinkwater and Bennett, 1991). Liver cancer susceptibility loci have been identified for the size and the multiplicity of liver tumors in different independent crosses and tumor induction protocols. Urethane induced liver tumors in males were studied in crosses with the susceptible strain C3H and different resistant strains. Six tumor size loci on different chromosomes were identified, *Hcs1-Hcs6* (Gariboldi *et al.*, 1993b; Manenti *et al.*, 1994). Four liver tumor susceptibility loci, *Hcr1*, *Hcr2*, *Hcf1* and *Hcf2*, were mapped which influence DEN induced liver tumor multiplicity in crosses between the susceptible strains DBA/2 and C57BR/cdJ and the resistant strain C57BL/6 (Lee *et al.*, 1995; Poole *et al.*, 1996). Two loci, *Hcs7* and *Hcf2*, map to an almost identical interval distal on chromosome 1 and could be identical, although mapped in different systems.

2.6 Lymphomas

The mouse strains AKR and SL/Kh are highly susceptible to spontaneous lymphomas. Lymphomas are caused by an endogenous virus and host genetic factors. Spontaneous lymphomas are rare in strain NFS/N that lacks endogenous ecotropic provirus. In segregating crosses involving these strains, four different loci were mapped *Tlsm1*, *Lyr2*, *Mmsr1*, and *Mmsr2* on chromosomes 4, 7, 17 and 18, respectively (Pataer *et al.*, 1996; Shisa *et al.*, 1996; Yamada *et al.*, 1994a). In female SWR × SJL intercross mice, three QTLs, *Gct1*, *2* and *4* were found for the susceptibility to DHEA induced ovarian granulosa cell tumors (Beamer *et al.*, 1998). Susceptibility to radiation-induced lymphomas shows a suggestive linkage to the distal end of chromosome 5 (Szymanska *et al.*, 1999); this part of the chromosome also exhibits frequent loss of heterozygosity in the lymphomas (Sitarz *et al.*, 2000). Two lymphomagenesis QTLs were published with suggestive linkages, *Lyr1* (Okumoto *et al.*, 1990) and *Foc1* (Yamada *et al.*, 1994b). *Lyr1* and *Foc1* map close to each other on chromosome 4 (*Figure 1*). This supports their mapping, even though the linkages were not significant (Lander and Kruglyak, 1995).

Murine plasmacytoma is a B-cell tumor that can be induced by i.p. injections of pristane or paraffin oils. Plasmacytomas form a mouse model for B-cell tumors such as Burkitt lymphomas. Whereas the strains BALB/c and NZB are susceptible to plasmacytoma, most other strains are resistant (Mock *et al.*, 1993). Three distinct plasmacytoma susceptibility loci have been mapped in crosses between the plasmacytoma susceptible strain BALB/c and the resistant strain DBA/s. Originally a single strong linkage (LOD=11.1) was found on chromosome 4 and a minor locus Pctm, distal on chromosome 1 (Mock *et al.*, 1993). Subsequent analysis of chromosome 4 congenics led to the identification of the two closely linked loci, *Pctr1* and *Pctr2*, distal on chromosome 4 (Potter *et al.*, 1994). Cdkn2a was proposed as a candidate for *Pctr1* (Zang *et al.*,1998). The fine-mapping of *Pctr2* led to an ambiguous mapping, two possible small intervals were found where *Pctr2* could be located (Mock *et al.*,1997).

2.7 Genetic interactions

An increasing number of genetic interactions is being found, many cancer QTLs are involved in reciprocal genetic interactions, or epistasis, where susceptibility at a locus depends on the genetic composition at the interacting locus. When two loci are involved in a counteracting interaction, their alleles confer either susceptibility or resistance, depending on a specific genetic combination; the same allele can have the opposite effect on a different genetic background. If we take the example of *Scc4* and *Scc5*, the BALB/c allele of *Scc4*, $Scc4^{BALB}$, is susceptible in combination with $Scc5^{STS/STS}$, but is resistant in combination with $Scc5^{BALB/BALB}$; $Scc5^{BALB}$ is resistant in combination with an $Scc4^{BALB}$ allele but susceptible in combination with $Scc4^{STS/STS}$ (van Wezel *et al.*, 1996). Consequently, in spite of the relatively large effects of these loci together, there is no effect if each of them is considered alone. The opposite effects of the same allele in combinations with the partner locus allele mask each other. At present, this type of interaction has been detected in many loci and with many biologic traits; four out of the 10 identified colon cancer loci and 13 of the 26 lung cancer loci are involved in two-way interactions (Fijneman, 1996, 1998; van Wezel, 1996, 1999). This indicates that interactions are an integral part of the genetic control of quantitative traits. These interactions probably reflect the involvement of the interacting loci in the same pathway, or in two interacting pathways. Potentially the analysis of such interactions may help to uncover the individual pathways along which the susceptibility genes operate.

2.8 Statistical criteria

In order to avoid large numbers of false positive QTLs, stringent statistical criteria for their detection are imperative. Lander and Kruglyak (1995) have defined a set of stringent criteria for QTL detection. A clear discrimination is made between significant and suggestive linkage. Only for significant QTLs should a locus name be provided. We propose to assign a temporary, working-draft, name for suggestive loci (w: Locusname). If an investigator finds five QTLs for a certain trait and the fifth linkage is suggestive it should be named w: QTL5. This linkage should be confirmed within three years, a period sufficient to produce and analyze new crosses.

3. Use of RCS for analysis of immune response

Immune response has two basic strategies: non specific (innate) and specific (acquired) response. They are not completely separate as the initial innate response to pathogens influences the latter specific responses. Antigen-induced activation of T-lymphocytes plays a central role in specific immune reactions which are involved in many important biological processes such as control of infection, defense against cancer, autoimmune diseases, allergy and allograft rejection. T-cells can be stimulated to produce cytokines and cytokine receptors which drive proliferation and differentiation, and to express cell surface and intra-cellular molecules involved in effector functions (Chambers and Allison, 1999). These processes include many pathways which are composed from many elements, both known and unknown. The type of response is influenced by the nature and amount of the antigen, way of presentation, cytokines (Lanzavecchia and Sallusto, 2000), hormones (Wilckens and De Rijk, 1997) and T-cell signaling pathways (Acuto and Cantrell, 2000). Mouse strains differ considerably in both specific and non specific immune response which indicates the important role of the genotype in these processes. Some of the loci and genes controlling autoim-munity (Encinas and Kuchroo, 2000; Vyse and Todd, 1996), asthma (De Sanctis and Drazen, 1997) and resistance to infections (Gruenheid and Gros, 2000; Dietrich, 2001; Quareshi et al., 1999) have been established but the majority of genes modifying immune functions remain to be identified. To map mouse genes which are responsible for differences in magnitude and quality of immune response we used the recombinant congenic strains (Stassen et al., 1996). We have analyzed in vitro the model of T-cell activation in which T-cell response was dissected into several parts reflecting the different steps of this process: specific stimulation by alloantigens, activation via CD3 complex and IL-2 receptor and production of cytokines. To dissect immune response to pathogen in vivo we analyzed Leishmania major infection (Figure 3).

3.1 Alloantigen response

The response to the alloantigens, which is estimated by the mixed lymphocyte response, depends on the genetic disparity between the donors of responding and stimulating cells. Differences in the MHC antigens and the Mls1 antigen encoded by the Mtv-7 provirus (Beuntner et al., 1992) induce the strongest response in mice. However, even with comparable incompatibilities in MHC and Mls antigens, some strains of genetically defined mice respond remarkably better than other strains (Holáň et al., 1996; Rychlíková et al., 1973). We have identified two Alan (**Alloan**tigen response) loci, Alan1 (Holáň et al., 2000) and Alan2 (Havelková et al., 2000), which control differences in magnitude in proliferative response to alloantigens in series CcS/Dem and OcB/Dem, respectively. Alan2 controlled response to three different alloantigens: C57BL/10, BALB/c and CBA. This finding might indicate, that at least two types of factors influencing alloreactivity could exist. The first type includes the structural differences in major and minor alloantigens. The second type, controlled by Alan loci, is most likely based on polymorphism in gene(s) which code factor(s) participating in the T-cell receptor signal transduction, or mediating co-stimulatory signals by antigen-presenting cells. It was suggested that allelic polymorphism in non MHC, non Mls loci influence graft-versus-host-disease (Allen, 2000) which might indicate that phenomena observed by us are more common in the alloresponse.

3.2 T-cell activation

In the further analysis of T-cell activation RCS were used to separate (Lipoldová *et al.*, 1995) and map multiple loci controlling T-cell proliferative response to CD3 antibodies (anti-CD3), interleukin (IL)-2 and production of Th2 cytokines IL-4 and IL-10 stimulated by Concanavalin A (ConA). These responses were studied in CcS/Dem series using several different concentrations. Multiple *Tria* (**T** cell **r**eceptor **i**nduced **a**ctivation), *Cinda* (**c**ytokine **ind**uced **a**ctivation), and *Cypr* (**Cy**tokine **pr**oduction) loci were identified that control lymphocyte activation by anti-CD3, IL-2 and ConA, respectively. Some of them, *Tria1–3* (Havelková *et al.*, 1999a, 1996), *Cypr1* (Kosařová *et al.*, 1999), operate in the certain genetic background in a wide dose region, whereas controlling functions of *Tria4,5* (Havelková *et al.*, 1999b), *Cypr2,3* (Kosařová *et al.*, 1999) and *Cinda1,2* (Krulová *et al.*, 1997) are limited only to some concentrations. These restricted effects may reflect induction of different types of ligand/receptor interactions, possibly by a differential activation of the high- and low-affinity receptors or by a different rate of cross-linking, resulting in triggering of distinct signaling pathways. Genes which operate in a wide range of concentrations might participate in some essential pathways; those which control response at only some concentrations, might participate in pathways which are preferentially activated at certain conditions. Another important finding is that production of two different Th2 cytokines IL-4 and IL-10 is controlled independently by non linked loci and that the relationship between the level of these two cytokines depends on the genotype at a locus *coral* (**cor**relation 1; Kosařová *et al.*, 1999). These studies complement biochemical analysis of signaling pathways and could help to find crucial points that are responsible for genetically influenced differences in T-cell activation.

3.3 Leishmania major

RCS proved to be especially useful in the analysis of relationships between different components of complex traits, such as susceptibility to infection by *Leishmania major*. Responses to *L. major* have been previously studied in recombinant inbred strains between susceptible strain BALB/c and resistant strain STS. Results suggested a single major gene *Scl1* on chromosome 11 which controls lesion size (Roberts *et al.*, 1993). This hypothesis has been further supported: (1) *in vitro* tests of differences in IL-12 responsiveness of BALB/c and B10.D2 cells; (2) susceptibility of BALB/c mice has been attributed to the *Irf1* gene closely linked to *Scl1* (Güler et al., 1996). *L. major* response has been analyzed using the CcS-series of the RCS, and our data indicated that the control of susceptibility of BALB/c mice is effectuated by multiple genes (Demant *et al.*, 1996). Several chromosomal regions were implicated in resistance to skin lesions in infected mice by Beebe *et al.* (1997), in serial backcrosses between susceptible BALB/c and resistant B10.D2 strains Two loci; *Lmr1* (**L**eishmania **m**ajor **r**esponse 1) and *Lmr2* controlling resistance to cutaneous lesions were mapped by Roberts *et al.* (1997), who studied F_2 hybrids between susceptible BALB/c and resistant C57BL/6 mice. As no indication about the function of these loci has been reported, we have been intrigued by the question of the relationship between the

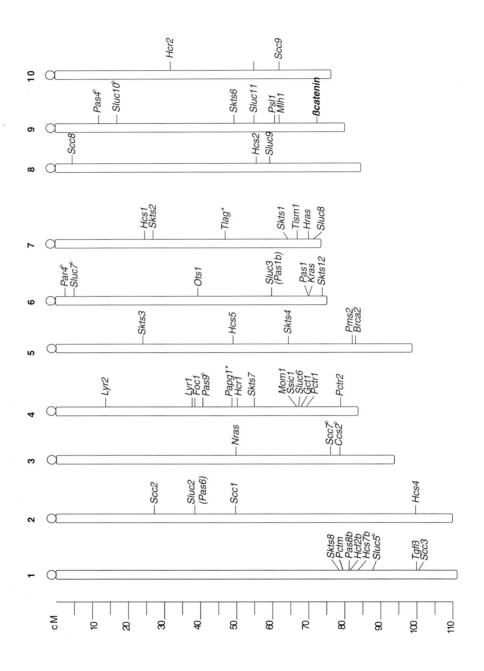

(a)

(b)

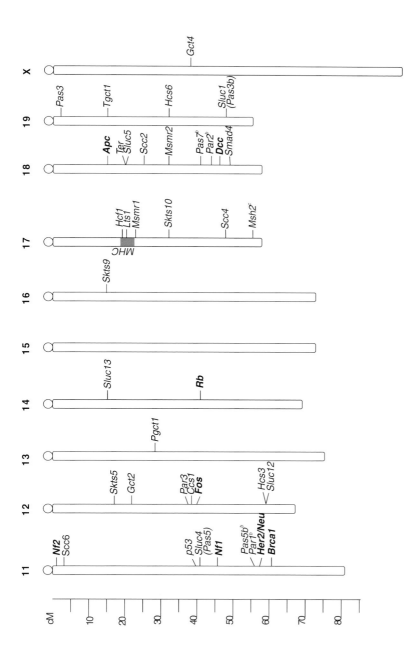

Figure 3. Segmentation of the 'donor' strain genome on the genetic background of the 'background' strain in recombinant congenic strains. Background strain: BALB/c; donor strain: STS/A recombinant congenic strains: CcS-1–CcS-20. Reproduced from Demant (1992), with permission.

multigenic control and the role of specific components of the immune response in development of the disease. In order to obtain insight into the mechanisms of the genetic susceptibility to leishmaniosis, we integrated the genetic linkage study of susceptibility with determination of several symptoms of disease and a number of immunologically relevant parameters (Lipoldová et al., 2000). By linkage analysis of F_2 hybrids between the resistant RC strain CcS-5 and the susceptible strain BALB/c we mapped five novel loci, Lmr3–7 each of which appears to be associated with a different combination of pathological symptoms and immunological reactions. Lmr3 on chromosome 5 controls serum IFN-γ and IgE levels, splenomegaly and possibly hepatomegaly. Lmr4 on chromosome 6 determines lesion size. Lmr5 on chromosome 10 influences lesion size, IgE, IL-12 and IFN-γ levels and possibly also splenomegaly, but not hepatomegaly. Loci Lmr4 and Lmr5 interact in determining IFN-γ level in serum. Lmr6 on chromosome 11 controls IL-4 level in serum. Lmr7 on chromosome 17 controls proliferation of lymphocytes from infected mice. Our data support earlier findings about organ differences in L. major response (Stenger et al., 1996), as skin lesions were controlled by Lmr4 and Lmr5, splenomegaly by Lmr3 and Lmr5, and hepatomegaly by Lmr3.

We did not detect any effect of the postulated major susceptibility gene Scl1 on chromosome 11 (Roberts et al., 1993) nor of Irf1, that has been suggested as a potential candidate (Güler et al., 1996). This part of chromosome 11 in the CcS-5 strain is of BALB/c origin (Stassen et al., 1996), as is the case with the region carrying Lmr2 on chromosome 9 (Roberts et al., 1997). The locus Lmr1 on chromosome 17 (Roberts et al., 1997) is located telomeric to Lmr7 on chromosome 17. Suggestive linkages with lesion development were reported (Beebe et al., 1997) to markers on chromosomes 6, 11 and 15. The significant linkage with Lmr6 coincides with the region of suggestive linkage on chromosome 11. Lmr4 on chromosome 6 is in a different location from that suggested by Beebe et al. (1997). In fact, both the regions on chromosomes 6 and 15 are of BALB/c origin and cannot be involved in our studies.

4. Summary

A wide variety of genetic loci – modifiers of quantitative traits – has been mapped in the mouse. We discussed only a selected part of them: the group of loci modifying development of various types of cancer, and our own work on genes modifying T-lymphocyte activation and resistance to infection. The data shows the wide interest in the definition of the modifier genes for various types of cancer and immunological disorders. The rapid progress of both the human and mouse genome projects make the sequence of regions containing individual QTLs accessible and will aid the identification of the pertinent candidate genes. At the same time, the development of new genotyping techniques will offer economic and rapid typing of a very large number of loci, thus permitting application of QTLs to characterization of individual patients. This will lead to a better definition of genetic heterogeneity of the disease, assessment of the prognosis, and in many cases to optimization of therapy.

References

Acuto, O. and Cantrell, D. (2000) T cell activation and the cytoskeleton. *Annu. Rev. Immunol.* **18**: 165–184.

Allen, R.D. (2000) The new genetics of bone marrow transplantation. *Genes Immun.* **1**: 316–320.

Angel, J.M., Beltran, L., Minda, K., Rupp, T. and DiGiovanni, J. (1997) Association of a murine chromosome 9 locus (Psl1) with susceptibility to mouse skin tumor promotion by 12-O-tetradecanoylphorbol-13-acetate. *Mol. Carcinog.* **20**: 162–167.

Angel, J.M., Popova, N., Lanko, N., Turusov, V.S. and DiGiovanni J. (2000) A locus that influences susceptibility to 1, 2-dimethylhydrazine-induced colon tumors maps to the distal end of mouse chromosome 3. *Mol. Carcinog.* **27**: 47–54.

Asada, Y., Varnum, D.S., Frankel, W.N. and Nadeau, J.H. (1994). A mutation in the Ter gene causing increased susceptibility to testicular teratomas maps to mouse chromosome 18. *Nat. Genet.* **6**: 363–368.

Balmain, A. and Nagase, H. (1998) Cancer resistance genes in mice: models for the study of tumour modifiers. *TIG* **14**: 139–144.

Beamer, W.G., Shultz, K.L., Tennent, B.J., Nadeau, J.H., Churchill, G.A. and Eicher, E.M. (1998) Multigenic and imprinting control of ovarian granulosa cell tumorigenesis in mice. *Cancer Res.* **58**: 3694–3699.

Bedell, M.A., Jenkins, N.A. and Copeland, N.G. (1997) Mouse models of human disease Part II Recent progress and future directions. *Genes Dev.* **11**: 11–43.

Beebe, A.M., Mauze., S, Schork, N.J. and Coffman, R.L. (1997) Serial backcross mapping of multiple loci associated with resistance to Leishmania major in mice. *Immunity* **6**: 551–557.

Beuntner, U., Frankel, W.N., Cote, M.S., Coffin, J.M. and Huber, B.T. (1992) Mls-1 is encoded by the long terminal repeat open reading frame of the mouse mammary tumor provirus Mtv-7. *Proc. Natl Acad. Sci. USA* **89**: 5432–5436.

Blake, J.A., Eppig, J.T., Richardson, J.E. and Davisson, M.T. The Mouse Genome Database Group. (2000) The Mouse Genome Database (MGD) : expanding genetic and genomic resources for the laboratory mouse. *Nucleic Acids Res.* **28**: 108–111. (URL: http://www.informatics.jax.org) .

Chambers, C.A. and Allison, J.P. (1999) Costimulatory regulation of T cell function. *Curr. Opin. Cell. Biol.* **11**: 203–210.

Collin, G.B., Asada, Y., Varnum, D.S and Nadeau, J.H. (1996) DNA pooling as a quick method for finding candidate linkages in multigenic trait analysis: an example involving susceptibility to germ cell tumors. *Mamm. Genome* **7**: 68–70.

Cormier, R.T., Bilger, A., Lillich, A.J., Halberg, R.B., Hong, K.H., Gould, K.A., Borenstein, N., Lander E.S. and Dove, W.F. (2000) The Mom1AKR intestinal tumor resistance region consists of Pla2g2a and a locus distal to D4Mit64. *Oncogene* **19**: 3182–3192.

Demant, P. (1992) Genetic Resolution of susceptibility to cancer – New Perspectives. *Sem. Cancer Biol.* **3**: 159–166.

Demant, P. and Hart, A.A.M. (1986) Recombinant congenic strains – a new tool for analyzing genetic traits determined by more than one gene. *Immunogenet.* **24**: 416–422.

Demant, P., Oomen, L.C.J.M. and Oudshoorn-Snoek, M. (1989) Genetics of tumor susceptibility in the mouse, MHC and non-MHC genes. *Adv. Cancer Res.* **53**: 117–179.

Demant P., Lipoldová, M. and Svobodová, M. (1996) Resistance to Leishmania major in mice. *Science* **274**: 1392 (technical comment).

De Sanctis, G.T. and Drazen, J.M. (1997) Genetics of native airway responsiveness in mice. *Am. J. Respir. Crit. Care Med.* **156**: S82–S88.

Devereux, T.R., Wiseman, R.W., Kaplan, N., Garren, S., Foley, J.F., White, C.M., Anna, C., Watson, M.A., Patel, A. and Jarchow,S. (1994). Assignment of a locus for mouse lung tumor susceptibility to proximal chromosome 19. *Mamm. Genome* **5**: 749–755.

Dietrich, W.F. (2001) Using mouse genetics to understand infectious disease pathogenesis. *Genome. Res.* **11**: 325–331.

Dietrich, W.F., Lander, E.S., Smith, J.S., Moser, A.R., Gould, K.A., Luongo, C., Borenstein, N. and Dove, W. (1993) Genetic identification of Mom-1, a major modifier locus affecting Min-induced intestinal neoplasia in the mouse. *Cell* **75**: 631–639.

Dietrich, W.F., J. Miller, R. Steen, M.A. *et al.* (1996) A comprehensive genetic map of the mouse genome. *Nature* **380**: 149–152.

Drinkwater, N.R. and Ginsler, J.J. (1986) Genetic control of hepatocarcinogenesis in C57BL/6J and C3H/HeJ inbred mice. *Carcinogenesis* **7**: 1701–1710.

Drinkwater, N.R. and Bennet, L.M. (1991) Genetic control of carcinogenesis in experimental animals. *Progr. Exper. Tumor Res.* **33**: 1–20.

Dux, A. and Demant, P. (1987). MHC-controlled susceptibility to C3H-MTV-induced mouse mammary tumors is predominantly systemic rather than local. *Int. J. Cancer* **40**: 372–377.

Encinas, J.A. and Kuchroo, V.K. (2000) Mapping and identification of autoimmunity genes. *Curr. Opin. Immunol.* **12**: 691–697.

Festing, M.F., Yang, A. and Malkinson, A.M. (1994) At least four genes and sex are associated with susceptibility to urethane-induced pulmonary adenomas. *Genet. Res.* **64**: 99–106.

Festing, M.F., Lin, L., Devereux, T.R., Gao, F., Yang, A., Anna, C.H., White, C.M., Malkinson, A.M. and You, M. (1998) At least four loci and gender are associated with susceptibility to the chemical induction of lung adenomas in A/J × BALB/c mice. *Genomics* **53**: 129–136.

Fijneman, R.J.A. and Demant, P. (1995a) A gene for susceptibility to small intestinal cnace, ssic1, maps to the distalpart of mouse chromosome 4. *Cancer Res.* **55**: 3179–3182.

Fijneman, R.J.A., Oomen, L.C. Snoek, M. and Demant, P. (1995b) A susceptibility gene for alveolar lung tumors in the mouse maps between Hsp70.3 and G7 within the H2 complex. *Immunogenetics* **41**: 106–109.

Fijneman, R.J.A., de Vries, S.S., Jansen, R.C. and Demant, P. (1996) Involvement of quantitative trait loci in a complex interaction: mapping of four new loci, Sluc1, Sluc2, Sluc3 and Sluc4, that influence susceptibility to lung cancer in the mouse. *Nat. Genet.* **13**: 465–467.

Fijneman, R.J.A., Jansen, R.C., van der Valk, M.A. and Demant, P. (1998) High frequency of interactions between lung cancer genes in the mouse: mapping of sluc5 to sluc14. *Cancer Res.* **58**: 4794–4798.

Fleiszer, D., Hilgers, J. and Skamene, E. (1988) Multigenic control of colon carcinogenesis in mice treated 1,2-dimethylhydrazine. *Curr. Top. Microbiol. Immunol.* **137**: 243–249.

Gariboldi, M., Manenti, G., Canzian, F., Falvella, F.S., Radice, M.T., Pierotti, M.A., Della Porta, G., Binelli, G. and Dragani, T.A. (1993a) A major susceptibility locus to murine lung carcinogenesis maps on chromosome 6. *Nat. Genet.* **3**: 132–136.

Gariboldi, M., Manenti, G., Canzian, F., Falvella, F.S., Pierotti, M.A., Della Porta, G., Binelli, G. and Dragani, T.A. (1993b) Chromosome mapping of murine susceptibility loci to liver carcinogenesis. *Cancer Res.* **53**: 209–211.

Gruenheid, S. and Gros, P. (2000) Genetic susceptibility to intracellular infections: *Nramp1*, macrophage function and divalent cation transport. *Curr. Opin. Immunol.* **3**: 43–48.

Güler, M.L., Gorham, J.D., Hsieh C.-S., Mackey, A.J., Steen, R.G., Dietrich, W.F. and Murphy, K.M. (1996) Genetic susceptibility to Leishmania: IL-12 responsiveness in Th1 development. *Science* **271**: 984–987.

Havelková, H., Krulová, M., Kosaová, M., Holáň, V., Hart, A.A.M., Demant, P. and Lipoldová, M. (1996) Genetic control of T-cell proliferative response in mice linked to chromosomes 11 and 15. *Immunogenetics* **44**: 475–477.

Havelková, H., Kosaová, M., Krulová, M., Holáň, V., Demant, P. and Lipoldová, M. (1999a) T-cell proliferative response is controlled by locus Tria3 on mouse chromosome 17. *Immunogenetics* **49**: 235–237.

Havelková, H., Kosaová, M., Krulová, M., Demant, P. and Lipoldová, M. (1999b) T-cell proliferative response is controlled by loci Tria4 and Tria5 on mouse chromosomes 7 and 9. *Mamm. Genome* **10**: 670–674.

Havelková, H., Badalová, J., Demant, P. and Lipoldová, M. (2000) A New Type of Genetic Regulation of Allogeneic Response. A Novel Locus on Mouse Chromosome 4, Alan2 Controls MLC Reactivity to Three Different Alloantigens: C57BL/10, BALB/c and CBA. *Genes Immun.* **1**: 483–487.

Holáň, V., Lipoldová, M. and Demant, P. (1996) Identical genetic control of MLC reactivity to different MHC incompatibilities, independent of production and response to IL-2. *Immunogenetics* **44**: 27–35.

Holáň, V., Havelková, H., Krulová, M., Demant, P. and Lipoldová, M. (2000) A novel alloreactivity controlling locus, Alan1, mapped to mouse chromosome 17. *Immunogenetics* **51**: 755–757.

Jacoby, R.F., Hohman, C., Marshall, D.J., Frick, T.J., Schlack, S., Broda, M., Smutko, J. and Elliott, R.W. (1994) Genetic analysis of colon cancer susceptibility in mice. *Genomics* 22: 381–387.

Kosařová, M., Havelková, H., Krulová, M., Demant, P. and Lipoldová, M. (1999) The production of two Th2 cytokines, interleukin-4 and interleukin-10 is controlled independently by a locus *Cypr1* and loci *Cypr2* and *Cypr3*, respectively. *Immunogenetics* 49: 134–141.

Krulová, M., Havelková, H., Kosařová, M., Holáň, V., Hart, A.A.M., Demant, P. and Lipoldová, M. (1997) IL-2 induced proliferative response is controlled by loci *Cinda1* and *Cinda2* on mouse chromosomes 11 and 12. A distinct control of the response induced by different IL-2 concentrations. *Genomics* 42: 11–15.

Lander, E. and Kruglyak, L. (1995) Genetic dissection of complex traits: guidelines for interpreting and reporting linkage results. *Nat. Genet.* 11: 241–247.

Lanzavecchia, A. and Sallusto, F. (2000) Dynamics of T lymphocyte responses: Intermediates, effectors, and memory cells. *Science* 290: 92–97.

Lee, G.H., Bennett, L.M., Carabeo, R.A. and Drinkwater, N.R. (1995) Identification of hepatocarcinogen-resistance genes in DBA/2 mice. *Genetics* 139: 387–395.

Lee, G.H., Bugni, J.M., Obata, M., Nishimori, H., Ogawa, K. and Drinkwater, N.R. (1997) Genetic dissection of susceptibility to murine ovarian teratomas that originate from parthenogenetic oocytes. *Cancer Res.* 57: 590–593.

Lichtenstein, P., Holm, N.V., Verkasalo, P.K., Iliadou, A., Kaprio, J., Koskenvuo, M., Pukkala, E., Skytthe, A. and Hemminki, K. (2000). Environmental and heritable factors in the causation of cancer – analyses of cohorts of twins from Sweden, Denmark, and Finland. *N. Engl. J. Med.* 343: 78–85.

Lipoldová, M., Kosařová, M., Zajícová, A., Holáň, V., Hart, A.A.M., Krulová, M. and Demant, P. (1995) Separation of multiple genes controlling the T-cell proliferative response to IL-2 and anti-CD3 using recombinant congenic strains. *Immunogenetics* 41: 301–311.

Lipoldová, M., Svobodová, M., Krulová, M. *et al.* (2000) Susceptibility to *Leishmania major* infection in mice: multiple loci and heterogeneity of immunopathological phenotypes. *Genes Immun.* 1: 200–206.

MacPhee, M., Chepenik, K.P., Liddell, R.A., Nelson, K.K., Siracusa L.D. and Buchberg A.M. (1995) The secretory phospholipase A2 gene is a candidate for the Mom1 locus, a major modifier of ApcMin-induced intestinal neoplasia. *Cell* 81: 957–966.

Manenti, G., Binelli, G., Gariboldi, M., Canzian, F., De Gregorio, L., Falvella, F.S., Dragani, T.A. and Pierotti, M.A. (1994) Multiple loci affect genetic predisposition to hepatocarcinogenesis in mice. *Genomics* 23: 118–124.

Manenti, G., Gariboldi, M., Fiorino, A., Zanesi, N., Pierotti, M.A. and Dragani, T.A. (1997) Genetic mapping of lung cancer modifier loci specifically affecting tumor initiation and progression. *Cancer Res.* 57: 4164–4166.

Matin, A., Collin, G.B., Asada, Y., Varnum, D. and Nadeau, J.H. (1999) Susceptibility to testicular germ-cell tumours in a 129.MOLF-Chr 19 chromosome substitution strain. *Nat. Genet.* 23: 237–240.

Mock, B.A., Krall, M.M. and Dosik, J.K. (1993) Genetic mapping of tumor susceptibility genes involved in mouse plasmacytomagenesis. *Proc. Natl Acad. Sci. USA* 90: 9499–9503.

Mock, B.A., Hartley, J., Le Tissier, P., Wax, J.S. and Potter, P. (1997) The plasmocytoma resistance gene, Ptcs2, delays the onset of tumorigenesis and resides in the telomeric region of chromosome 4. *Blood* 90: 4092–4098.

Mock, B.A., Lowry, D.T., Rehman, I., Padlan, C., Yuspa, S.H. and Hennings, H. (1998) Multigenic control of skin tumor susceptibility in SENCARA/Pt mice *Carcinogenesis* 19: 1109–1115.

Moen, C.J., van der Valk, M.A., Snoek, M., van Zutphen, B.F., von Deimling, O., Hart, A.A. and Demant, P. (1991). The recombinant congenic strains – a novel genetic tool applied to the study of colon tumor development in the mouse. *Mamm. Genome* 1: 217–227.

Moen, C.J.A, Snoek, M., Hart, A.A.M. and Demant, P. (1992) *Scc1*, a novel colon cancer susceptibility gene in the mouse: linkage to CD44 (Ly-24, Pgp1) on chromosome 2. *Oncogene* 7: 563–566.

Moen, C.J.A, Groot, P.C., Hart, A.A.M. Snoek, M. and Demant, P. (1996) Fine mapping of colon tumor susceptibility (Scc) genes in the mouse, different from the genes known to be somatically mutated in colon cancer. *Proc. Natl Acad. Sci. USA* 93: 1082–1086.

Muller, A.J., Terensky, A.K. and Levine, A.J. (2000) A male germ cell tumor-susceptibility-determining locus, pgct1, identified on mouse chromosome 13. *Proc. Natl Acad. Sci. USA* 97: 8421–8426.

Murphy, E.D. (1966) Characteristic tumors. In: Green, E.L. (ed), *Biology of the Laboratory Mouse*, McGraw-Hill, pp. 521–570.

Nagase, H., Bryson, S., Cordell, H., Kemp, C.J., Fee, F. and Balmain, A. (1995) Distinct genetic control of benign and malignant skin tumors in mice. *Nat. Genet.* **10**: 424–429.

Nagase, H., Mao, J.H. and Balmain, A.A. (1999) subset of skin tumor modifier loci determines survival time of tumor-bearing mice. *Proc. Natl Acad. Sci. USA* **96**: 15032–15037.

Noguchi, T. and Noguchi, M. (1985). A recessive mutation (ter) causing germ cell deficiency and a high incidence of congenital testicular teratomas in 129/Sv-ter mice. *J. Natl. Cancer Inst.* **75**: 385–392.

Obata, M., Nishimori, H., Ogawa, K. and Lee, G.H. (1996) Identification of the Par2 (Pulmonary adenoma resistance) locus on mouse chromosome 18, a major genetic determinant for lung carcinogen resistance in BALB/cByJ mice. *Oncogene* **13**: 1599–1604.

Okumoto, M. Nishikawa, R. Imai, S. and Hilgers, J. (1990) Genetic analysis of resistance to radiation lymphomagenesis with recombinant inbred strains of mice. *Cancer Res.* **50**: 3848–3850.

Oomen, L.C.J.M., Demant, P., Hart, A.A.M. and Emmelot, P. (1983) Multiple genes in the H2 complex affect differently the number and growth rate of transplacentally induced lung tumors in mice. *Int. J. Cancer* **31**: 447–454.

Pataer, A., Kamoto, T., Lu, L.M., Yamada Y. and Hiai, H. (1996) Two dominant host resistance genes to pre-B lymphoma in wild-derived inbred mouse strain MSM/Ms. *Cancer Res.* **56**: 3716–3720.

Peto, J. (1980) Genetic predisposition to cancer. In: Cairns, J., Lyon, J.L. and Skolnick, M. (eds), *Cancer Incidence in Defined Populations (Banbury Report 4)*, Cold Spring Harbor Lab. Press, Cold Spring Harbor, pp. 203–213.

Ponder, B.A.J. (1990) Inherited disposition to cancer. *Trends Genet.* **6**: 213–218.

Poole, T.M. and Drinkwater, N.R. (1996). Strain dependent effects of sex hormones on hepatocarcinogenesis in mice. *Carcinogenesis* **17**: 191–196.

Potter, M., Mushinski, E.B., Wax, J.S., Hartley, J. and Mock, B.A. (1994) Identification of two genes on chromosome 4 that determine resistance to plasmacytoma induction in mice. *Cancer Res.* **54**: 969–975.

Quareshi, S.T., Gros, P. and Malo, D. (1999) Host resistance to infection: genetic control of lipopolysaccharide responsiveness by TOLL-like receptor genes. *Trends Genet.* **15**: 291–294.

Risch, N. and Merikangas, K. (1996) The future of genetic studies of complex human diseases. *Science* **273**: 1516–1517.

Roberts, L.J., Baldwin, T.M., Curtis, J.M., Handman, E. and Foote, S.J. (1997) Resistance to *Leishmania major* is linked to H2 region on chromosome 17 and to chromosome 9. *J. Exp. Med.* **9**: 1705–1710.

Roberts, M., Mock, B.A and Blackwell, J.M. (1993) Mapping of genes controlling *Leishmania major* infection in CXS recombinant inbred mice. *Eur. J. Immunogenet.* **20**: 349–362.

Ryan, J., Barker, P.E., Nesbitt, M.N. and Ruddle, F.H. (1987) Kras2 as a genetic marker for lung tumor susceptibility in inbred mice. *J. Natl Cancer Inst.* **79**: 1351–1357.

Rychlíková, M., Demant, P. and Ivanyi, P. (1973) The mixed lymphocyte reaction in H-2K, H-2D, and non-H-2 incompatibility. *Biomedicine* **18**: 401–407.

Shisa, H., Yamada, Y., Kawarai, A., Terada, N., Kawai, M., Matsushiro, H. and Hiai, H. (1996) Genetic and epigenetic resistance of SL/Ni mice to lymphomas. *Jpn J. Cancer Res.* **87**: 258–262.

Sitarz, M., Wirth-Dzieciolowska, E. and Demant, P. (2000). Loss of heterozygosity on chromosome 5 in vicinity of the telomere in gamma-radiation-induced thymic lymphomas in mice. *Neoplasma* **47**: 148–150.

Stassen, A.P.M., Groot, P.C., Eppig, J.T. and Demant, P. (1996) Genetic composition of the recombinant congenic strains. *Mamm. Genome* **7**: 55–58.

Stenger, S., Donhauser N., Thüring, H., Röllinghoff, M. and Bogdan, C. (1996) Reactivation of latent leishmaniasis by inhibition of inducible nitric oxide synthase. *J. Exp. Med.* **183**: 1501–1514.

Su, L.K., Kinzler, K.W., Vogelstein, B., Preisinger, A.C., Moser, A.R., Luongo, C., Gould, K.A. and Dove, W.F. (1992) Multiple intestinal neoplasia caused by a mutation in the murine homolog of the Apc gene. *Science* **256**: 668–670.

Szymanska, H., Sitarz, M., Krysiak, E., Piskorowska, J., Czarnomska, A., Skurzak, H., Hart, A.A., de Jong, D. and Demant, P. (1999). Genetics of susceptibility to radiation-induced lymphomas, leukemias and lung tumors studied in recombinant congenic strains. *Int. J. Cancer* **83**: 674–678.

Talbot, C.J., Nicod, A., Cherny, S.S., Fulker, D.W., Collins, A.C. and Flint, J. (1999) High-resolution mapping of quantitative trait loci in outbred mice. *Nature Genet*. **21**: 305–308.

van der Valk, M.A. (1981) Survival, tumor incidence and gross pathology in 33 mouse strains. In: Hilgers, J. and Sluyser, M. (eds), *Mammary Tumors in Mice*. Elsevier, Amsterdam, pp. 46–115.

van Wezel, T., Ruivenkamp, C.A., Stassen, A.P., Moen, C.J. and Demant, P. (1999) Four new colon cancer susceptibility loci, *Scc6* to *Scc9* in the mouse. *Cancer Res*. **59**: 4216–4218.

van Wezel, T., Stassen, A.P.M., Moen, C.J.A., Hart, A.A.M., van der Valk, M.A. and Demant, P. (1996) Gene interaction and single gene effects in colon tumour susceptibility in mice. *Nature Genet*. **13**: 468–471.

Vyse, T.J. and Todd, J.A. (1996) Genetic analysis of autoimmune disease. *Cell* **85**: 311–318.

Wang D.G., Fan, J.B., Siao, C.J. *et al*. (1998) Large-scale identification, mapping, and genotyping of single-nucleotide polymorphisms in the human genome. *Science* **280**: 1077–1082.

Wilckens, T. and De Rijk, R. (1997) Glucocorticoids and immune function: unknown dimensions and new frontiers. *Immunol. Today* **18**: 418–424.

Yamada, Y., Shisa, H., Matsushiro, H., Kamoto, T., Kobayashi, Y., Kawarai, A. and Hiai, H. (1994a) T lymphomagenesis is determined by a dominant host gene Thymic Lymphoma susceptible mouse-1 (Tlsm1) in mouse models. *J. Exp. Med.* **180**: 2155–2162.

Yamada, Y., Matsushiro, H., Ogawa, M.S., Okamoto, K., Nakakuki, Y., Toyokuni, S., Fukumoto, M. and Hiai, H. (1994b) Genetic predisposition to pre-B lymphomas in SL/Kh strain mice. *Cancer Res.* **54**: 403–407.

Zhang, S., Ramsay, E.S. and Mock, B.A. (1998) Cdkn2a, the cyclin-dependent kinase inhibitor encoding p16INK4a and p19ARF, is a candidate for the plasmacytoma susceptibility locus, Pctr1. *Proc. Natl Acad. Sci. USA* **95**: 2429–2434.

8

The *GNAS1* gene

D.T. Bonthron

1. Introduction

Recent studies of *GNAS1*, a gene first cloned in 1988, have revealed an unexpected degree of regulatory complexity. These new aspects of the gene's function are not only relevant to the pathogenesis of the disorders that result from *GNAS1* mutations, but also promise to improve our understanding of the epigenetic processes that underlie genomic imprinting. This chapter therefore focuses particularly on the imprinting of the various *GNAS1* gene products, the mutational spectrum of the gene having recently been comprehensively reviewed elsewhere (Aldred and Trembath, 2000).

2. Pseudohypoparathyroidism (PHP) and Albright hereditary osteodystrophy (AHO)

The name of the Boston endocrinologist Fuller Albright (1900–1969) is now usually associated with two syndromic endocrine disorders; Albright hereditary osteodystrophy (AHO) and McCune Albright syndrome (MAS). By a strange coincidence, both of these phenotypically quite dissimilar conditions have turned out to result from genotypic abnormalities in the same gene, *GNAS1*. AHO is caused by heterozygous germline inactivating mutations, and MAS by mosaic somatic mutations that constitutively activate the *GNAS1* gene product, $G_s\alpha$.

2.1 Clinical features

PHP is defined by biochemical features of hypocalcemia and/or hyperphosphatemia but in the presence of a normal or elevated parathyroid hormone (PTH) level. The syndrome of PHP was originally described by Albright *et al.* (1942). Several subtypes of PHP are now defined on the basis of endocrinological parameters. The commonest of these, PHP-Ia, has the clinical features of: short stature with an osteodystrophy (AHO) affecting particularly the metacarpals and metatarsals (brachydactyly type E; *Figure 1*); frequently subcutaneous or intracerebral calcification; round facies; and mild mental retardation in some patients.

Genotype to Phenotype second edition, edited by S. Malcolm and J. Goodship.
© 2001 BIOS Scientific Publishers Ltd, Oxford.

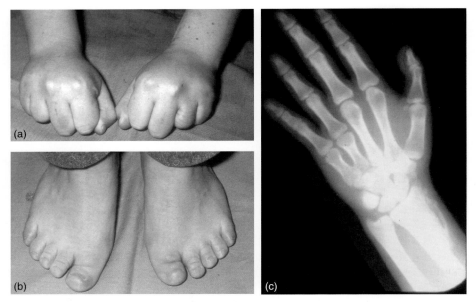

Figure 1. Skeletal features of Albright osteodystrophy. (a) Brachydactyly type E; there is often disproportionate shortening of metacarpals IV and V, as revealed when the patient makes a fist. (b) Brachydactyly resulting from shortening of all the metatarsals. (c) Striking shortening of metacarpals IV and V. Also visible is an area of abnormal subcutaneous calcification along the distal radius. Courtesy of Dr D.G.D. Barr, Royal Hospital for Sick Children, Edinburgh, UK.

2.2 Endocrinological abnormalities

PHP-Ia is characterized by end-organ resistance to the action of various hormones that act as ligands for cell-surface receptors of the seven transmembrane domain class. These hormones include PTH, resulting in hypocalcemia, thyroid stimulating hormone (TSH), resulting in hypothyroidism, gonadotropins, resulting in gonadal dysfunction, and glucagon. The distinct receptors for these hormones have in common the fact that they increase intracellular levels of the second messenger cyclic AMP. All these receptors belong to the so-called 'serpentine' class, which have seven transmembrane domains. Receptor activation is coupled to adenylate cyclase activation through the complex activity of the stimulatory heterotrimeric G protein, G_s. Some of the factors modulating the interaction between serpentine receptor and G protein have been reviewed in detail (Bourne, 1997).

2.3 Hormone receptor-adenylate cyclase coupling by G_s

The cycle of G_s activation is illustrated in simple form in *Figure 2*. Receptor activation results in exchange of GTP for GDP and dissociation of the $G_s\alpha$ subunit from the β and γ subunits. The free activated GTP-bound $G_s\alpha$ subunit then activates adenylate cyclase. This stimulatory activity is terminated by the $G_s\alpha$ subunit's intrinsic GTPase activity, GDP-bound $G_s\alpha$ being inactive and rebinding to the β and γ subunits. The activation of $G_s\alpha$ is also accompanied by a shift from

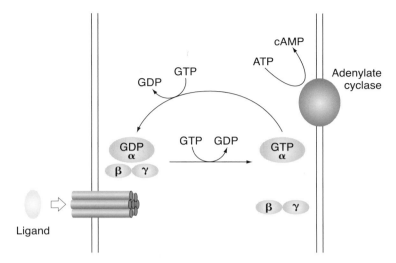

Figure 2. Hormone signalling *via* G$_s$. In the basal state, G$_s$ is a membrane-associated αβγ heterotrimer with GDP bound to the α subunit. Hormone activation of a G$_s$-coupled seven transmembrane domain receptor promotes exchange of GTP for GDP and G$_s$α dissociation from the βγ subunits. G$_s$α-GTP activates adenylate cyclase, resulting in an increase in intracellular cAMP. The signalling is terminated after hydrolysis of G$_s$α-GTP to G$_s$α-GDP, mediated by an intrinsic GTPase activity of the G$_s$α subunit, following which G$_s$α-GDP reassociates with βγ subunits. The oncogenic *gsp* mutations at G$_s$α codons 201 and 227 constitutively activate the signalling pathway by removing the intrinsic GTPase activity (see text).

membrane to cytosol (Wedegaertner *et al.*, 1996). (As a consequence of this, constitutively activated G$_s$α mutants (see below) are cytoplasmic in distribution.) This intracellular translocation is mediated at least in part through depalmitoylation of G$_s$α (Wedegaertner and Bourne, 1994). The two modification cycles of G$_s$α depalmitoylation and βγ dissociation appear to potentiate each other in determining G$_s$α translocation between membrane and cytosol (Iiri *et al.*, 1996).

2.4 Classification

Patients with PHP may be subclassified into types I and II. In type I, but not type II, the normal rise in urinary cyclic AMP following infusion of a PTH analog is blunted or absent (Chase *et al.*, 1969). PHP-Ib patients are distinguished from PHP-Ia by having isolated renal PTH resistance (with no other endocrine disturbance) and no features of AHO.

Some individuals, including relatives of PHP-Ia patients, may have the skeletal features of AHO, but with no evidence for endocrine or biochemical disturbance. These individuals are sometimes referred to as having pseudopseudohypoparathyroidism (PPHP). PPHP patients carry the same *GNAS1* mutations as their relatives with PHP-Ia (see below), and the basis for such phenotypic variability was unclear until the recognition that it resulted from parental (genomic) imprinting of *GNAS1*.

2.5 Deficiency of $G_s\alpha$ in PHP

The recognition that the hormones to which there is end-organ resistance in PHP-Ia are adenylate cyclase-coupled, led to the biochemical demonstration of deficient receptor–cyclase coupling activity in patients' erythrocyte membranes (Farfel *et al.*, 1980). (The coupling factor, originally termed 'N', is now known as G_s.) A roughly 50% deficiency of G_s is found both in patients with PHP-Ia and their relatives with PPHP (Levine *et al.*, 1986; Schuster *et al.*, 1993). The multiple hormone resistance seen in PHP-Ia thus results from deficiency of a common coupling factor (G_s) that mediates cyclase activation in response to several different serpentine receptors.

2.6 Molecular genetics

Isolation of genomic clones containing *GNAS1*, the human gene encoding the α subunit of G_s, revealed that $G_s\alpha$ is encoded by 13 exons, spanning some 19 kb (Kozasa *et al.*, 1988). However, alternative splicing patterns result in the generation of four alternative isoforms that result (a) from inclusion or exclusion of exon 3 in the transcript and (b) from selection of alternative intron 3 splice acceptor sites, separated by 3 nt (see *Figure 3*). The functional differences between these protein isoforms are not known.

The first molecular lesion identified in PHP-Ia was an ATG→GTG mutation of the $G_s\alpha$ initiation codon, within *GNAS1* exon 1 (Patten *et al.*, 1990). This resulted in an abnormally migrating protein that was immunoreactive with antisera against the C- but not the N-terminus of $G_s\alpha$. In this family, $G_s\alpha$ mRNA levels

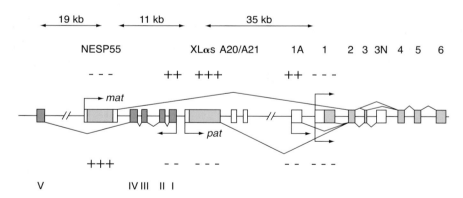

Figure 3. Schematic diagram of the components of the *GNAS1* gene. The diagram is not to scale: approximate distances between major landmarks are indicated at the top. In the middle, exons are indicated as boxes. (Exons 7–13 have been omitted.) Coding regions are shaded light gray. The five exons of the antisense transcript are dark gray and numbered I–V. The sense exons are identified above. Only the major splicing patterns are shown. Exons A20, A21 are included between XLαs and exon 2 in a minor mRNA species. The allelic origin of transcripts from the various promoters is shown by the grey arrows above the gene (maternal) or below it (paternal). For exon 1, transcription is biallelic in most tissues but maternal in a few. Differential methylation of maternal (above) and paternal (below) alleles is indicated by (+) methylated or (–) unmethylated.

were normal, but other studies had shown reduced $G_s\alpha$ mRNA levels in some families (Carter *et al.*, 1987). It soon became clear that there was considerable allelic heterogeneity in PHP-Ia, the next two lesions identified being a splice donor mutation and a single nucleotide deletion within the coding region resulting in a frameshift (Weinstein *et al.*, 1990).

3. Anomalous inheritance of PHP-Ia – imprinting of *GNAS1*

The deficiency of cases of male-to-male transmission of PHP led some to speculate that the condition might be an X-linked dominant disorder. However, with the demonstration of $G_s\alpha$ deficiency in PHP and the mapping of *GNAS1* to chromosome 20, autosomal dominant inheritance became established. The likely explanation for this anomaly was provided by Davies and Hughes (1993), who drew attention to the differing phenotypes of cases that had been maternally or paternally transmitted. All 60 published cases of the former had a PHP-Ia phenotype with hormone resistance, whereas all three of the latter had only the skeletal features of AHO (i.e. had PPHP). The implication was that the maternal allele of *GNAS1* normally provides the function that is deficient in PHP-Ia. *GNAS1* was thus predicted to be an imprinted gene, expressed from the maternal allele. Such imprinting would have to be tissue-restricted, as the 50% reduction in erythrocyte $G_s\alpha$ levels seen in PHP patients implies biallelic expression in erythropoietic tissue.

Despite individual reports of families not fitting the imprinting hypothesis (Schuster *et al.*, 1994), most subsequent literature has lent strong support. The pedigrees published in a large multicenter study, for example (Marguet *et al.*, 1997), strikingly illustrate this maternal inheritance bias, despite the authors' comments to the contrary. PPHP patients, like PHP-Ia patients, have 50% reduction in $G_s\alpha$ activity, but do not show loss of renal cAMP excretion, implying that unlike PHP-Ia patients, they have considerable residual $G_s\alpha$ activity in the proximal renal tubule (Fischer *et al.*, 1998).

3.1 Molecular studies

Initial attempts to demonstrate directly that *GNAS1* is monoallelically expressed used a single-nucleotide polymorphism (SNP) in exon 5 to distinguish the maternal and paternal alleles in human fetal RNA from a range of tissues (Campbell *et al.*, 1994). No allelic imbalance could be detected in any samples. While this finding did not exclude monoallelic expression in selected populations of cells (e.g. PTH target cells in the proximal renal tubule), it was clear that as in blood, *GNAS1* expression was biallelic in most tissues.

A paternally-expressed G protein encoded by GNAS1. Clear demonstration at a molecular level that *GNAS1* is imprinted eventually came serendipitously. Using restriction landmark genomic scanning (RLGS), a genome-wide screen for novel differentially methylated regions (DMRs) was performed. Two DMRs were identified. One proved to lie upstream of the *ZAC* tumor suppressor gene on 6q24, which is paternally expressed (Kamiya *et al.*, 2000). The other DMR, named A20, identified a database expressed sequence tag representing a novel transcript from *GNAS1* (Hayward *et al*., 1998). The A20 DMR was shown to lie about 35 kb upstream of exon 1. A large exon located at this position splices to the acceptor site of exon 2 to generate a mRNA encoding the human homolog of an 'extra-large' G protein, XLαs, originally identified in rat PC12 cells (Kehlenbach *et al*., 1994). XLαs has a novel N-terminal sequence continuous with the $G_s\alpha$ open reading frame contained in exons 2–13. The human XLαs-containing transcripts were shown to be exclusively derived from the paternal allele in all tissues (Hayward *et al*., 1998a). A large region of differential methylation including the XLαs exon is methylated on the maternal allele only. Thus, while $G_s\alpha$, the first recognized *GNAS1* product, is mostly biallelically derived, another G protein encoded at the same locus, XLαs, is paternally derived. As the two G proteins share sequences encoded by exons 2–13, determination of the physiological function of XLαs might have implications for the pathophysiology of PHP/PPHP that result from mutations in these exons (see below).

Maternal transcripts from GNAS1. A third class of protein-coding *GNAS1* transcript was identified by 5′-RACE experiments aimed at identifying additional promoters (Hayward *et al*., 1998b). Again, a large single exon, 14 kb upstream of XLαs, was shown to be included in transcripts that splice to the exon 2 acceptor site. This exon encodes the human homolog of the bovine neuroendocrine secretory protein NESP55 (Ischia *et al*., 1997). Unlike the XLαs exon, the NESP55 exon includes both initiation and termination codons, so that *GNAS1* exons 2–13 comprise part of the 3′ untranslated region. NESP55 thus shares no amino acid sequence homology with XLαs or $G_s\alpha$; it is a protein of unknown function, originally isolated from adrenal medulla chromaffin cell secretory granules.

A CpG island that includes part of the NESP55 exon shows the opposite pattern of differential methylation to XLαs, being methylated on the paternal allele. Correspondingly, NESP55 transcripts are derived exclusively from the maternal allele in all tissues (Hayward *et al*., 1998b).

A second paternally active promoter. Variant $G_s\alpha$ transcripts lacking exon 1 but instead containing sequences derived from an exon now known as 1a have been reported in the dog (Ishikawa *et al*., 1990), human (Swaroop *et al*., 1991) and mouse (Liu *et al*., 2000). Exon 1a lies ~2.5 kb upstream of exon 1 in these species. Exon 1a-containing transcripts lack the $G_s\alpha$ initiator codon and so are probably not translated. Nonetheless, in some tissues they comprise a large proportion of all *GNAS1* transcripts (Swaroop *et al*., 1991). In both mouse and human, exon 1a lies within a DMR that is methylated on the maternal allele. Exon 1a-containing transcripts are derived from the paternal allele (Liu *et al*., 2000a,b). Their precise function is unclear, but they may be involved in the tissue-specific imprinting of the downstream exon 1 promoter.

Alternative 3′ exon. Further variability in *GNAS1* transcript type results from the presence of an exon variously referred to as N1 or 3N, lying within intron 3 of the gene and including an alternative stop codon and polyadenylation site. Transcripts of forms 1–2–3–3N (Crawford *et al*., 1993), 1a–2–3–3N (Ishikawa *et al*.,

1990), XL-A20-A21–2–3–3N, XL-2–3–3N (Hayward *et al.*, 1998a; Pasolli *et al.*, 2000) and NESP-2–3–3N (Weiss *et al.*, 2000) have all been described. It is not known whether any of the proteins encoded by these truncated transcripts have independent functional significance. All such proteins, other than that encoded by NESP-2–3–3N, would be C-terminally truncated.

These complex patterns of alternative promoter usage and splicing of *GNAS1* are summarized in *Figure 3*.

3.2 Functions of novel GNAS1 products

XL αs. The biological function of XLαs remains rather unclear at present, and it will probably need mouse knockout experiments for its physiological role to be fully revealed. In the rat, both XLαs and its C-truncated variant XLαN1 (corresponding to exons XL-2–3–3N) are expressed mostly in neuroendocrine tissues such as adrenal, endocrine pancreas, and at highest level, in the pituitary, especially the pars intermedia of the posterior pituitary (Pasolli *et al.*, 2000). Although originally reported to be associated predominantly with the trans-Golgi network, recent immunolocalization studies have shown XLαs to be almost exclusively plasma-membrane associated (Pasolli *et al.*, 2000). Unlike $G_s\alpha$, XLαs does not appear to be activated by known $G_s\alpha$-coupled receptors; nonetheless, it does associate with G protein $\beta\gamma$ subunits, and activated XLαs (either GTPγS-bound or in the shape of the GTPase-deficient mutant Q548L, analogous to the Q227L $G_s\alpha$ mutant) strongly activates adenylate cyclase (Klemke *et al.*, 2000). Thus, although the signal transduction role of XLαs in neuroendocrine cells remains unclear, it does appear likely that both it and $G_s\alpha$ act through the common effector pathway of adenylate cyclase.

NESP55. There is even less information available on the biochemical properties of NESP55. Although this protein was originally suggested to be the source of biologically active peptides as a result of proteolytic processing (Ischia *et al.*, 1997), there is as yet no direct evidence pointing to a specific physiological role. The NESP55 protein is located within large dense-core neurosecretory vesicles (Leitner *et al.*, 1999). Bovine NESP55 is found in decreasing order of abundance within the adrenal medulla, anterior and posterior pituitary, brain and intestine (Lovisetti-Scamihorn *et al.*, 1999); within the adrenal medulla, adrenergic cells contain much higher levels of NESP55 than noradrenergic cells (Bauer *et al.*, 1999b). Within the rat brain, its mRNA is confined to neuronal subpopulations in a distribution overlapping that of adrenergic, noradrenergic and serotonergic systems (Bauer *et al.*, 1999a). NESP55 protein is postranslationally modified by the addition of glycosaminoglycan (Weiss *et al.*, 2000) and differentially cleaved to a variety of peptide products in various tissues (Lovisetti-Scamihorn *et al.*, 1999).

3.3 Regulation of GNAS1 *imprinting*

Comparative sequence analysis of the 20 kb imprinted NESP55-XLαs region of mouse and human *Gnas/GNAS1* showed the most highly conserved region to lie about 3 kb upstream of XLαs (Hayward and Bonthron, 2000). This region also corresponds in the human gene to the site of initiation for a spliced, paternally expressed antisense transcript that terminates several kb upstream of NESP55. In other genes, such antisense transcripts may act to suppress the sense transcripts' promoter on the same allele (Wutz et al., 1997). It is possible, therefore, that silencing of the paternal NESP55 promoter depends on antisense transcription across the region. Direct demonstration of such a role, however, will require site-directed manipulation of the putative control elements. This might be achieved either (a) by targeting the endogenous mouse gene or (b) by altering such elements in the context of a large transgene whose ability to imprint at a heterologous locus in the mouse genome can be examined.

Pointers from the mouse. In the mouse, the genomic region corresponding to distal 20q13 is the distal part of chromosome 2. This region has been recognized to be imprinted on the basis of the contrasting phenotypes that result from paternal or maternal uniparental disomy (Williamson et al., 1998).

The molecular details of imprinting of the mouse *Gnas* gene appear mostly to be closely similar to those of *GNAS1*. DMRs associated with the murine *Nesp* and *Gnas-xl* regions were independently isolated by methylation-sensitive representational difference analysis (Kelsey et al., 1999; Peters et al., 1999). The corresponding mRNA species are maternally and paternally expressed, respectively. In addition, two potential regulatory features are conserved between *Gnas* and *GNAS1*: (a) an antisense transcript, similar in at least some respects to the *GNAS1* antisense transcript, traverses the *Nesp* region (Li et al., 2000; Wroe et al., 2000); (b) exon 1a, whose paternal transcripts probably lack coding potential, is maternally methylated in both species (Liu et al., 2000a,b). In the mouse, the exon 1a DMR is established at a time that makes it a plausible candidate site for the primary gametic imprint.

The availability of a murine *Gnas* knockout model has enabled some other aspects of imprinting to be analyzed. In particular, the prediction of cell-type specific monoallelic expression of $G_s\alpha$ transcripts has been partly verified. $G_s\alpha$ is predominantly maternally derived in proximal renal tubule (Weinstein et al., 2000; Yu et al., 1998) and in brown and white adipose tissue (Yu et al., 2000). Mice with a knockout of the maternal *Gnas* allele are obese and have reduced metabolic rates; this may provide some useful insight into the origin of the obesity of PHP-Ia patients (who have maternal *GNAS1* mutations). However, unlike humans with paternal *GNAS1* mutations, mice with a knockout of the paternal *Gnas* allele are hypermetabolic and have decreased adipose tissue lipid (Yu et al., 2000). In the mouse, at least some of these effects on energy metabolism appear to be due to decreased or increased sympathetic nervous system activity, respectively. It is not known if such alterations occur in PHP-Ia patients.

If the predominantly maternal expression of $G_s\alpha$ in murine renal cortex can be extrapolated to humans, it may go some way towards explaining the parental origin effect on hormone resistance in PHP-Ia. However, re-examination of $G_s\alpha$ imprinting in various hormonal target tissues in humans is still required. At present, we know for sure only that $G_s\alpha$ is normally predominantly maternally derived in the pituitary (see below; Hayward et al., 2001).

4. Mutational spectrum of *GNAS1*

With rare exceptions, no mutation-type specific genotype–phenotype correlation has been observed for *GNAS1* mutations. There is a spectrum of missense, frameshift, and splice site mutations in both PHP-Ia and PPHP. These mutations are presumably null, given the nature of some of them and the 50% loss of $G_s\alpha$ activity in patients. Nonetheless, there is wide phenotypic variability within families as well as between families with different mutations, and the reasons for this, apart from the imprinting effects discussed earlier, are not known. Ectopic calcification is a particularly unpredictable clinical feature; occasional patients present in infancy with relentlessly progressive subcutaneous calcification while others with the same mutation (even within the same family) may have much more typical PHP-Ia with little or no calcification. A comprehensive review of the *GNAS1* mutational spectrum observed to date has recently been given (Aldred and Trembath, 2000). The only recurrent mutation seen in independent families is a 4-bp deletion of exon 7 (Weinstein *et al.*, 1992), which has been described in 11 families, including two in which a *de novo* mutation was seen (Aldred and Trembath, 2000; Yu *et al.*, 1995). This recurrent mutation results in a frameshift and a premature termination of the open reading frame at codon 202. However, the mutant mRNA is also virtually undetectable in lymphoblastoid cell RNA, presumably as a result of instability. The same may be true for other *GNAS1* mutations, as in some early studies reduced levels of $G_s\alpha$ mRNA were seen in a substantial proportion of patients (Carter *et al.*, 1987; Schuster *et al.*, 1993).

One unusual mutation, A366S, caused a combination of PHP-Ia and testotoxicosis in two unrelated patients (Iiri, 1994). Luteinizing hormone (LH) normally activates testosterone production via the G_s-coupled LH receptor. The A366S mutation enhances GDP release from $G_s\alpha$ and constitutively activates the pathway. However, this activation only has an effect at testis temperature; at 37°C, the mutant protein is rapidly degraded, accounting for the PHP-Ia phenotype in other tissues.

4.1 Is XLαs involved in the phenotype?

Mutations of exons 2–13 have the possibility not only of affecting $G_s\alpha$ function, but also that of XLαs; this would not seem unlikely, given that the biochemical activity of XLαs appears to be, like that of $G_s\alpha$, that of a stimulatory G protein (Klemke *et al.*, 2000). PHP-Ia patients with mutations in exon 1 (which is unique to the $G_s\alpha$ transcript) have a phenotype indistinguishable to that of patients with mutations in exons 2–13 (which are common to $G_s\alpha$ and XLαs; Fischer *et al.*, 1998; Patten *et al.*, 1990). However, as PHP-Ia patients carry their mutation on the maternal allele, XLαs function, which is paternally derived, would be expected to be normal in both groups. Nonetheless, there is little or no evidence of a specific phenotypic effect on XLαs from clinical observations of families with *GNAS1* mutations. Individuals with paternally-derived mutations typically have PPHP, and do not have additional clinical features that are not also seen in PHP-Ia patients with maternally-derived mutations.

5. Activating mutations of *GNAS1*

5.1 Pituitary tumors

The intrinsic GTPase activity that normally terminates the activated state of GTP-bound $G_s\alpha$ is lost as a result of specific mutations of codons 201 and 227. These mutations occur in approximately 40% of growth-hormone secreting pituitary tumors (Landis *et al.*, 1989, 1990; Lyons *et al.*, 1990; Yang *et al.*, 1996). Less frequently, the same mutations have also been identified in thyroid adenomas and carcinomas (Suarez *et al.*, 1991), Leydig cell tumors of the ovary and testis (Fragoso *et al.*, 1998), corticotroph adenomas, pheochromocytomas and parathyroid adenomas (Williamson *et al.*, 1995a,b).

Levels of $G_s\alpha$ are generally lower in *gsp*⁺ than *gsp*⁻ somatotroph tumors. This may be the result of increased degradation of the mutant $G_s\alpha$ protein (Ballaré *et al.*, 1998). However, others have found that $G_s\alpha$ mRNA levels are higher in *gsp*⁻ tumors (Barlier *et al.*, 1999), suggesting negative transcriptional feedback of the *gsp* oncogene on *GNAS1* expression. Tumors resulting from the *gsp* oncogene occur in older patients than *gsp*⁻ tumors, and respond much better to treatment with the somatostatin analog octreotide than do *gsp*⁻ tumors (Barlier *et al.*, 1998; Yang *et al.*, 1996). Codon 201 and 227 mutation testing of somatotroph adenomas is therefore of prognostic significance.

5.2 Maternal origin of oncogenic mutations

By mutation analysis of RT-PCR products representing the allele-specific NESP55 and XLαs transcripts, we have recently demonstrated that *gsp* mutations in somatotroph adenomas occur virtually exclusively on the maternal allele (Hayward *et al.*, 2001). This bias could perhaps result from cellular lethality of a paternal *gsp* mutation. (In the pituitary, such a mutation, to judge from recent experimental studies (Klemke *et al.*, 2000), would be expected to constitutively activate all intracellular XLαs and thereby massively stimulate adenylate cyclase.) However, from our own studies, it also appears that in contrast to the situation in most tissues, $G_s\alpha$ transcripts in normal human pituitary are maternally derived. It is therefore also possible that the maternal mutational bias reflects the fact that paternal *gsp* mutations simply do not have any effect on $G_s\alpha$ activity in this tissue. It is not yet known whether the same parental bias in origin of the somatic mutation is also present in McCune Albright syndrome (see below), but in view of the tissue-specific variations in *GNAS1* imprinting, such a comparison might be quite illuminating. Further studies indicate that in both *gsp*⁺ and *gsp*⁻ somatotroph tumors, the maternal monoallelic $G_s\alpha$ transcription pattern is frequently lost, with both alleles becoming more or less equally expressed. Thus, as has been seen for some other imprinted growth control genes, relaxation of $G_s\alpha$ imprinting may play a role in the tumorigenic process.

5.3 McCune Albright syndrome

The R201H, R201C and R201G mutations are also found to underlie the McCune Albright syndrome (MAS). MAS is a non inherited disorder in which patchy skin hyperpigmentation, fibrous lesions of various bones (polyostotic fibrous dysplasia) and endocrine hyperfunction occur. The latter manifests most frequently in girls, as precocious puberty. Partial forms of the syndrome may occur, presumably reflecting variable distribution of the mosaic codon 201 mutation. The mutation may or may not be demonstrable in blood, but is generally present in the bony lesions, even if absent from the unaffected periosteum in the same bone (Kitoh *et al.*, 1999). These lesions are composed of disorganized fibrous tissue, and appear to result from upregulation of G_s function within osteogenic precursor cells (Riminucci *et al.*, 1997). Interestingly, however, and unlike somatotroph tumors bearing the same mutations, the individual lesions of fibrous dysplasia appear not to be clonal in origin. In an experimental model of MAS, in which marrow stromal cells from the fibrous dysplastic lesions are allowed to reform bone in immuno-compromised mice, neither the mutation-bearing nor normal progenitors alone are capable of recapitulating the development of fibrous dysplasia; however a mixture of normal and mutated progenitors can do so (Bianco *et al.*, 1998). Fibrous dysplasia in MAS patients may therefore be the consequence of specific abnormal interactions between stromal progenitors carrying the activating *GNAS1* mutation and their normal counterparts.

6. PHP-Ib; a tissue-specific *GNAS1* imprinting defect

PHP-Ib is characterized, like PHP-Ia, by hormone resistance, but this is restricted to the renal action of PTH; multiple hormone resistance is not seen. Patients have PTH-resistant hypocalcemia and/or hyperphosphatemia, and defective renal phosphaturic and cAMP response to PTH. However, they have normal levels of erythrocyte $G_s\alpha$ and do not show evidence of AHO or other features of PHP-Ia. For this reason, PHP-Ib used to be regarded as genetically completely distinct from PHP-Ia, and perhaps due to PTH receptor mutations. However, it did not prove possible to demonstrate PTHR mutations in this condition. Linkage analysis of four PHP-Ib kindreds was used to localize a PHP-Ib gene to a region of 20q13.3 immediately centromeric to *GNAS1* (Juppner *et al.*, 1998). Only a single recombinant in one family appeared to exclude *GNAS1* (or at least part of the gene) from the PHP-Ib interval. Furthermore, as for PHP-Ia, the pedigrees analyzed again indicated a requirement for maternal inheritance of the PHP-Ib mutation, paternally inherited cases being non penetrant. These unexpected findings raise the possibility that a mutation within or closely linked to *GNAS1*, that specifically ablates $G_s\alpha$ function only in the renal tubule, might underlie PHP-Ib. The precise nature of such mutation has not at time of writing been defined, but in a large proportion of PHP-Ib patients, abnormal methylation of the exon 1a DMR occurs, such that this region has a paternal epigenotype on both alleles (M. Bastepe *et al.*, 2001; Liu *et al.*, 2000a). As discussed above, there is circumstantial evidence for believing that in proximal renal tubule, the main PTH target tissue, $G_s\alpha$ is expressed only from the maternal allele. Probably, therefore, this target tissue is uniquely susceptible to the presence of a paternal epigenotype on the maternal *GNAS1* allele; this imprinting abnormality would greatly reduce $G_s\alpha$ function in the proximal tubule, while leaving it intact in other tissues in which $G_s\alpha$ is biallelically expressed. In some but not all PHP-Ib patients, abnormal methylation of the NESP55 and XLαs regions is also seen, but the resulting loss of NESP55 expression does not appear to correlate with any additional phenotypic features (Liu *et al.*, 2000a).

7. Mechanistic heterogeneity in PHP-Ib

Imprinting mutations appear not to underlie all cases of PHP-Ib. In one family, the mutation ΔI382, removing one residue from the C-terminal region of $G_s\alpha$, was recently described (Wu *et al.*, 2000). This finding nicely exemplifies the subtle allelic heterogeneity that underlies the various *GNAS1* endocrine syndromes. Co-transfection experiments showed that the ΔI382 mutation selectively impairs $G_s\alpha$ coupling to the PTH receptor, but not to the other $G_s\alpha$-coupled receptors for TSH and LH. As AHO was not present in this PHP-Ib family, the mutation also provides genetic evidence that AHO is probably not the result of skeletal PTH resistance, but rather due to some other defect of G_s signaling.

8. Summary

Many details remain to be elucidated about the normal function of the *GNAS1* gene. The most important of these relate to tissue-specific imprinting of the $G_s\alpha$-encoding transcripts, since ultimately it is this biological feature that determines the hormone resistance seen in PHP-Ia patients. However, the complex transcription and splicing patterns and allele-specific regulation of this locus also make it an attractive model system in which to study control of an imprinted locus. Unlike some large imprinted domains that span megabase-sized regions, the complex *GNAS1* locus, at <100 kb in size, may prove amenable to experimental manipulations that will allow its regulation to be finely dissected.

References

Albright, F., Burnett, C.H., Smith, P.H. and Parson, W. (1942) Pseudo-hypoparathyroidism – an example of 'Seabright-Bantam syndrome': report of three cases. *Endocrinology* 30: 922–932.

Aldred, M.A. and Trembath, R.C. (2000) Activating and inactivating mutations in the human GNAS1 gene. *Hum. Mutat.* 16: 183–189.

Ballaré, E., Mantovani, S., Lania, A., Di Blasio, A.M., Vallar, L. and Spada, A. (1998) Activating mutations of the Gs alpha gene are associated with low levels of Gs alpha protein in growth hormone-secreting tumors. *J. Clin. Endocrinol. Metab.* 83: 4386–4390.

Barlier, A., Gunz, G., Zamora, A.J., Morange-Ramos, I., Figarella-Branger, D., Dufour, H., Enjalbert, A. and Jaquet, P. (1998) Prognostic and therapeutic consequences of Gs alpha mutations in somatotroph adenomas. *J. Clin. Endocrinol. Metab.* 83: 1604–1610.

Barlier, A., Pellegrini-Bouiller, I., Gunz, G., Zamora, A.J., Jaquet, P. and Enjalbert, A. (1999) Impact of gsp oncogene on the expression of genes coding for Gsalpha, Pit-1, Gi2alpha, and somatostatin receptor 2 in human somatotroph adenomas: involvement in octreotide sensitivity. *J. Clin. Endocrinol. Metab.* 84: 2759–2765.

Bastepe, M., Pincus, J.E., Sugimoto, T., Tojo, K., Kanatani, M., Azuma, Y., Kruse, K., Rosenbloom, A.L., Koshiyama, H. and Juppner, H. (2001) Positional dissociation between the genetic mutation responsible for pseudohypoparathyroidism type Ib and the associated methylation defect at exon A/B: evidence for a long-range regulatory element within the imprinted GNAS1 locus. *Hum. Mol. Genet.* **10**: 1231–1241.

Bauer, R., Ischia, R., Marksteiner, J., Kapeller, I. and Fischer-Colbrie, R. (1999a) Localization of neuroendocrine secretory protein 55 messenger RNA in the rat brain. *Neuroscience* **91**: 685–694.

Bauer, R., Weiss, C., Marksteiner, J., Doblinger, A., Fischer-Colbrie, R. and Laslop, A. (1999b) The new chromogranin-like protein NESP55 is preferentially localized in adrenaline-synthesizing cells of the bovine and rat adrenal medulla. *Neurosci. Lett.* **263**: 13–16.

Bianco, P., Kuznetsov, S.A., Riminucci, M., Fisher, L.W., Spiegel, A.M. and Robey, P.G. (1998) Reproduction of human fibrous dysplasia of bone in immunocompromised mice by transplanted mosaics of normal and Gsalpha-mutated skeletal progenitor cells. *J. Clin. Invest.* **101**: 1737–1744.

Bourne, H.R. (1997) How receptors talk to trimeric G proteins. *Curr. Opin. Cell. Biol.* **9**: 134–142.

Campbell, R., Gosden, C.M. and Bonthron, D.T. (1994) Parental origin of transcription from the human *GNAS1* gene. *J. Med. Genet.* **31**: 607–614.

Carter, A., Bardin, C., Collins, R., Simons, C., Bray, P. and Spiegel, A. (1987) Reduced expression of multiple forms of the alpha subunit of the stimulatory GTP-binding protein in pseudohypoparathyroidism type Ia. *Proc. Natl Acad. Sci. USA* **84**: 7266–7269.

Chase, L.R., Melson, G.L. and Aurbach, G.D. (1969) Pseudohypoparathyroidism: defective excretion of 3′,5′-AMP in response to parathyroid hormone. *J. Clin. Invest.* **48**: 1832–1844.

Crawford, J.A., Mutchler, K.J., Sullivan, B.E., Lanigan, T.M., Clark, M.S. and Russo, A.F. (1993) Neural expression of a novel alternatively spliced and polyadenylated Gs alpha transcript. *J. Biol. Chem.* **268**: 9879–9885.

Davies, S.J. and Hughes, H.E. (1993) Imprinting in Albright's hereditary osteodystrophy. *J. Med. Genet.* **30**: 101–103.

Farfel, Z., Brickman, A.S., Kaslow, H.R., Brothers, V.M. and Bourne, H.R. (1980) Defect of receptor-cyclase coupling protein in pseudohypoparathyroidism. *N. Engl. J. Med.* **303**: 237–242.

Fischer, J.A., Egert, F., Werder, E. and Born, W. (1998) An inherited mutation associated with functional deficiency of the alpha-subunit of the guanine nucleotide-binding protein Gs in pseudo- and pseudopseudohypoparathyroidism. *J. Clin. Endocrinol. Metab.* **83**: 935–938.

Fragoso, M.C., Latronico, A.C., Carvalho, F.M. *et al.* (1998) Activating mutation of the stimulatory G protein (gsp) as a putative cause of ovarian and testicular human stromal Leydig cell tumors. *J. Clin. Endocrinol. Metab.* **83**: 2074–2078.

Hayward, B.E., Barlier, A., Korbonits, M., Grossman, A.B., Enjalbert, A., Jacquet, P. and Bonthron, D.T. (2001) Imprinting of the Gsα gene in the pathogenesis of acromegaly. *J. Clin. Invest.* **107**: R31–R36.

Hayward, B.E. and Bonthron, D.T. (2000) An imprinted antisense transcript at the human GNAS1 locus. *Hum. Mol. Genet.* **9**: 835–841.

Hayward, B.E., Kamiya, M., Strain, L., Moran, V., Campbell, R., Hayashizaki, Y. and Bonthron, D.T. (1998a) The human *GNAS1* gene is imprinted and encodes distinct paternally and biallelically expressed G proteins. *Proc. Natl Acad. Sci. USA* **95**: 10038–10043.

Hayward, B.E., Moran, V., Strain, L. and Bonthron, D.T. (1998b) Bidirectional imprinting of a single gene: GNAS1 encodes maternally, paternally, and biallelically derived proteins. *Proc. Natl Acad. Sci. USA* **95**: 15475–15480.

Iiri, T., Backlund, P.S., Jr., Jones, T.L., Wedegaertner, P.B. and Bourne, H.R. (1996) Reciprocal regulation of Gs alpha by palmitate and the beta gamma subunit. *Proc. Natl Acad. Sci. USA* **93**: 14592–14597.

Ischia, R., Lovisetti-Scamihorn, P., Hog-Angeletti, R., Wolkersdorfer, M., Winkler, H. and Fischer-Colbrie, R. (1997) Molecular cloning and characterization of NESP55, a novel chromogranin-like precursor of a peptide with 5-HT1B receptor antagonist activity. *J. Biol. Chem.* **272**: 11657–11662.

Ishikawa, Y., Bianchi, C., Nadal-Ginard, B. and Homcy, C.J. (1990) Alternative promoter and 5′ exon generate a novel Gs alpha mRNA. *J. Biol. Chem.* **265**: 8458–8462.

Juppner, H., Schipani, E., Bastepe, M. *et al.* (1998) The gene responsible for pseudohypoparathyroidism type Ib is paternally imprinted and maps in four unrelated kindreds to chromosome 20q13.3. *Proc. Natl Acad. Sci. USA* **95**: 11798–11803.

Kamiya, M., Judson, H., Okazaki, Y. *et al.* (2000) The cell cycle control gene ZAC/PLAGL1 is imprinted – a strong candidate gene for transient neonatal diabetes. *Hum. Mol. Genet.* **9**: 453–460.

Kehlenbach, R.H., Matthey, J. and Huttner, W.B. (1994) XL alpha s is a new type of G protein [published erratum appears in *Nature* 1995, 375: 253]. *Nature* **372**: 804–809.

Kelsey, G., Bodle, D., Miller, H.J., Beechey, C.V., Coombes, C., Peters, J. and Williamson, C.M. (1999) Identification of imprinted loci by methylation-sensitive representational difference analysis: application to mouse distal chromosome 2. *Genomics* **62**: 129–138.

Kitoh, H., Yamada, Y. and Nogami, H. (1999) Different genotype of periosteal and endosteal cells of a patient with polyostotic fibrous dysplasia. *J. Med. Genet.* **36**: 724–725.

Klemke, M., Pasolli, H., Kehlenbach, R., Offermans, S., Schultz, G. and Huttner, W. (2000) Characterization of the extra-large G-protein α-subunit XLαs. II. Signal transduction properties. *J. Biol. Chem.* **275**: 33633–33640.

Kozasa, T., Itoh, H., Tsukamoto, T. and Kaziro, Y. (1988) Isolation and characterization of the human Gs alpha gene. *Proc. Natl Acad. Sci. USA* **85**: 2081–2085.

Landis, C.A., Masters, S.B., Spada, A., Pace, A.M., Bourne, H.R. and Vallar, L. (1989) GTPase inhibiting mutations activate the alpha chain of Gs and stimulate adenylyl cyclase in human pituitary tumours. *Nature* **340**: 692–696.

Landis, C.A., Harsh, G., Lyons, J., Davis, R.L., McCormick, F. and Bourne, H.R. (1990) Clinical characteristics of acromegalic patients whose pituitary tumors contain mutant Gs protein. *J. Clin. Endocrinol. Metab.* **71**: 1416–1420.

Leitner, B., Lovisetti-Scamihorn, P., Heilmann, J., Striessnig, J., Blakely, R.D., Eiden, L.E. and Winkler, H. (1999) Subcellular localization of chromogranins, calcium channels, amine carriers, and proteins of the exocytotic machinery in bovine splenic nerve. *J. Neurochem.* **72**: 1110–1116.

Levine, M.A., Jap, T.S., Mauseth, R.S., Downs, R.W. and Spiegel, A.M. (1986) Activity of the stimulatory guanine nucleotide-binding protein is reduced in erythrocytes from patients with pseudohypoparathyroidism and pseudopseudohypoparathyroidism: biochemical, endocrine, and genetic analysis of Albright's hereditary osteodystrophy in six kindreds. *J. Clin. Endocrinol. Metab.* **62**: 497–502.

Li, T., Vu, T.H., Zeng, Z.L., Nguyen, B.T., Hayward, B.E., Bonthron, D.T., Hu, J.F. and Hoffman, A.R. (2000) Tissue-specific expression of antisense and sense transcripts at the imprinted Gnas locus. *Genomics* **69**: 295–304.

Liu, J., Litman, D., Rosenberg, M.J., Yu, S., Biesecker, L.G. and Weinstein, L.S. (2000a) A GNAS1 imprinting defect in pseudohypoparathyroidism type IB. *J. Clin. Invest.* **106**: 1167–1174.

Liu, J., Yu, S., Litman, D., Chen, W. and Weinstein, L.S. (2000b) Identification of a methylation imprint mark within the mouse *Gnas* locus. *Mol. Cell. Biol.* **20**: 5808–5817.

Lovisetti-Scamihorn, P., Fischer-Colbrie, R., Leitner, B., Scherzer, G. and Winkler, H. (1999) Relative amounts and molecular forms of NESP55 in various bovine tissues. *Brain Res.* **829**: 99–106.

Lyons, J., Landis, C.A., Harsh, G. *et al.* (1990) Two G protein oncogenes in human endocrine tumors. *Science* **249**: 655–659.

Marguet, C., Mallet, E., Basuyau, J.P., Martin, D., Leroy, M. and Brunelle, P. (1997) Clinical and biological heterogeneity in pseudohypoparathyroidism syndrome. *Hormone Res.* **48**: 120–130.

Pasolli, H., Klemke, M., Kehlenbach, R., Wang, Y. and Huttner, W. (2000) Characterization of the extra-large G-protein α-subunit XLαs. I. Tissue distribution and subcellular localization. *J. Biol. Chem.* **275**: 33622–33632.

Patten, J.L., Johns, D.R., Valle, D., Eil, C., Gruppuso, P.A., Steele, G., Smallwood, P.M. and Levine, M.A. (1990) Mutation in the gene encoding the stimulatory G protein of adenylate cyclase in Albright's hereditary osteodystrophy. *N. Engl. J. Med.* **322**: 1412–1419.

Peters, J., Wroe, S.F., Wells, C.A., Miller, H.J., Bodle, D., Beechey, C.V., Williamson, C.M. and Kelsey, G. (1999) A cluster of oppositely imprinted transcripts at the Gnas locus in the distal imprinting region of mouse chromosome 2. *Proc. Natl Acad. Sci. USA* **96**: 3830–3835.

Riminucci, M., Fisher, L.W., Shenker, A., Spiegel, A.M., Bianco, P. and Gehron Robey, P. (1997) Fibrous dysplasia of bone in the McCune–Albright syndrome: abnormalities in bone formation. *Am. J. Pathol.* **151**: 1587–1600.

Schuster, V., Eschenhagen, T., Kruse, K., Gierschik, P. and Kreth, H.W. (1993) Endocrine and molecular biological studies in a German family with Albright hereditary osteodystrophy. *Eur. J. Pediatr.* **152**: 185–189.

Schuster, V., Kress, W. and Kruse, K. (1994) Paternal and maternal transmission of pseudohypoparathyroidism type Ia in a family with Albright hereditary osteodystrophy: no evidence of genomic imprinting. *J. Med. Genet.* **31**: 84.

Suarez, H.G., du Villard, J.A., Caillou, B., Schlumberger, M., Parmentier, C. and Monier, R. (1991) gsp mutations in human thyroid tumours. *Oncogene* **6**: 677–679.

Swaroop, A., Agarwal, N., Gruen, J.R., Bick, D. and Weissman, S.M. (1991) Differential expression of novel Gs alpha signal transduction protein cDNA species. *Nucleic Acids Res.* **19**: 4725–4729.

Wedegaertner, P.B. and Bourne, H.R. (1994) Activation and depalmitoylation of Gs alpha. *Cell* **77**: 1063–1070.

Wedegaertner, P.B., Bourne, H.R. and von Zastrow, M. (1996) Activation-induced subcellular redistribution of Gs alpha. *Mol. Biol. Cell.* **7**: 1225–1233.

Weinstein, L.S., Gejman, P.V., de Mazancourt, P., American, N. and Spiegel, A.M. (1992) A heterozygous 4-bp deletion mutation in the Gs alpha gene (GNAS1) in a patient with Albright hereditary osteodystrophy. *Genomics* **13**: 1319–1321.

Weinstein, L.S., Gejman, P.V., Friedman, E., Kadowaki, T., Collins, R.M., Gershon, E.S. and Spiegel, A.M. (1990) Mutations of the Gs alpha-subunit gene in Albright hereditary osteodystrophy detected by denaturing gradient gel electrophoresis. *Proc. Natl Acad. Sci. USA* **87**: 8287–8290.

Weinstein, L.S., Yu, S. and Ecelbarger, C.A. (2000) Variable imprinting of the heterotrimeric G protein G(s) alpha-subunit within different segments of the nephron. *Am. J. Physiol. Renal. Physiol.* **278**: F507–514.

Weiss, U., Ischia, R., Eder, S., Lovisetti-Scamihorn, P., Bauer, R. and Fischer-Colbrie, R. (2000) Neuroendocrine secretory protein 55 (NESP55): alternative splicing onto transcripts of the GNAS gene and posttranslational processing of a maternally expressed protein. *Neuroendocrinology* **71**: 177–186.

Williamson, C.M., Beechey, C.V., Papworth, D., Wroe, S.F., Wells, C.A., Cobb, L. and Peters, J. (1998) Imprinting of distal mouse chromosome 2 is associated with phenotypic anomalies in utero. *Genet. Res.* **72**: 255–265.

Williamson, E.A., Ince, P.G., Harrison, D., Kendall-Taylor, P. and Harris, P.E. (1995a) G-protein mutations in human pituitary adrenocorticotrophic hormone-secreting adenomas. *Eur. J. Clin. Invest.* **25**: 128–131.

Williamson, E.A., Johnson, S.J., Foster, S., Kendall-Taylor, P. and Harris, P.E. (1995b) G protein gene mutations in patients with multiple endocrinopathies. *J. Clin. Endocrinol. Metab.* **80**: 1702–1705.

Wroe, S.F., Kelsey, G., Skinner, J.A., Bodle, D., Ball, S.T., Beechey, C.V., Peters, J. and Williamson, C.M. (2000) An imprinted transcript, antisense to Nesp, adds complexity to the cluster of imprinted genes at the mouse Gnas locus. *Proc. Natl Acad. Sci. USA* **97**: 3342–3346.

Wu, W.I., Schwindinger, W.F., Aparicio, L.F. and Levine, M.A. (2001) Selective resistance to parathyroid hormone caused by a novel uncoupling mutation in the carboxyl terminus of Gαs: A cause of pseudohypoparathyroidism Type Ib. *J. Biol. Chem.* **276**: 165–171.

Wutz, A., Smrzka, O.W., Schweifer, N., Schellander, K., Wagner, E.F. and Barlow, D.P. (1997) Imprinted expression of the Igf2r gene depends on an intronic CpG island. *Nature* **389**: 745–749.

Yang, I., Park, S., Ryu, M., Woo, J., Kim, S., Kim, J., Kim, Y. and Choi, Y. (1996) Characteristics of gsp-positive growth hormone-secreting pituitary tumors in Korean acromegalic patients. *Eur. J. Endocrinol.* **134**: 720–726.

Yu, S., Yu, D., Hainline, B.E., Brener, J.L., Wilson, K.A., Wilson, L.C., Oude-Luttikhuis, M.E., Trembath, R.C. and Weinstein, L.S. (1995) A deletion hot-spot in exon 7 of the Gs alpha gene (GNAS1) in patients with Albright hereditary osteodystrophy. *Hum. Mol. Genet.* **4**: 2001–2002.

Yu, S., Yu, D., Lee, E., Eckhaus, M., Lee, R., Corria, Z., Accili, D., Westphal, H. and Weinstein, L.S. (1998) Variable and tissue-specific hormone resistance in heterotrimeric Gs protein alpha-subunit (Gsalpha) knockout mice is due to tissue-specific imprinting of the gsalpha gene. *Proc. Natl Acad. Sci. USA* **95**: 8715–8720.

Yu, S., Gavrilova, O., Chen, H. *et al.* (2000) Paternal versus maternal transmission of a stimulatory G-protein alpha subunit knockout produces opposite effects on energy metabolism. *J. Clin. Invest.* **105**: 615–623.

9

Genomic disorders

Susan L. Christian and David H. Ledbetter

1. Introduction

Chromosomal rearrangements that cause deletions, duplications, inversions or marker chromosome formation often result in human disease due to alteration in the dosage of a gene or genes that leads to the phenotypic effects observed. The human diseases that result from these rearrangements are referred to as genomic disorders (Lupski, 1998a). The most common recurring chromosomal rearrangements are mediated by low-copy, chromosome specific repeated DNA sequences present at breakpoint 'hotspots' referred to as duplicons (Eichler, 1998). These duplicons range in size from 3 to 500 kb and include small duplications containing exons of single genes up to 400 kb genomic duplications containing multiple genes. Misalignment of duplicons during meiosis followed by double stranded breaks in the DNA is the probable mechanism of rearrangements causing genomic disorders.

Several excellent reviews have been published recently describing some of these disorders (Ji *et al.*, 2000a; Lupski, 1998a; Mazzarella and Schlessinger, 1998). In this review, genomic disorders have been classified into two groups based on the number of genes affected by the chromosomal rearrangement (*Table 1*). Examples will be presented of the different types of duplicons observed in the human genome followed by a description of several mechanisms of DNA double–strand break repair.

2. Type I: single gene disorders

This first type of genomic disorder is comprised of single gene disorders that are associated with both point mutations and recurrent chromosomal rearrangements that alter expression of the functional copy of a single gene. In these cases the disease gene is present either within or between a duplicated genomic sequence (i.e. duplicon). This type of genomic disorder is characterized by smaller duplicons, usually <50 kb, and the size of the rearrangements ranges from ~4 kb -1.5 Mb (*Table 1*).

2.1 α thalassemia

Alpha thalassemia (OMIM 141800) is caused by reduced (α+) or complete (α−) loss of expression of the α globin subunits of the hemoglobin molecule. The α+

Genotype to Phenotype second edition, edited by S. Malcolm and J. Goodship.
© 2001 BIOS Scientific Publishers Ltd, Oxford.

Table 1. Example of genomic disorders

Genomic disorder	Type	Location	Duplicon size	Orientation	Chr. Abn.	Size of rearrangement
alpha thalassemia	I	16p13.3	5 kb	direct	del	3.7–4.2 kb
red–green color blindness	I	Xq28	39 kb	direct	del/dup	39 kb
Hunter syndrome	I	Xq28	3 kb	inverted	inv	20 kb
Hemophilia A	I	Xq28	9.5 kb	inverted	inv	550 kb
Juvenile nephronophithisis	I	2q13	45 kb	direct	del	290 kb
CMT1a	I	17p12	24 kb	direct	dup	1.5 Mb
HNPP	"	"	"	"	del	"
SMS	II	17p11.2	>200 kb	direct	del	~5 Mb
PWS/AS	II	15q11-q13	~400 kb	inverted	del	~4 Mb
Dup15q11-q13 (autism)	"	"	"	"	dup, marker	4–10 Mb
DGS/VCFS	II	22q11.2	200 kb	?	del	3.0 Mb /~1.5 Mb
dup22q11.2	"	"	"	"	dup	3.0 Mb /~1.5 Mb
Cat-eye syndrome	II	22q11.2	200 kb	?	marker	~500 kb – 2 Mb
Williams syndrome	II	7q11.2	>300 kb	direct	del	1.6–2.0 Mb
Nerurofibromatosis type I	II	17q11.2	15–100 kb	direct	del	1.5 Mb

disease is common in the Mediterranean, Middle East, Africa, India and Melanesia with frequencies ranging from 5% to 80% while the more severe α-condition is localized to Southeast Asia with a frequency of 5–15% (Weatherall *et al.*, 1995). The hemoglobin molecule is comprised of two α and two β globin gene subunits. The α globin genes are arranged in a tandem array at chromosome 16p13.3 and are comprised of two active genes (α_1 and α_2), two pseudogenes ($\psi\alpha_1$ and $\psi\alpha_2$), one α-like gene (ζ_2) and its pseudogene ($\psi\zeta_1$) (Higgs *et al.*, 1989). The α_1 and α_2 genes are located within ~4 kb genomic duplicons containing three subregions of homology termed X, Y, and Z (Higgs *et al.*, 1989; *Figure 1a*). The X boxes are 4.2 kb apart and the Z boxes are 3.7 kb apart. The most common molecular defects causing α+ thalassemia arise through recombination between either the X boxes producing a 4.2 kb deletion or the Z boxes producing a 3.7 kb deletion (Higgs *et al.*, 1989).

2.2 Red–green color blindness

Red–green color blindness (OMIM 303900, 303800) are X-linked disorders that affect ~8% of Caucasian, ~5% of Asian and ~3% of African males (Nathans, 1999). The genes affecting red and green color vision are located on chromosome Xq28 in tandem sets of 1–4 red and 1–7 green opsin genes (Nathans, 1999; Neitz and Neitz, 2000). The 39 kb duplicons in this region are comprised of one 15 kb opsin gene plus 24 kb of flanking sequence (Nathans, 1999; *Figure 1b*). Unequal homologous recombination within the flanking sequence produces changes in the number of the red–green opsin genes, while intergenic recombination produces hybrid red–green genes (Nathans, 1999). Red–green color vision defects occur when there are no intact red and/or green opsin genes on the X chromosome (Nathans, 1999; Neitz and Neitz, 2000).

2.3 Hunter syndrome

Hunter syndrome or mucopolysaccharidosis type II (MPSII) (OMIM 309900) is an X-linked disorder with an incidence of 1/132 000 live male births (Lagerstedt *et al.*, 1997). This lysosomal storage disorder is caused by a defect in the iduronate-2-sulfatase (*IDS*) gene leading to accumulation of heparan and dermatan sulfate in various tissues (Lagerstedt *et al.*, 1997). Excess accumulation of these glucosaminoglycans causes a range of clinical phenotypes including progressive mental retardation, short stature, coarse facial features and skeletal abnormalities with the severe forms causing death before age 15. The *IDS* gene is ~24 kb in size with a 3 kb pseudogene, *IDS2*, located ~20 kb distal in an inverted orientation on Xq28 (Bondeson *et al.*, 1995a, Timms *et al.*, 1995; *Figure 1c*). Approximately 13% of Hunter syndrome patients have a paracentric inversion between IDS and IDS2 causing loss of function of the *IDS* gene (Bondeson *et al.*, 1995b). Cloning of the inversion breakpoints in six patients determined that all recombinations occurred within a 1 kb region with sequence identity >98% (Lagerstedt *et al.*, 1997).

2.4 Hemophilia A

Hemophilia A (OMIM 306700) is a common X-linked bleeding disorder with an incidence of ~1/5000 live male births caused by loss of function of the coagulation factor VIII gene located on chromosome Xq28. Approximately 45% of mutations in the factor VIII gene are due to a paracentric inversion involving two copies of 'gene A' (Lakich *et al.*, 1993). One copy of this gene is located within the 35 kb intron 22 of the factor VIII gene, while two additional copies of gene A are located ~500 kb distal (Naylor *et al.*, 1993, 1995; *Figure 1d*). A foldback pairing mechanism for the telomeric portion of Xq has been proposed as an intermediate structure. A subsequent crossover event between two gene A copies causes an inversion that disrupts the factor VIII gene by separating exons 1–22 from exons 23–26 (Lakich *et al.*, 1993; Naylor *et al.*, 1993). A similar mechanism has also been proposed for Hunter syndrome.

2.5 Juvenile nephronophthisis

Familial juvenile nephronophthisis (NPH) (OMIM 256100) is an autosomal recessive disorder responsible for 6–10% of chronic renal failure in children. Mutations in the *NPHP1* gene, located on chromosome 2q13, account for ~85% of the purely renal form of the disease (Konrad *et al.*, 1996). In ~80% of these patients a homozygous deletion of 290 kb causes loss of the *NPHP1* gene (Saunier

Single Gene Disorders

(a) α thalassemia

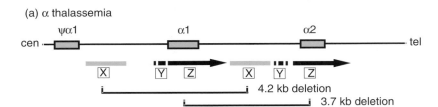

(b) Red green colorblindness

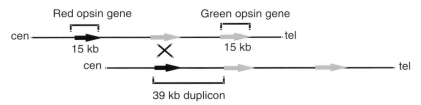

(c) Hunter syndrome

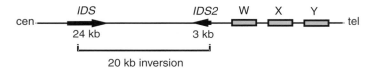

(d) Hemophilia A

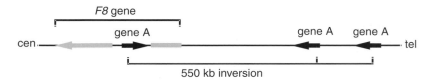

(e) Juvenile nephronophthisis

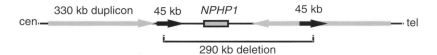

(f) CMT1A/HNPP

◀ **Figure 1.** Genomic organization of duplicons in single gene disorders. (a) The alpha globin gene cluster including two alpha globin genes (α1 and α2) and a α1 pseudogene ($\psi\alpha_1$) are represented by the dark striped rectangles. The duplicated X, Y, and Z boxes are presented below the line. The region between the X boxes and between the Z boxes indicates the common rearrangements observed in alpha thalassemia. (b) The red opsin genes are indicated with a black arrow and the green opsin genes with a light arrow. The 39 kb duplicon that contains a 15 kb opsin gene plus 24 kb of flanking sequence is indicated between two opsin genes. In red–green color blindness misalignment of these duplicons followed by homologous recombination causes gain or loss of the red and green opsin genes. (c) The iduronate-2-sulfatase gene (*IDS*) is indicated with a long black arrow and the *IDS2* pseudogene with a short black arrow. Additional genes W, X, and Y are present as light rectangles distal to *IDS2*. The line between the *IDS* gene and the *IDS2* pseudogene indicates the site of the common inversion. (d) The factor VIII (*F8*) gene is indicated by a long light arrow. The three copies of gene A are indicated by short black arrows. A 550 kb inversion occurs between the proximal copy of gene A, located within intron 22 of the *F8* gene, and one of the distal copies of gene A. (e) The *NPHP1* gene is indicated by a light colored box between two duplicons. The 330 kb duplicons are indicated by light arrows and the 45 kb duplicons by black arrows. A 290 kb deletion occurs by homologous recombination between the 45 kb duplicons. (f) The *PMP22* dosage sensitive gene is indicated by the light colored box. The 24 kb CMT1A-REPs are indicated by black arrows and the *COX10* gene by a light striped box. Homologous recombination between the CMT1A-REPs causes the duplications and deletions associated with CMT1A and HNPP, respectively. cen, centromere; tel, telomere.

---

et al., 2000). The deleted region is flanked by two ~330 kb inverted repeats and two 45 kb direct repeats (Saunier *et al.*, 2000; *Figure 1e*). Unequal homologous recombination between the 45 kb repeats causes loss of the *NPHP1* gene while recombination between the 330 kb repeats causes a nonpathologic paracentric inversion (Saunier *et al.*, 2000).

2.6 CMT1a/HNPP

Charcot-Marie-Tooth syndrome, type 1A (CMT1A) (OMIM 118220) and hereditary neuropathy with liability to pressure palsies (HNPP) (OMIM 162500) are disorders producing peripheral neuropathies related to dosage of the *PMP22* gene (Lupski, 1998b). CMT1 has an estimated prevalence of 1/2500 individuals with 70–90% showing a duplication of the *PMP22* gene while 82% of familial and 86% of sporadic cases of HNPP are the result of deletion of the *PMP22* gene (Lupski, 1998b; Nelis *et al.*, 1996). The *PMP22* gene is located on chromosome 17p12 within a 1.5 Mb region containing multiple genes that are flanked by two 24 kb duplicons, termed CMT1A-REPs, that are arranged in a direct orientation (Lupski, 1998b; *Figure 1f*). Unequal homologous recombination events produce either duplication (CMT1A) or deletion (HNPP) of a 1.5 Mb region containing 30–50 genes (Lupski, 1998b). However, only *PMP22* is thought to be dosage sensitive (Lupski, 1998b).

The proximal CMT1A-REP contains exon VI of the *COX10* gene while the distal CMT1A-REP is located within the ancestral copy of the *COX10* gene indicating an evolutionary duplication of this region (Reiter *et al.*, 1997; *Figure 1f*).

Although the two CMT1A-REP duplicons contain 98.7% sequence identity, a 557 base pair hotspot of recombination was identified in 21/23 HNPP patients (Reiter *et al.*, 1998). Located near this hotspot is a *mariner* insect transposon-like element (MITE) that may provide a target for double-strand breaks and stimulate homologous recombination (Lupski, 1998b; Reiter *et al.*, 1996).

Many other examples of single gene disorders that may arise through chromosomal rearrangements have been observed including spinal muscular atrophy (SMA) caused by a ~500 kb inverted duplication in 5q13 in 18% of cases, duplication of the *PLP* gene in Pelizaeus–Merzbacher disease, β thalassemia, familial isolated growth-hormone deficiency (GH1), congenital adrenal hyperplasia III (CAH, 21-hydroxylase deficiency), glucocorticoid-remediable aldosteronism (GRA), Bartter syndrome type III, and Gaucher disease (Lupski, 1998a; Mazzarella and Schlessinger, 1998).

3. Type II: contiguous gene syndromes

This type of genomic disorder is characterized by large chromosomal rearrangements with complex phenotypes associated with the dosage effects of multiple, unrelated genes and is referred to as a contiguous gene syndrome (Ledbetter and Ballabio, 1995). Although many genomic disorders with recurrent regions of chromosomal breakage are known, the mechanisms of rearrangement are only now being characterized at a molecular level. The best understood mechanisms of rearrangement are facilitated by larger (up to 400 kb) duplicons that flank large genomic regions containing multiple genes (*Table 1*).

3.1 Chromosome 17p11.2

Deletion 17p11.2 [Smith–Magenis syndrome (SMS)]. SMS (OMIM 182290) is a moderate to severe mental retardation syndrome with a prevalence of approximately 1/25 000 live births (Greenberg *et al.*, 1991). Additional clinical findings include developmental delay, self-injurious behavior, peripheral neuropathy, craniofacial and skeletal anomalies, and sleep disturbances (Greenberg *et al.*, 1996). Greater than 80% of SMS cases are the result of a 5 Mb deletion on 17p11.2 (Juyal *et al.*, 1996; Lupski *et al.*, 1998a). Located centromeric to the CMT1A/HNPP deletion/duplication region, the SMS deletion region is flanked by duplicons termed SMS-REPs (Chen *et al.*, 1997; *Figure 2a*). Partial cosmid contigs

▶ **Figure 2.** Genomic organization of duplicons in contiguous gene disorders. The overall genomic organization of the contiguous gene syndromes is very similar. In all illustrations the black arrows or rectangles represent duplicons, and the orientation of the duplicons is indicated with the arrows, if known. Genes located either within or between the duplicons are indicated with small vertical lines. The locations and sizes of the rearrangements observed in these disorders are indicated below the duplicons. (a) SMS, Smith Magenis syndrome; dup, duplication; (b) PWS, Prader–Willi syndrome; AS, Angelman syndrome; BP, breakpoint; (c) DGS, DiGeorge syndrome; VCFS, Velo-cardio-facial syndrome; CES, Cat-eye syndrome; LCR, low copy repeat; (d) Williams syndrome; (e) NF1, Neurofibromatosis type 1.

(a) SMS/dup17p11.2

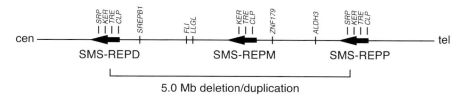

5.0 Mb deletion/duplication

(b) PWS/AS/dup15q11-q13

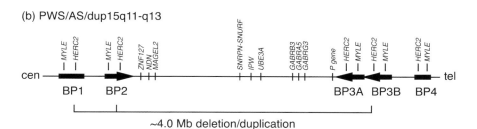

~4.0 Mb deletion/duplication

(c) DGS/VCFS/CES

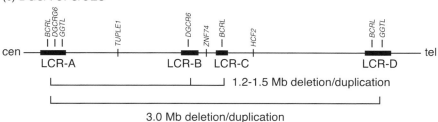

1.2-1.5 Mb deletion/duplication

3.0 Mb deletion/duplication

(d) Williams syndrome

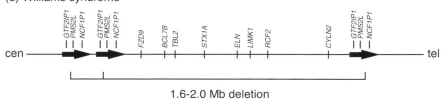

1.6-2.0 Mb deletion

(e) NF1 Microdeletion syndrome

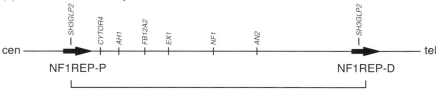

1.5 Mb deletion

have been developed across the SMS-REPs with a size estimated to be >200 kb (Lupski, 1998b). In addition, a third smaller duplicon is located within the commonly deleted region (Chen *et al.*, 1997; *Figure 2a*). Four genes, *SRP*, *TRE*, *KER*, and *CLP*, are located within both the proximal and distal SMS-REP duplicons, and analysis of SMS deletion patients using the *CLP* cDNA probe identified a 1.2 Mb novel junction fragment in 29/31 patients analyzed suggesting a common mechanism of rearrangement (Chen *et al.*, 1997).

Duplication 17p11.2. The unequal homologous recombination event that causes SMS also produces a reciprocal chromosomal duplication with an abnormal phenotype (Brown *et al.*, 1996; Chen *et al.*, 1997, Potocki *et al.*, 1999, 2000). Although only a few cases have been reported thus far, some of the features of this duplication include mild mental retardation, developmental delay, short stature, dental abnormalities and behavior problems including attention deficit, hyperactivity, obsessive compulsive disorder or autism (Brown *et al.*, 1996; Potocki, *et al.*, 1999, 2000).

3.2 Chromosome 15q11-q13

Another region producing large genomic rearrangements is located on chromosome 15q11-q13 where multiple structural abnormalities including deletions, duplications, triplications, and supernumerary marker chromosomes are observed.

Deletion 15q11-q13 [Prader–Willi syndrome(PWS)/Angelman syndrome(AS)]. The best understood genomic disorders in this region are PWS (OMIM 176279) and AS (OMIM 105830) that each occur with a frequency of ~1/10 000–1/15 000 individuals (Cassidy *et al.*, 2000). The most common cause of PWS and AS is a ~4 Mb deletion within this region that accounts for ~70% of all cases (Jiang *et al.*, 1999; Nicholls, 1999). Although the deletions causing PWS and AS are identical in size, the presence of genomic imprinting within 15q11-q13 produces different phenotypes depending on the parental origin of the deletion. A paternal deletion results in PWS with mild mental retardation, short stature, small hands and feet and obesity, while a maternal deletion produces AS, a much more severe mental retardation disorder that also includes seizures, absent speech, ataxic gait and inappropriate laughter (Cassidy *et al.*, 2000). Although most cases of AS are associated with a 4 Mb deletion, AS is also caused by loss of the maternally expressed gene, *UBE3A*, through point mutations in the gene, microdeletions within the imprinting control center, paternal uniparental disomy, unbalanced translocations or other unidentified molecular abnormalities (Jiang *et al.*, 1999). Several paternally expressed genes including *SNRPN–SNURF*, *ZNF127*, *NDN*, *MAGEL2* and *IPW* have been identified within the commonly deleted region. Although a direct association between a specific gene and a feature of PWS has not been established there is some evidence to suggest that the SNRPN–SNURF gene may encode a paternally expressed postnatal growth factor (Nicholls 1999; *Figure 2b*).

The common deletion region is flanked by two proximal duplicons (Amos-Landgraf *et al.*, 1999; Christian *et al.*, 1995; Knoll *et al.*, 1990) and two distal duplicons (Amos-Landgraf *et al.*, 1999; Christian *et al.*, 1999; Kuwano *et al.*, 1992;

Figure 2b). Recently, a ~400 kb BAC contig was developed across the duplicon region and found to contain two genes, *HERC2* and *MYLE*, with at least five additional ESTs present (Amos-Landgraf *et al.*, 1999; Christian *et al.*, 1999; Ji *et al.*, 2000b). Fine mapping indicates that the distal duplicon is present in two copies in a direct orientation and, unlike most duplicons identified thus far, the proximal and distal duplicons are arranged in an inverted orientation relative to each other (Christian *et al.*, 1999).

Duplication/triplication 15q11-q13. Chromosomal rearrangements that produce additional copies of the ~4 Mb deleted in PWS and AS have also been observed in several cases (reviewed in Mao *et al.*, 2000). The phenotype of maternal duplications producing trisomy for the 4 Mb region includes developmental delay, seizures, minor facial anomalies and autism while the phenotype associated with paternal duplications is unclear but may range from phenotypically normal to developmental delay and behavior problems (Mao *et al.*, 2000). A more complex rearrangement involving the 15q11-q13 duplicons has been predicted as the mechanism associated with interstitial triplications of the 4 Mb region producing an overall tetrasomic condition. An abnormal phenotype is observed whether inherited from the mother or father that is more severe than observed in the duplications providing additional support for the presence of dosage sensitive genes in this region (Cassidy *et al.*, 1996; Long *et al.*, 1998; Schinzel *et al.*, 1994; Ungaro *et al.*, 2001).

Supernumerary marker formation 15q11 (small) or 15q11-q13 (large). In addition to deletions and duplications, 15q11-q13 rearrangements may also produce an extra or supernumerary marker chromosome (also referred to as inv dup (15) or dicentric chromosome) (Crolla *et al.*, 1995). Rearrangements involving the proximal duplicons produce small marker chromosomes with no phenotypic effects (Huang *et al.*, 1997) while maternal rearrangements involving the distal duplicons are associated with an abnormal phenotype that includes autism (Flejter *et al.*, 1996; Robinson *et al.*, 1998; Wandstrat *et al.*, 1998, Wolpert *et al.*, 2000). The effect of these maternal large marker chromosomes is to produce either trisomy or tetrasomy for the 4 Mb region increasing the copy number of dosage sensitive genes in a manner similar to duplications and triplications.

3.3 Chromosome 22q11-q12

22q11 deletion syndrome. Interstitial deletion of chromosome 22q11 is one of the most frequent chromosomal rearrangements observed in humans with an incidence of 1/4000 live births (Scambler, 2000). The deletion has been associated with >80 phenotypic features occurring in many combinations resulting in several syndromes associated with this deletion including Velo-cardio-facial (or Shprintzen) syndrome (VCFS; OMIM 192430), DiGeorge syndrome (DGS; OMIM 188400), conotruncal heart malformations (OMIM 217095), and Opitz G/BBB syndrome (OMIM 145410; Scambler, 2000). The best characterized syndromes are VCFS, that presents with conotruncal heart defects, cleft palate, learning disabilities and a characteristic facial appearance, and DGS that includes the anomalies of VCFS with the additional features of thymic aplasia and severe hypocalcemia (Swillen *et al.*, 2000). Additionally, schizophrenia, schizoaffective disorder or bipolar disorder have been recognized in ~30% of individuals with 22q11 deletions (Swillen *et al.*, 2000).

Molecular analysis of 151 VCFS patients indicated that 83% have a deletion of chromosome 22q11 (Carlson *et al.*, 1997). Flanking these deletions are duplicons termed either VCFS-REP (Edelmann *et al.*, 1999a) or LCRs (low copy repeats; Shaikh *et al.*, 2000; *Figure 2c*). A single proximal duplicon and three distal duplicons are involved in ~98% of all deletions (Shaikh *et al.*, 2000; *Figure 2c*). The most common rearrangement produces a 3.0 Mb deletion in ~87% of cases while two smaller internal duplicons mediate deletions of ~1.2 Mb (8%) and 1.5 Mb (2%; Shaikh *et al.*, 2000). The larger outer duplicons are ~250–350 kb in size while the smaller internal duplicons are ~40–135 kb consistent with the higher frequency of rearrangements involving the larger duplicons (Shaikh *et al.*, 2000). Within the duplicons are smaller repeated regions that are arranged in both direct and inverted orientation and contain many genes including *GGT, BCRL, V7-rel, GGT-Rel, POM121L-1*, and *SMA1* (Edelmann *et al.*, 1999a; Shaikh *et al.*, 2000). The deleted region also includes many genes, however the correlation between haploinsufficiency of individual genes and phenotypic effects has not been delineated (Swillen *et al.*, 2000).

Duplication 22q11-q12. The reciprocal duplication product of the 22q11 deletion has been reported in several cases (Edelmann *et al.*, 1999b; Fujimoto and Lin, 1996; Knoll *et al.*, 1995; Lindsay *et al.*, 1995). The phenotypic features range from normal (Edelmann *et al.*, 1999b) to multiple abnormalities that include failure to thrive, developmental delay, hypotonia, preauricular ear pits or tags, high arched palate, sleep apnea, and seizure-like episodes (Edelmann *et al.*, 1999b; Fujimoto and Lin, 1996; Knoll *et al.*, 1995; Lindsay *et al.*, 1995).

Supernumerary marker chromosome 22q11-q12 [Cat-eye syndrome (CES)]. Supernumerary marker chromosomes containing extra copies of the 22q11 region often result in a disorder termed CES (OMIM 115479). CES has a variable phenotype that includes ocular coloboma, mental retardation, anal atresia, heart defects, urogenital defects, and dysmorphic facies (Mears *et al.*, 1994). The breakpoints for CES are clustered in the same regions as the 22q11 deletions (McTaggart *et al.*, 1998; Mears *et al.*, 1994). Type I marker chromosomes are small and have a rearrangement involving the proximal duplicon while type II chromosomes are larger and rearrange at one of the distal duplicons (McTaggart *et al.*, 1998). Small markers have been reported with a normal phenotype while large markers have a phenotype ranging from normal to individuals with one or more features of CES (Crolla *et al.*, 1997).

3.4 Williams–Beuren syndrome (WBS)

WBS (OMIM 194050) is a mental retardation disorder with a frequency of approximately 1/20 000 – 1/50 000 (Greenberg, 1990). WBS is characterized by short stature, a friendly outgoing personality, infantile hypercalcemia, distinctive 'elfin' facial features, dental anomalies and cardiac anomalies including supravalvular aortic stenosis (Francke, 1999). More than 90% of WBS cases are associated with a ~2.0 Mb deletion of 7q11.23 (Nickerson *et al.*, 1995; Pérez Jurado *et al.*, 1996; Robinson *et al.*, 1996). This region contains at least 16 genes affecting the overall phenotype including the elastin gene, whose mutations are known to be associated with valvular weakness leading to supravalvular aortic stenosis (DeSilva *et al.*, 1999; Francke, 1999; Hockenhull *et al.*, 1999; Peoples *et al.*, 2000).

Three duplicons of >300 kb flank the deletion region, two proximal and one distal (DeSilva *et al.*, 1999; Peoples *et al.*, 2000; *Figure 2d*). Multiple genes are present within the duplicons including *PMS2L*, *IB291*, *GTF2I*, *NCF1*, *ZP3* and *STAG3-related* genes (Francke, 1999; Pezzi *et al.*, 2000). Rearrangement within the duplicon region was confirmed by identification of a novel >3 Mb junction fragment using a cDNA probe for *IB291*, a gene located within the duplicons (Perez Jurado *et al.*, 1998). Currently, the reciprocal duplication product has not been identified for this rearrangement.

3.5 Neurofibromatosis type I (NF1). NFI (OMIM 162200) is an autosomal dominant disorder with a prevalence of 1/4000 that predisposes to the development of benign and malignant tumors (López-Correa *et al.*, 2000). In ~5–10% of cases disruption in the *NF1* gene is caused by a microdeletion of ~1.5 Mb on chromosome 17q11.2 (López-Correa *et al.*, 2000). The deleted region contains at least 11 genes in addition to the *NF1* gene producing a more severe phenotype than mutation of the *NF1* gene alone. These additional features include earlier onset and excessive development of neurofibromas, developmental delay, dysmorphic features and learning disability (Jenne *et al.*, 2000; López-Correa et al., 2000; Riva *et al.*, 2000; Upadhyyaya *et al.*, 1998). Duplicons, termed NF1-REPs, of ~15–100 kb flank the deleted region in a direct orientation and mediate the rearrangements in most cases (Dorschner *et al.*, 2000; *Figure 2e*). The duplicons contain at least four ESTs and an expressed SH3GL pseudogene (Dorschner *et al.*, 2000).

4. Mechanism of rearrangement

Chromosomal rearrangements are often initiated by double-stranded DNA breaks (DSBs) that are repaired improperly (Morgan *et al.*, 1998). These rearrangements may be either intra- or interchromosomal and are most likely to occur during meiotic recombination (Haber, 2000). Evidence to support the presence of both intra- and interchromosomal rearrangements in genomic disorders has been obtained by analysis of polymorphic markers flanking deletions or duplications in CMT1a/HNPP (Lopes *et al.*, 1998), PWS/AS (Carrozzo *et al.*, 1997; Robinson *et al.*, 1998), DGS/VCFS (Baumer *et al.*, 1998; Edelmann *et al.*, 1999a), WS (Urbán *et al.*, 1996; Baumer *et al.*, 1998) and NF1 (López-Correa *et al.*, 2000).

The presence of duplicons within the human genome thus creates genomic instability that predisposes these regions to misalignment and DSBs resulting in these common genomic disorders. Two pathways for DNA repair are recognized that predisopose to chromosomal rearrangements termed homologous recombination and nonhomologous DNA end-joining (or illegitimate recombination;

Figure 3). These DNA repair pathways together with the orientation of the duplicons produce the chromosomal rearrangements commonly observed.

4.1 Direct orientation

Homologous recombination involving misaligned duplicons in a tandem or direct orientation on sister chromatids or homologous chromosomes causes formation of reciprocal deletions and duplications (*Figure 3a*). These rearrangements are produced following DSBs where regions of high homology allow one broken end to 'invade' either the sister chromatid or allelic homolog (Haber, 2000). The resulting crossover of DNA is resolved producing a deletion and a reciprocal duplication. Examples of this type of rearrangement are observed in many genomic disorders including CMT1A/HNPP, SMS/dup 17p11.2, and VCFS/dup 22q11.2.

Nonhomologous DNA end-joining involving misaligned duplicons in a direct orientation causes formation of a supernumerary marker chromosome plus a linear acentric fragment that is subsequently lost during successive rounds of mitosis (*Figure 3b*). In this mechanism of rearrangement the broken DNA strands that result from breakage across homologous regions of DNA attach to DNA ends that share little or no homology. The strands undergo DNA synthesis and ligation yielding a junction with little or no homology to the original strand (Tsukamoto and Ikeda, 1998). An example of this rearrangement is the marker 22 chromosome that causes CES.

Another mechanism of homologous recombination observed in genomic disorders in a direct orientation is gene conversion where a broken DNA strand copies the sequence from the highly homologous segment while leaving the original locus unchanged (Bishop and Schiestl, 2000). This type of recombination has been observed with α thalassemia and red–green color blindness (Higgs *et al.*, 1989; Zhou and Li, 1996).

4.2 Inverted orientation

Although the same repair pathways are utilized in rearrangements involving duplicons in an inverted orientation, the resulting products of the rearrangements differ. Homologous recombination on a single chromatid mediated by misaligned duplicons in an inverted orientation produces an inversion as observed in Hunter syndrome or Hemophilia A. Other inversions between duplicons probably exist within the human genome that have not been detected using standard cytogenetic G-banding techniques. One example of a nonpathologic inversion mediated by homologous recombination is the rearrangement between the 330 kb duplicons in the juvenile nephronophithisis region on 2q13 (Saunier *et al.*, 2000). Homologous recombination between sister chromatids or homologous chromosomes would produce a dicentric and acentric chromosome similar to the products in *Figure 3b*. An example of this type of dicentric chromosome is observed in the inv dup (15) chromosomes associated with autism.

Nonhomologous DNA end-joining between inverted repeats within a single chromatid produces a deletion, as observed in PWS/AS, plus a circular acentric fragment that is lost (*Figure 3c*). DNA end-joining between sister chromatids or homologous chromosomes would produce reciprocal deletion and duplication products similar to the products in *Figure 3a*.

(a) Direct repeats – homologous recombination

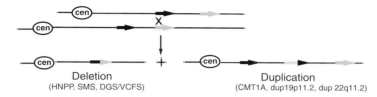

Deletion
(HNPP, SMS, DGS/VCFS)

Duplication
(CMT1A, dup19p11.2, dup 22q11.2)

(b) Direct repeats – nonhomologous end-joining

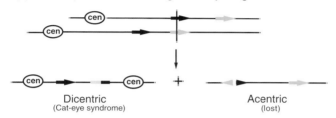

Dicentric
(Cat-eye syndrome)

Acentric
(lost)

(c) Inverted repeats

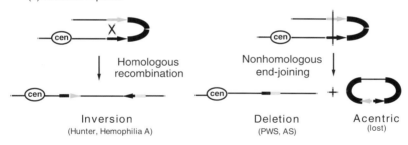

Homologous
recombination

Nonhomologous
end-joining

Inversion
(Hunter, Hemophilia A)

Deletion
(PWS, AS)

Acentric
(lost)

Figure 3. Mechanisms of chromosomal rearrangement. In all examples duplicons are represented as either dark or light arrows with the orientation indicated by the direction of the arrows. The sites of DSBs are indicated as an X for the crossover events observed in homologous recombination and a vertical line for nonhomologous end-joining. cen, centromere (a) Homologous recombination between duplicons in a direct orientation is represented where misalignment may cause double stranded breaks. DNA repair mediated by a crossover event causes formation of a deletion and a reciprocal duplication. The same products would be predicted for rearrangements involving either homologous chromosomes or sister chromatids. (b) Nonhomologous DNA end-joining in homologous chromosomes or sister chromatids in a direct orientation causes formation of one chromosome with two centromeres (dicentric) and one chromosome with no centromeres (acentric). (c) Misalignment between duplicons in an inverted orientation in single chromatids is mediated through a fold-back mechanism. Homologous recombination causes formation of an inversion while nonhomologous DNA end-joining would result in a deletion and a circular acentric fragment.

6. Conclusions

Genomic disorders that arise through rearrangement of genomic duplicons are emerging as an important mechanism of human genetic disease. Sequence analysis of other organisms including the single celled yeast, *Saccharomyces cerevisiae* (Johnston, 2000), and the model diploid plant, *Arabidopsis thaliana* (Blanc *et al.*, 2000) has detected a surprising degree of genomic duplication. The presence of duplicons within the human genome indicates that genomic duplication is also a frequent event in humans. As sequencing of the human genome nears completion, the identification of previously undiscovered genomic disorders is not only possible but probable. The identification of the chromosomal rearrangements that underlie these human genetic diseases will allow for development of molecular testing to confirm the diagnosis and also characterizastion of genes responsible for specific features of the genomic disorder.

References

Amos-Landgraf, J.M., Ji, Y., Gottlieb, T. *et al.* (1999) Chromosome breakage in the Prader-Willi and Angelman syndromes involves recombination between large, transcribed repeats at proximal and distal breakpoints. *Am. J. Hum. Genet.* 65: 370–386.

Baumer, A.F., Dutly, D., Balmer, M., Riegel, T., Tukel, M., Krajewska-Walaseck, M. and Schinzel, A.A. (1998) High level of unequal meiotic crossovers at the origin of the 22q11.2 and 7q11.23 deletions. *Hum. Mol. Genet.* 7: 887–894.

Bishop A.J.R. and Schiestl, R.H. (2000) Homologous recombination as a mechanism for genome rearrangements: environmental and genetic effects. *Hum. Molec. Genet.* 9: 2427–2434.

Blanc, G., Barakat, A., Guyot, R., Cooke, R. and Delseny, M. (2000) Extensive duplication and reshuffling in the *Arabidopsis* genome. *Plant Cell* 12: 1093–1101.

Bondeson, M.L., Malmgren, H., Dahl, N., Carlberg, B.M. and Pettersson, U. (1995a) Presence of an IDS-related locus (IDS2) in Xq28 complicates the mutational analysis of Hunter syndrome. *Eur. J. Hum. Genet.* 3: 219–227.

Bondeson, M.L., Dahl, N., Malmgren, H., Kleijer, W.J., Tonnesen, T., Carlberg, B.M. and Pettersson, U. (1995b) Inversion of the IDS gene resulting from recombination with IDS-related sequences is a common cause of the Hunter syndrome. *Hum. Molec. Genet.* 4: 615–621.

Brown, A., Phelan, M.C., Patil, S., Crawford, E., Rogers, R.C. and Schwartz, C. (1996) Two patients with duplication of 17p11.2: the reciprocal of the Smith-Magenis syndrome deletion? *Am. J. Med. Genet.* 63: 373–377.

Carlson, C., Sirotkin, H., Pandita, R. *et al.* (1997) Molecular definition of 22q11 deletions in 151 velo-cardio-facial syndrome patients. *Am. J. Hum. Genet.* 61: 620–629.

Carrozzo, R., Rossi, E., Christian, S.L. *et al.* (1997) Inter- and intrachromosomal rearrangements are both involved in the origin of 15q11-q13 deletions in Prader–Willi syndrome. *Am. J. Hum. Genet.* 61: 228–231.

Cassidy, S.B., Conroy, J., Becker, L. and Schwartz, S. (1996) Paternal triplication of 15q11-q13 in a hypotonic, developmentally delayed child without Prader–Willi or Angelman syndrome. *Am. J. Med. Genet.* 62: 206(A4).

Cassidy, S.B., Dykens, E. and Williams, C.A. (2000) Prader–Willi and Angelman syndromes: sister imprinted disorders. *Am. J. Med. Genet. (Semin. Med. Genet.)* 97: 136–146.

Chen, K.S., Manian, P., Koeuth, T., Potocki, L., Zhao, Q., Chinault, A.C., Lee, C.C. and Lupski, J.R. (1997) Homologous recombination of a flanking repeat gene cluster is a mechanism for a common contiguous gene deletion syndrome. *Nat. Genet.* 17: 154–163.

Christian, S.L., Robinson, W.P., Huang, B., Mutirangura, A., Line, M.R., Nakao, M., Surti, U., Chakravarti, A. and Ledbetter, D.H. (1995) Molecular characterization of two proximal deletion breakpoint regions in both Prader–Willi and Angelman syndrome patients. *Am. J. Hum. Genet.* 57: 40–48.

Christian, S.L., Fantes, J.A. Mewborn, S.K., Huang, B. and Ledbetter, D.H. (1999) Large genomic duplicons map to sites of instability in the Prader–Willi/Angelman syndrome chromosome region (15q11-q13). *Hum. Molec. Genet.* **8**: 1025–1027.

Crolla, J.A., Harvey, J.F., Sitch, F.L. and Dennis, N.R. (1995) Supernumerary marker 15 chromosomes: a clinical, molecular and FISH approach to diagnosis and prognosis. *Hum. Genet.* **95**: 161–170.

Crolla, J.A., Howard, P., Mitchell, C., Long, F.L. and Dennis, N.R. (1997) A molecular and FISH approach to determining karyotype and phenotype correlations in six patients with supernumerary marker (22) chromosomes. *Am. J. Med. Genet.* **72**: 440–447.

DeSilva, U., Massa, H., Trask, B.J. and Green, E.D. (1999) Comparative mapping of the region of human chromosome 7 deleted in Williams syndrome. *Genome Res.* **9**: 428–436.

Dorschner, M.O., Sybert, V.P., Weaver, M., Pletcher, B.A. and Stephens, K. (2000) NF1 microdeletion breakpoints are clustered at flanking repetitive sequences. *Hum. Molec. Genet* **9**: 35–46.

Edelmann, L., Pandita, R.K. and Morrow, B. E (1999a) Low-copy repeats mediate the common 3-Mb deletion in patients with Velo-cardio-facial syndrome. *Am. J. Hum. Genet.* **64**: 1076–1086.

Edelmann, L. Pandita, R.K., Spiteri, E. *et al.* (1999b) A common molecular basis for rearrangement disorders on chromosome 22q11. *Hum. Molec. Genet.* **8**: 1157–1167.

Eichler, E.E. (1998) Masquerading repeats: paralogous pitfalls of the human genome. *Genome Res.* **8**: 758–762.

Flejter, W.L., Bennett-Baker, P.E., Ghaziuddin, M., McDonald, M., Sheldon, S. and Gorski, J.L. (1996) Cytogenetic and molecular analysis of inv dup(15) chromosomes observed in two patients with autistic disorder and mental retardation. *Am. J. Med. Genet.* **61**: 182–187.

Francke, U. (1999) Williams-Beuren syndrome: genes and mechanisms. *Hum. Molec. Genet.* **8**: 1947–1954.

Fujimoto, A. and Lin, M.S. (1996) *De novo* direct duplication of chromosome segment 22q11.2-q13.1. *Am. J. Med. Genet.* **62**: 300–301.

Greenberg, F. (1990) Williams syndrome professional symposium. *Am. J. Med. Genet.* **6** (Suppl.): 85–88.

Greenberg, F. Guzzetta, V., Montes de Oca-Luna, R. *et al.* (1991) Molecular analysis of the Smith-Magenis syndrome: a possible contiguous-gene syndrome associated with del(17)(p11.2). *Am. J. Hum. Genet.* **49**: 1207–1218.

Greenberg, F., Lewis, R.A., Potocki, L. *et al.* (1996) Multi-disciplinary clinical study of Smith-Magenis syndrome. *Am. J. Med. Genet.* **62**: 247–254.

Haber, J.E. (2000) Partners and pathways, repairing a double-strand break. *Trends Genet.* **16**: 259–264.

Higgs, D.R., Vickers, M.A., Wilkie, A.O., Pretorius, I.M., Jarman, A.P. and Weatherall, D.J. (1989) A review of the molecular genetics of the human alpha-globin gene cluster. *Blood* **73**: 1081–1084.

Hockenhull, E.L., Carette, M.J., Metcalfe, K., Donnai, D., Read, A.P. and Tassabehji, M. (1999) A complete physical contig and partial transcript map of the Williams syndrome critical region. *Genomics* **58**: 138–145.

Huang, B. Crolla, J.A., Christian, S.L., Wolf-Ledbetter, M.E., Macha, M.E. Papenhausen, P.N. and Ledbetter, D.H. (1997) Refined molecular characterization of the breakpoints in small inv dup(15) chromosomes. *Hum. Genet.* **99**: 7–11.

Jenne, D.E., Tinschert, S., Stegmann, E., Reimann, H., Nurnberg, P., Horn, D., Naumann, I., Buske, A. and Thiel, G. (2000) A common set of at least 11 functional genes is lost in the majority of NF1 patients with gross deletions. *Genomics* **66**: 93–97.

Ji, Y., Eichler, E.E., Schwartz, S. and Nicholls, R.D. (2000a). Structure of chromosomal duplicons and their role in mediating human genomic disorders. *Genome Res.* **10**: 597–610.

Ji, Y., Rebert, N.A., Joslin, J.M., Higgins, M.J., Schultz, R.A. and Nicholls, R.D. (2000b) Structure of the highly conserved *HERC2* gene and of multiple partially duplicated paralogs in human. *Genome Res.* **10**: 319–329.

Jiang, Y.-H., Lev-Lehman, E., Bressler, J., Tsai, R.F. and Beaudet A.L. (1999) Genetics of Angelman syndrome. *Am. J. Hum. Genet.* **65**: 1–6.

Johnston, M. (2000) The yeast genome: on the road to the Golden Age. *Curr. Opin. Genet. Dev.* **10**: 617–623.

Juyal, R.C., Figuera, L.E., Hauge, X., Elsea, S.H., Lupski, J.R., Greenberg, F., Baldini, A. and Patel, P.I. (1996) Molecular analyses of 17p11.2 deletions in 62 Smith-Magenis syndrome patients. *Am. J. Hum. Genet.* **58**: 998–1007.

Knoll, J.H., Nicholls, R.D., Magenis, R.E., Glatt, K., Graham, J.M.Jr., Kaplan, L. and Lalande, M. (1990) Angelman syndrome: three molecular classes identified with chromosome 15q11q13-specific DNA markers. *Am. J. Hum. Genet.* **47**: 149–155.

Knoll, J.H.M., Asamoah, A., Pletcher, B.A. and Wagstaff, J. (1995) Interstitial duplication of proximal 22q: phenotypic overlap with cat eye syndrome. *Am.J. Med. Genet.* **55**: 221–224.

Konrad, M., Saunier, S., Heidet, L. *et al.* (1996) Large homozygous deletions of the 2q13 region are a major cause of juvenile nephronophthisis. *Hum. Molec. Genet.* **5**: 367–371.

Kuwano, A., Mutirangura, A., Dittrich, B. *et al.* (1992) Molecular dissection of the Prader–Willi/Angelman syndrome region (15q11–13) by YAC cloning and FISH analysis. *Hum. Molec. Genet.* **1**: 417–425.

Lagerstedt, K., Karsten, S.L., Carlberg, B.-M., Kleijer, W.J., Tonnesen, T., Pettersson, U. and Bondeson, M.-L. (1997) Double-strand breaks may initiate the inversion mutation causing the Hunter syndrome. *Hum. Molec. Genet.* **6**: 627–633.

Lakich, D., Kazazian, H.H.Jr., Antonarakis, S.E. and Gitschier, J. (1993) Inversions disrupting the factor VIII gene are a common cause of severe haemophilia A. *Nat. Genet.* **5**: 236–241.

Ledbetter, D.H. and Ballabio, A. (1995) Molecular cytogenetics of contiguous gene syndromes: mechanisms and consequences of gene dosage imbalance. In: Scriver, C.R., Beaudet, A.L., Sly, W.S. and Valle, D. (eds), *The Metabolic and Molecular Bases of Inherited Disease*. McGraw-Hill, Inc. New York, Vol. 1, pp. 811–839.

Lindsay, E.A., Shaffer, L.G., Carrozzo, R., Greenberg, F. and Baldini, A. (1995) *De novo* tandem duplication of chromosome segment 22q11-q12: clinical, cytogenetic, and molecular characterization. *Am. J. Med. Genet.* **56**: 296–299.

Long, F.L., Duckett, D.P., Billam, L.J., Williams, D.K. and Crolla, J.A. (1998) Triplication of 15q11-q13 with inv dup(15) in a female with developmental delay. *J. Med. Genet.* **35**: 425–428.

Lopes, J., Ravise, N., Vandenberghe, A. *et al.* (1998) Fine mapping of de novo CMT1A and HNPP rearrangements within CMT1A-REPs evidences two distinct sex-dependent mechanisms and candidate sequences involved in recombination. *Hum. Molec. Genet.* **7**: 141–148.

López-Correa, C., Brems, H., Lázaro, C., Marynen, P. and Legius, E. (2000) Unequal meiotic crossover: a frequent cause of NF1 microdeletions. *Am. J. Hum. Genet.* **66**: 1969–1974.

Lupski, J. (1998a) Genomic disorders: structural features of the genome can lead to DNA rearrangements and human disease traits. *Trends Genet.* **14**: 417–422.

Lupski, J.R. (1998b) Charcot-Marie-Tooth disease: lessons in genetic mechanisms. *Mol. Med.* **4**: 3–11.

McTaggert, K.E., Budarf, M.L., Driscoll, D.A., Emanuel, B.S., Ferreira, P. and McDermid, H.E. (1998) Cat eye syndrome chromosome breakpoint clustering: identification of two intervals also associated with 22q11 deletion syndrome breakpoints. *Cytogenet. Cell Genet.* **81**: 222–228.

Mao, R., Jalal, S.M., Snow, K., Michels, V.V., Szabo, S.M. and Babovic-Vuksanovic, D. (2000) Characteristics of two cases with dup(15)(q11.2-q12): one of maternal and one of paternal origin. *Genet. Med.* **2**: 131–135.

Mazzarella, R. and Schlessinger, D. (1998) Pathological consequences of sequence duplications in the human genome. *Genome Res.* **8**: 1007–1021.

Mears, A.J., Duncan, A.M.V., Budarf, M.L., Emanuel, B.S., Sellinger, B., Siegel-Bartelt, J., Greenberg, C.R. and McDermid, H.E. (1994) Molecular characterization of the marker chromosome associated with Cat Eye Syndrome. *Am. J. Hum. Genet.* **55**: 134–142.

Morgan, W.F., Corcoran, J., Hartmann, A., Kaplan, M.I., Limoli, C.L. and Ponnaiya, B. (1998) DNA double-strand breaks, chromosomal rearrangements, and genomic instability. *Mut. Res.* **404**: 125–128.

Nathans, J. (1999) The evolution and physiology of human color vision: insights from molecular genetics of visual pigments. *Neuron* **24**: 299–312.

Naylor, J., Brinke, A., Hassock. S., Green, P.M. and Giannelli, F. (1993) Characteristic mRNA abnormality found in half the patients with severe haemophilia A is due to large DNA inversions. *Hum. Molec. Genet.* **2**: 1773–1778.

Naylor, J.A., Buck, D., Green, P.M., Williamson, H., Bentley, D. and Giannelli, F. (1995) Investigation of the factor VIII intron 22 repeated region (*int22h*) and the associated inversion junctions. *Hum. Molec. Genet.* **4**: 1217–1224.

Neitz, M. and Neitz. J. (2000) Molecular genetics of color vision and color vision defects. *Arch. Ophthalmol.* **118**: 691–700.

Nelis E. and Van Broeckhoven, C. (1996) Estimation of the mutation frequencies in Charcot-Marie-Tooth disease type 1 and hereditary neuropathy with liability to pressure palsies: A European collaborative study. *Eur. J. Hum. Genet.* **4**: 25–33.

Nicholls, R.D., Ohta, T. and Gray, T.A. (1999) Genetic abnormalities in Prader–Willi syndrome and lessons from mouse models. *Acta Paediatr. Suppl.* **88**: 99–104.

Nickerson, E., Greenberg, F., Keating, M.T., McCaskill, C. and Shaffer, L.G. (1995) Deletions of the elastin gene at 7q11.23 occur in approximately 90% of patients with Williams syndrome. *Am. J. Hum. Genet.* **56**: 1156–1161.

Peoples, R., Franke, Y., Wang, Y.K., Perez-Jurado, L., Paperna, T., Cisco, M. and Francke, U. (2000) A physical map, including a BAC/PAC clone contig, of the Williams-Beuren syndrome-deletion region at 7q11.23. *Am. J. Hum. Genet.* **66**: 47–68.

Pérez Jurado, L.A., Peoples, R., Kaplan, P., Hamel, B.C. and Francke, U. (1996) Molecular definition of the chromosome 7 deletion in Williams syndrome and parent-of-origin effects on growth. *Am. J. Hum. Genet.* **59**: 781–792.

Pérez Jurado, L.A., Wang, Y.K., Peoples, R., Coloma, A., Cruces, J. and Francke, U. (1998) A duplicated gene in the breakpoint regions of the 7q11.23 Williams-Beuren syndrome deletion encodes the initiator binding protein TFII-I and BAP-135, a phosphorylation target of BTK. *Hum. Molec. Genet.* **7**: 325–334.

Pezzi, N., Prieto, I., Kremer, L., Pérez Jurado, L.A., Valero, C., Del Mazo, J., Martínez-A.C. and Barbero, J.L. (2000) *STAG3*, a novel gene encoding a protein involved in meiotic chromosome pairing and location of *STAG-3*-related genes flanking the Williams-Beuren syndrome deletion. *FASEB J.* **14**: 581–592.

Potocki, L., Chen, K.-S., Koeuth, T. *et al.* (1999) DNA rearrangements on both homologues of chromosome 17 in a mildly delayed individual with a family history of autosomal dominant carpal tunnel syndrome. *Am. J. Hum. Genet.* **64**: 471–478.

Potocki, L., Chen Park, S-S., Osterholm, D.E. *et al.* (2000) Molecular mechanism for duplication 17p11.2-the homologous recombination reciprocal of the Smith-Magenis microdeletion. *Nature Genet.* **24**: 84–87.

Reiter, L.T., Murakami, T., Koeuth, T., Pentao, L., Muzny, D.M., Gibbs, R.A. and Lupski, J.R. (1996) A recombinational hotspot responsible for two inherited peripheral neuropathies is located near a *mariner* transposon-like element. *Nature Genet.* **12**: 288–297.

Reiter, L.T., Murakami, R., Koeuth, T., Gibbs, R.A. and Lupski, J.R. (1997) The human *COX10* gene is disrupted during homologous recombination between the 24 kb proximal and distal CMT1A-REPs. *Hum. Molec. Genet.* **6**: 1595–1603.

Reiter, L.T., Hastings, P.J., Nelis, E., De Jonghe, P., Van Broeckhoven, D. and Lupski, J.R. (1998) Human meiotic recombination products revealed by sequencing a hotspot for homologous strand exchange in multiple HNPP deletion patients. *Am. J. Hum. Genet.* **62**: 1023–1033.

Riva, P., Corrado, L., Natacci, F. *et al.* (2000) NF1 microdeletion syndrome: refined FISH characterization of sporadic and familial deletions with locus-specific probes. *Am. J. Hum. Genet.* **66**: 100–109.

Robinson, W.P., Waslynka, J., Bernasconi, F., Wang, M., Clark, S. Kotzot, D. and Schinzel, A. (1996) Delineation of 7q11.2 deletions associated with Williams-Beuren syndrome and mapping of a repetitive sequence to within and to either side of the common deletion. *Genomics* **34**: 17–23.

Robinson, W.P., Dutly, F., Nicholls, R.D., Bernasconi, F., Penaherrera, M., Michaelis, R.C., Abeliovich, D. and Schinzel, A.A. (1998) The mechanisms involved in formation of deletions and duplications of 15q11-q13. *J. Med. Genet.* **35**: 130–136.

Saunier, S., Calado, J., Benessy, F., Silbermann, F., Heilig, R., Weissenbach, J. and Antignac, C. (2000) Characterization of the *NPHP1* locus: Mutational mechanism involved in deletions in familial juvenile nephronophthisis. *Am. J. Hum. Genet.* **66**: 778–789.

Schinzel, A.A., Brecevic, L., Bernasconi, F., Binkert, F., Berthet, F., Wuilloud, A. and Robinson, W.P. (1994) Intrachromosomal triplication of 15q11-q13. *J. Med. Genet.* **31**: 798–803.

Scambler, P.J. (2000) The 22q11 deletion syndromes. *Hum. Molec. Genet.* **9**: 2421–2426.

Shaikh, T.H., Kurahashi, H., Saitta, S.C. *et al.* (2000) Chromosome 22-specific low copy repeats and the 22q11.2 deletion syndrome: Genomic organization and deletion endpoint analysis. *Hum. Molec. Genet.* **9**: 489–501.

Swillen, A., Vogels, A., Devriendt, K. and Fryns, J.P. (2000) Chromosome 22q11 deletion syndrome: update and review of the clinical features, cognitive-behavioral spectrum, and psychiatric complications. *Am. J. Med. Genet. (Semin. Med. Genet.)* **97**: 128–135.

Timms, K.M., Lu, F., Shen, Y., Pierson, C.A., Muzny, D.M., Gu, Y., Nelson, D.L. and Gibbs, R.A. (1995) 130 kb of DNA sequence reveals two new genes and a regional duplication distal to the human iduronate-2-sulfate sulfatase locus. *Genome Res.* **5**: 71–78.

Tsukamoto, Y. and Ikeda, H. (1998) Double-strand break repair mediated by DNA end-joining. *Genes Cells* **3**: 135–144.

Ungaro, P., Christian, S.L., Fantes, J.A. Mutirangura, A., Black, S., Reynolds, J., Malcolm, S., Dobyns, W.B. and Ledbetter, D.H. (2001) Molecular characterization of four cases of intrachromosomal triplication of chromosome 15q11-q14. *J. Med. Genet.* **38**: 26–34.

Upadhyaya, M., Ruggieri, M., Maynard, J. *et al.* (1998) Gross deletions of the neurofibromatosis type 1 (*NF1*) gene are predominantly of maternal origin and commonly associated with a learning disability, dysmorphic features and developmental delay. *Hum. Genet.* **102**: 591–597.

Urbán, Z., Helms, C., Fekete, G., Csiszár, K., Bonnet D., Munnich, A., Donis-Keller, H. and Boyd, C.D. (1996) 7q11.23 deletions in Williams syndrome arise as a consequence of unequal meiotic crossover. *Am. J. Hum. Genet.* **59**: 958–962.

Wandstrat, A.E., Leana-Cox, J., Jenkins, L. and Schwartz, S. (1998) Molecular cytogenetic evidence for a common breakpoint in the largest inverted duplications of chromosome 15. *Am. J. Hum. Genet.* **62**: 925–936.

Weatherall, D.J., Clegg, J.B., Higgs, D.R. and Wood, W.G. (1995) The hemoglobinopathies. In Scriver, C.R., Beaudet, A.L., Sly, W.S. and Valle, D. (eds), *The Metabolic and Molecular Bases of Inherited Disease.* McGraw-Hill, Inc. New York, Vol. 3, pp. 3417–3484.

Wolpert, C.M., Menold, M.M., Bass, M.P. *et al.* (2000) Three probands with autistic disorder and isodicentric chromosome 15. *Am. J. Med. Genet.* **96**: 365–372.

Zhou, Y.H. and Li, W.H. (1996) Gene conversion and natural selection in the evolution of X-linked color vision genes in higher primates. *Mol. Biol. Evol.* **13**: 780–783.

10

Genotype to phenotype in the spinocerebellar ataxias

Paul F. Worth and Nicholas W. Wood

1. Introduction

The spinocerebellar ataxias (SCAs) are a group of inherited neurological disorders which are clinically and genetically heterogeneous. The relative rarity of each of the individual SCAs belies the fact that the nature of the mutation and the pathological process involved, appear to be common not only to at least five of the SCAs, but also to a further three neurodegenerative disorders; Huntington's disease (HD), X-linked spinobulbar muscular atrophy (SBMA), and dentatorubral pallidoluysian atrophy (DRPLA). The common endpoint in all these conditions is death of neurones, and knowledge of the pathophysiology of these diseases may have far-reaching consequences for the understanding of wide range of other neurodegenerative disorders.

Before the advent of genetic testing, the classification of the hereditary ataxias was purely on the basis of the clinical features. In 1893, Pierre Marie (Marie, 1893) observed that certain individuals with hereditary ataxia were distinct from those described in Friedreich's series of papers between 1863 and 1877. Marie noted that among his patients, the age of onset was later, tendon reflexes were increased, abnormal eye movements were frequent and inheritance was autosomal dominant. However, it is now apparent that his cases were heterogeneous, and many other clinical and pathological studies over the next 90 years sought to distinguish reliably between the various subtypes.

The most commonly used clinical classification was proposed in 1982, when Harding divided the autosomal dominant cerebellar ataxias (ADCA) into three subtypes, based on the different clinical features of 11 families (Harding, 1982; *Table 1*). The presence or absence of clinical features over and above cerebellar ataxia such as spasticity, extrapyramidal signs, ophthalmoplegia and maculopathy determines the ADCA type.

Although Harding's classification remains a useful clinical tool, providing an index of prognosis in many cases, each of the three clinical subtypes has been found

Genotype to Phenotype second edition, edited by S. Malcolm and J. Goodship.
© 2001 BIOS Scientific Publishers Ltd, Oxford.

Table 1. Autosomal dominant cerebellar ataxia: clinico-genetic classification. The table shows the major clinical signs which characterise each type of ADCA and the genetic loci, and their chromosomal locations, which account for each. Genes which have been cloned are shown in italics, together with the range of repeat sizes associated with normal and pathological alleles; note that for *SCA8* this is the number of combined CTA/CTG repeats. *SCA13* is not shown on this table as the clinical features do not fit any of the types described by Harding.

ADCA type	Clinical features	Genetic loci and chromosomal location		Normal allele	Pathological allele
ADCA I	Cerebellar syndrome plus: pyramidal signs supranuclear ophthalmoplegia extrapyramidal signs peripheral neuropathy dementia	*SCA1*	6p22-23	6–44	39–83
		SCA2	12q23-24.1	13–33	32–77
		SCA3	14q32.1	12–40	54–89
		SCA4	16q24-ter	–	–
		SCA8	13q21	16–91	110–130
		SCA12	5q31-33	7–28	66–93
ADCA II	Cerebellar syndrome plus: Pigmentary maculopathy Other signs as ADCA I	*SCAS7*	3p12-21.1	4–35	36–306
ADCA III	'Pure' cerebellear syndrome Mild pyramidal signs	*SCA5*	Cent 11	–	–
		SCA6	19p13	4–18	20–33
		SCA10	22q13	–	–
		SCA11	15q15.1-21.1	–	–
		SCA14	19q13.4-ter	–	–

to be genetically heterogeneous. There are at least 14 loci which have been shown to account for the three ADCA subtypes (*Table 1*). These loci have each been assigned the notation SCA (**S**pino**C**erebellar **A**taxia) followed by a number referring to the order in which the loci were identified. To date, the genes at seven of these loci have been cloned, and in five of these, *SCA*1 (Orr *et al.*, 1993), *SCA*2 (Imbert *et al.*, 1996; Pulst *et al.*, 1996; Sanpei *et al.*, 1996), *SCA*3 (Cancel *et al.*, 1995; Kawaguchi *et al.*, 1994; Schols *et al.*, 1995), *SCA*6 (Zhuchenko *et al.*, 1997) and *SCA*7 (David *et al.*, 1997; Del-Favero *et al.*, 1998; Koob *et al.*, 1998), affected individuals have expansions in a CAG trinucleotide repeat in the coding region of at least one copy of the relevant gene, analogous to the mutational mechanism in HD (1993), SBMA (La Spada *et al.*, 1991, 1992) and DRPLA (Nagafuchi *et al.*, 1994). All these CAG repeat disorders share common properties: (1) they are generally adult onset, although juvenile cases are observed, especially when fathers transmit the disease; (2) the disease course is progressive and death usually occurs after 10–30 years duration; (3) the clinical symptoms appear above a threshold number of CAG repeats; (4) there is an inverse correlation between the number of CAG repeats and the age at onset; (5) the trinucleotide repeat exhibits variable degrees of meiotic instability resulting in genetic anticipation, with the exception of *SCA*6; (6) the gene is expressed ubiquitously; (7) the pathological protein contains an expanded polyglutamine domain which is believed to confer a toxic gain of function on the respective protein, accumulation of which occurs in neuronal intranuclear or cytoplasmic inclusions in affected and unaffected brain structures.

Many detailed clinical studies of each of the SCAs have now been reported and it is clear that there are important, though sometimes subtle phenotypic differences between SCAs that together constitute a given ADCA type (see *Table 1*). However, SCA1, *SCA2, SCA3, SCA6* and *SCA7* together account for only around 25–60% of all ADCA families (A. Durr, personal communication) and therefore more genes and mutations remain to be identified, hence the complexity of the clinico-genetic classification of the dominant ataxias will continue to increase.

There are two SCA genes which have been recently identified, *SCA8* and *SCA12*, the molecular characteristics of which display interesting departures from the other polyglutamine-encoding SCAs. These will be discussed in subsequent sections.

The overall prevalence of ADCA is not known but it appears to be less than 1:10 000 (Hirayama *et al.*, 1994). The relative prevalence of the SCA mutations varies throughout the world; for example, in Portugal, *SCA3* accounts for 80% of ADCA families (Silveira *et al.*, 1998), but only 43% in Japan (Takano *et al.*, 1998) and is virtually unheard of in Italy. Similarly, *SCA6* is relatively frequent in Japan (30%; Matsuyama *et al.*, 1997, Takano *et al.*, 1998), Germany (13%; Schols *et al.*, 1998) and the United Kingdom (N. Wood *et al.*, unpublished data) but rare in France (Stevanin *et al.*, 1997) and Portugal.

This chapter is concerned with the way in which genotype–phenotype correlation in the SCAS and the study of the molecular characteristics of the pathogenic trinucleotide repeat expansions have led to important advances in the understanding of the mechanisms which may underlie the process of cell death in these and other neurodegenerative disorders.

2. Molecular characteristics of CAG repeats

2.1 Size distribution of normal and pathological alleles

Normal *SCA1, SCA2, SCA3* and *SCA7* alleles carry a variable number of CAG repeats but the degree of polymorphism varies according to the locus. Normal *SCA1, SCA3* and *SCA6* alleles are highly polymorphic; approximately 80% of control individuals are heterozygous for normal alleles. In contrast, only 24 and 35% of controls are heterozygous for normal *SCA2* and *SCA7* alleles, respectively; alleles with 22 and 10 CAG repeats account for about 80% of normal alleles at the *SCA2* and *SCA7* locus, respectively (Cancel *et al.*, 1997; David *et al.*, 1997).

The number of CAG repeats which represents the pathological threshold, and the range of the number of CAG repeats found in both normal and pathological alleles, varies between the different SCAs (*Table 1*). For example, for *SCA3* the normal range is 12–40 repeats, while pathological alleles contain 54–89 repeats. In contrast, for *SCA6*, the normal range is 4–18, while abnormal is 20–31. The gap between the ranges of repeat size of normal and pathological alleles also varies from two repeats in *SCA6*, to 14 in *SCA3*. The implication of these differences and why, for example, a repeat size of less than 30 is pathological in *SCA6* but not in any of the other SCAs, is unclear.

2.2 Instability of trinucleotide repeats

A notable feature of expanded CAG trinucleotide repeats is their tendency to exhibit both meiotic and mitotic instability, which is observed to varying degrees in *SCA1*, *SCA2*, *SCA3* and *SCA7*. Meiotic instability results in either an increase or a decrease in CAG repeat size during vertical transmission, and may account for the clinical phenomenon of anticipation. Meiotic instability may occur as a result of slippage during DNA replication or from the formation of stable hairpin structures. This latter mechanism would be predicted to result in large expansions or contractions depending on their location on the leading or lagging strand, respectively (Eichler *et al.*, 1994; Kang *et al.*, 1995; Wells *et al.*, 1998). Epigenetic factors, such as the position and orientation with regard to the origin of replication, can modify repeat instability in *Escherichia coli* (Kang *et al.*, 1995) and *Saccharomices cerevisiae* (Freudenreich *et al.*, 1997; Maurer *et al.*, 1996; Schweitzer and Livingston, 1998). The size of the repeat required to form stable structures may be important as demonstrated in HD and *SCA7* (David *et al.*, 1998; Ranen *et al.*, 1995). In *SCA3*/MJD, the analysis of polymorphisms located close to the CAG repeat showed that they can act both in *trans* and in *cis* (Igarashi *et al.*, 1996; Takiyama *et al.*, 1999). With the exception of *SCA7* (David *et al.*, 1998) there appears to be no relationship between instability and repeat length of pathological alleles, i.e. large alleles are not necessarily more unstable.

In contrast to expanded alleles, normal *SCA1*, *SCA2*, *SCA3*, *SCA6* and *SCA7* alleles show little or no meiotic instability, although the presence of a range of normal repeat lengths illustrates the inherent modest instability of these repeats. In normal *SCA1* and most normal *SCA2* alleles, the CAG repeat is interrupted by CAT and CAA interruptions, respectively (Cancel *et al.*, 1997; Chung *et al.*, 1993; Quan *et al.*, 1995). These interruptions, which also code for glutamines, are usually absent in the expanded allele, and it has been suggested that they may confer stability to the normal allele. Interrupted alleles with repeat numbers which fall within the expected pathological range, but which are apparently not associated with disease have been found in both *SCA1* and *SCA2*. No interruptions have been found in the normal alleles of the other SCAs, and therefore the length of the uninterrupted repeat alone may be the critical factor.

2.3 Paternal transmission effects

Increases in repeat size of pathological *SCA1*, *SCA2*, *SCA3* and *SCA7* alleles tend to occur more frequently in paternal transmissions than in maternal transmissions. In addition, pathological *SCA3* alleles are much less unstable than those of *SCA7*. Paternally transmitted *SCA7* alleles have been shown to undergo large increases (*Figure 1*) of up to 166 repeats (David *et al.*, 1998; Giunti *et al.*, 1999). The reason for this apparent paternal bias toward instability may be that ovogenesis stops early in prenatal development, but spermatogenesis is a lifelong process, involving many more cell divisions, and therefore more 'mistakes' in replication may occur. In contrast to *SCA7*, *SCA6* expansions show little or no meiotic instability; there is only one report in which a paternally-derived, 24-repeat *SCA6* allele underwent expansion by two repeats (Matsuyama *et al.*, 1997). Single sperm analyses of patients with *SCA1* (Chong *et al.*, 1995; Koefoed *et al.*, 1998), *SCA2* (Cancel *et al.*, 1997), *SCA3* (Watanabe *et al.*, 1996), *SCA7* (David *et al.*, 1998) and DRPLA (Takiyama *et al.*, 1999) have shown varying degrees of gonadal mosaicism. For example, in the sperm of one *SCA7* patient, alleles with repeats of

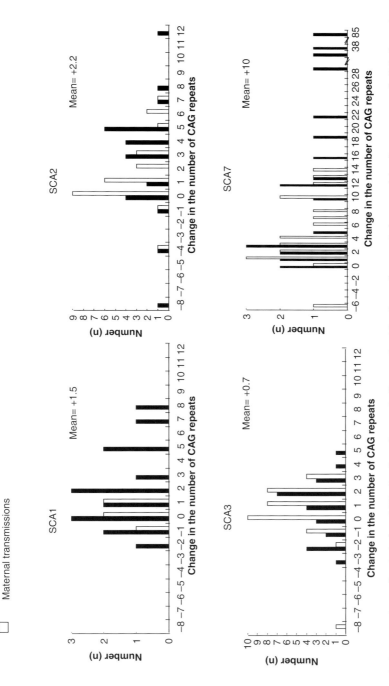

Figure 1. A comparison of meiotic instability in the spinocerebellar ataxias. Numbers of transmissions resulting in instability in the spinocerebellar ataxias. Numbers of transmissions resulting in a contracted allele (negative change in repeat number) and expanded allele (positive change in repeat number) observed in the different SCAs. Mean overall change in repeat number is given. Instability is low in SCA3, and most marked in SCA7. Most large increases in repeat number occur during paternal transmissions.

42 to >155 were found. In contrast, single sperm analysis of *SCA*6 patients revealed no gonadal mosaicism (Shizuka *et al.*, 1998).

2.4 Mitotic instability and somatic mosaicism

In addition to exhibiting meiotic instability, pathological *SCA*1, *SCA*2, *SCA*3 and *SCA*7 alleles show mitotic instability which results in somatic mosaicism. For a given SCA, meiotic and mitotic instability tend to be correlated, such that in *SCA*7, a high degree of somatic mosaicism is observed in blood leucocytes (*Figure 2*), whereas in *SCA*6, no such mosaicism has been detected. Somatic mosaicism is, however, always less marked than gonadal mosaicism for a given SCA (Cancel *et al.*, 1998; Chong *et al.*, 1995; Lopes-Cendes *et al.*, 1996; Takano *et al.*, 1996; Ueno *et al.*, 1995). To date, there is no good evidence of a significant effect of somatic mosaicism on the clinical phenotype.

2.5 De novo mutations

While expanded pathological alleles may undergo further expansion or contraction during meiosis, it has now been established for many of these disorders that there is a bimodal distribution of normal alleles with a small minority of alleles in the 'high-normal' range. It has been shown that these 'high-normal' alleles including intermediate alleles (IA), may also undergo expansion into the abnormal pathological range during meiosis. In so doing they result in *de novo*, or new mutations. Evidence for *de novo* pathological expansion has been shown in HD (Goldberg *et al.*, 1995), *SCA*2 (Schols *et al.*, 1997)and in *SCA*7 (Giunti *et al.*, 1999); in SCA7, IAs have been observed to undergo expansion into the pathological range (*Figure 3*). One study (Takano *et al.*, 1998) recently showed that the different prevalence of the dominantly inherited ataxias and DRPLA in separate Caucasian and Japanese populations was related to the frequency of the larger normal alleles of the respective gene. They found that the relative prevalence of *SCA1* and *SCA2* was higher in 177 Caucasian pedigrees (15% and 14%, respectively) than in 202 Japanese pedigrees (3% and 5%, respectively), and that

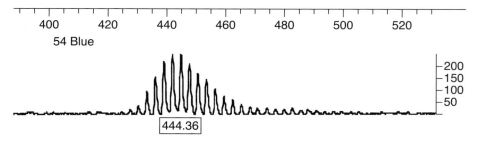

Figure 2. An electropherogram of a patient with an expansion in the *SCA7* gene. Mitotic instability present in peripheral blood leucocytes produces this characteristic 'hedgehog' appearance of the electropherogram, with the multiple peaks each representing alleles with different numbers of repeats, widely scattered around the mode, in this case approximately 445 base pairs. Normal, unexpanded alleles produce a single peak.

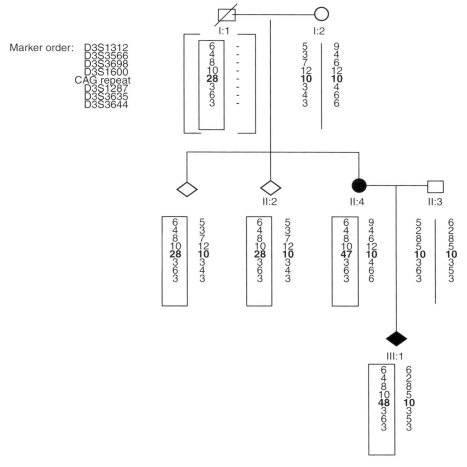

Marker order: D3S1312
                D3S3566
                D3S3698
                D3S1600
                CAG repeat
                D3S1287
                D3S3635
                D3S3644

Figure 3. An *SCA7* pedigree demonstrating a *de novo* mutation. The birth order has been changed and sex of three individuals, indicated by diamonds, has been concealed to preserve anonymity. The haplotype of individual I: 1 has been reconstructed from that of his offspring. Paternal transmission of a 28 repeat *SCA7* allele from I:1, who was unaffected, to two of his children, resulted in an unchanged number of repeats, but this same allele underwent expansion into the pathological range (47 repeats) when transmitted to individual II:4, who was affected. This allele was then relatively stable, undergoing expansion by only 1 repeat during maternal transmission to her child, III:1, who was also affected.

accordingly, the frequencies of the large normal *SCA1* alleles (>30 repeats) and *SCA2* alleles (>22 repeats) were greater in Caucasians. Conversely, *SCA3*, *SCA6* and DRPLA were more common in Japanese pedigrees (43%, 11% and 20%) than in Caucasian pedigrees (30%, 5% and 0%) and the frequencies of the large normal *SCA3* (>27 repeats), *SCA6* (>13 repeats) and DRPLA (>17 repeats) alleles was higher in the Japanese. This close correlation suggests that large normal alleles provide a reservoir for the appearance of *de novo* cases by undergoing pathological expansion. What is the origin of these IA? There are two possibilities: normal alleles may undergo *de novo* expansion, or unstable pathological expansions could undergo contraction into the intermediate range; in the latter case, the disease could appear to skip at least one generation. As yet there is no documented case of either of these possibilities, but the mechanism may be disease-specific.

Several outcomes of this instability are possible over many generations, the mean size of all pathological alleles may increase, decrease or remain unchanged. If the size of all expanded alleles tends to a net increase over time, the prevalence of the disease might be expected to fall; as pathological alleles undergo progressive expansion, resulting in larger repeat numbers, higher infantile or juvenile incidence would be predicted, with subsequent pre-reproductive death and loss of the expanded allele. Conversely, if the size of the pathological alleles tends to a mean decrease, the disease prevalence may increase. If the prevalence is to remain steady, *de novo* mutations must occur to replace progressively expanded, eventually un-transmitted, pathological alleles, or there must be no net increase in the size of pathological expansions. There is currently no evidence that the prevalence of *SCA7* is changing, and consequently it seems likely that new mutations probably occur relatively frequently or that expanded alleles contract as often as they undergo expansion. Haplotype analysis suggests multiple founders among existing *SCA7* patients and this finding favors the former hypothesis. In marked contrast, in *SCA6*, meiotic instability is low or zero, and as strong linkage disequilibrium is observed in unrelated German kindreds (Dichgans *et al.*, 1999), suggesting a single ancestral founder, a low *de novo* mutation rate in this disorder is likely.

2.6 SCA8 and SCA12 as exceptions to the 'rule'

While the foregoing discussion relates to *SCA1*, *SCA2*, *SCA3*, *SCA6* and *SCA7*, there are two SCAs which diverge from the common patterns described above. Koob *et al.* (1998) demonstrated linkage of an expansion in the CTG portion of a $(CTA)_n (CTG)_n$ trinucleotide repeat on chromosome 13q21 (*SCA8*) to the disease phenotype in a large family with cerebellar ataxia. Alleles of affected individuals contained 110–130 combined CTA/CTG repeats (CR), while in 1200 control chromosomes, alleles ranged in size from 16 to 92 CRs. Interesting departures from the usual molecular features of trinucleotide repeat disorders were noted: the repeat tract is untranslated, <100% penetrance, and maternal inheritance bias, with paternally and maternally transmitted alleles often undergoing large contractions and expansions, respectively. In this study, eight other ataxia families that were too small for linkage analysis were shown to harbor expanded *SCA8* alleles with repeat sizes ranging from c. 90 to 250 CR; these were termed 'potentially pathogenic' alleles. Thus, for *SCA8* distinct normal and pathological ranges could not be determined. Other studies have subsequently identified *SCA8* as a relatively common cause of cerebellar ataxia (Ikeda *et al.*, 2000; Silveira *et al.*, 2000). However, a number of other studies have recently cast doubt on the validity of the *SCA8* gene as a cause of cerebellar ataxia (Stevanin *et al.*, 2000; Vincent *et al.*, 2000; Worth *et al.*, 2000). In one study (Worth *et al.*, 2000), expanded alleles were segregating in two ataxia pedigrees, but alleles of 174, 133, 103, 101 and 100 CR's were also found in five different control subjects. Vincent *et al.* (2000) showed that large alleles between 100 and 1300 repeats were present in 14 patients with psychosis and in five control subjects. Stevanin *et al.* (2000) found alleles with 107, 111 and 123 CR among control subjects and alleles with 92–111 CR in a family with Lafora body disease. In addition, disease and expansions did not co-segregate in two families with ADCA.

It has been suggested that either additional factors other than expanded *SCA8* alleles are necessary for the development of ataxia (Moseley *et al.*, 2000), or that there is a critical maximum number of repeats above which ataxia does not develop. Alternatively, *de novo* expansion in *SCA8* may occur frequently in the normal population and these expanded alleles exist as polymorphisms in linkage disequilibrium with 'true' causative mutations in a gene for cerebellar ataxia on chromosome 13q21. Until a pathological mechanism and explanation for non penetrance can be established, we have advised against testing individuals with cerebellar ataxia for the presence of the *SCA8* expansion (Worth *et al.*, 2000).

The molecular features of *SCA12* are again different to those of the other SCAs, and have generated much recent interest. Holmes *et al.* (1999) identified a large pedigree of German descent. Age at onset ranges from 8–55 and presenting symptoms include upper extremity tremor, progressing slowly to a cerebellar syndrome, hyperreflexia, hypokinesia, abnormal eye movements, and in some cases dementia. Affected subjects have expansions containing 66–78 CAG repeats in the presumed 5'-untranslated region of a subunit of the protein phosphatase *PP2A* gene, while controls have 7–28 repeats. The *SCA12* repeat was reported to be only mildly unstable with an average contraction of –0.4 CAG repeats in four vertical transmissions. We have not found this mutation in screening over 390 ataxic patients. Although the authors acknowledged that the *SCA12* repeat could be in linkage disequilibrium with a second causative mutation, no positive controls have been identified.

3. Genotype–phenotype correlation

3.1 Clinical features of patients with SCAs

While there are differences between both the molecular characteristics and the clinical phenotypes of the different SCAs, there are also important similarities. The only clinical sign that is specific for a single locus is decreased visual acuity leading to blindness due to progressive macular dystrophy in most *SCA7* patients. No other clinical sign is specifically associated with a given genotype. However, group differences in the frequency of several signs, and their characteristic combination in several family members makes prediction of the genotype possible in some cases.

Part of the variability in phenotype for a given SCA has been explained by a bias resulting from clinical evaluation of patients with different disease durations. The frequencies of decreased vibration sense, Babinski sign, ophthalmoplegia, amyotrophy and sphincter disturbances are positively correlated with disease duration in ADCA I families (Durr *et al.*, 1993). Clinical signs such as dysphagia or sphincter disturbances increase with disease duration in *SCA2* (Cancel *et al.*,

1997), *SCA3*/MJD (Durr *et al.*, 1996) and *SCA7* (David *et al.*, 1998), as does dysarthria in SCA6 patients (Stevanin *et al.*, 1997).

An early decrease in saccade velocity and reduced tendon reflexes without extrapyramidal signs is suggestive of *SCA2* (Burk *et al.*, 1997; Cancel *et al.*, 1997; Schols *et al.*, 1997; Wadia *et al.*, 1998). *SCA3*/MJD and *SCA6* patients present frequently with cerebellar oculomotor signs such as saccadic smooth pursuit, gaze-evoked nystagmus and diplopia. Extrapyramidal signs, myokimia and bulging eyes have been reported to characterize Machado–Joseph disease (Lima and Coutinho, 1980) but are not frequent in non Portuguese Western European *SCA3*/MJD patients and might, therefore, be related to ethnic background (Durr *et al.*, 1996; Schols *et al.*, 1995, 1996; Stevanin *et al.*, 1994). *SCA3*/MJD patients also frequently have ophthalmoplegia or amyotrophy (Durr *et al.*, 1996; Giunti *et al.*, 1995; Maciel *et al.*, 1995; Maruyama *et al.*, 1995; Matilla *et al.*, 1995; Schols *et al.*, 1996; Takiyama *et al.*, 1995). The phenotype of ADCA III as described by Harding is that of a relatively 'pure' late onset cerebellar ataxia, without prominent features attributable to degeneration of other parts of the central nervous system. However, although most SCA6 patients have a relatively pure phenotype, there are a small minority who have additional clinical features more in keeping with ADCA type I, e.g. positive Babinski and hyperreflexia. These patients may simply have had an earlier age at onset, and therefore longer disease duration, in order for these signs to become manifest. Episodic ataxia has been described as the presenting sign in some *SCA6* patients (Jodice *et al.*, 1997). The clinical signs associated with the *SCA1* mutation are in general broader and homogeneous, and the patients have usually a pyramidal syndrome, often with hyperreflexia (Burk *et al.*, 1996).

3.2 Factors affecting age at onset

To varying degrees, patients with *SCA1*, *SCA2*, *SCA3*, *SCA6* and *SCA7* all display the phenomenon of anticipation, that is, the tendency towards earlier age at onset, increased disease severity and increased rate of disease progression with successive generations. In general, and for a given SCA, individuals may be expected to develop the disease earlier, and exhibit a more severe phenotype, including a more rapid rate of disease progression, with increasing repeat number. *Figures 4a* and *4b* show plots of age at onset against repeat number for the different SCAs, and a significant inverse correlation has been demonstrated in each case. It will be noted that the slope of the correlation curve is different for each SCA; the slope of the curve for *SCA1* is steeper than that of *SCA7*, suggesting that the protein context of the polyglutamine domain is important. Despite this close correlation, it is not possible to give an accurate prediction of the age at onset from knowledge of the repeat number, owing to considerable data scatter, and because repeat number accounts for only 50–80% of the variability in age at onset. Other factors may have an important influence, including a possible gender effect in *SCA3* (Kawakami *et al.*, 1995) and the number of CAG repeats of the normal allele has also been reported to influence age at onset (Durr *et al.*, 1996). Individuals homozygous for expansions in *SCA2* (Sanpei *et al.*, 1996), *SCA3* (Kawakami *et al.*, 1995; Lerer *et al.*, 1996; Sobue *et al.*, 1996) and *SCA6* (Geschwind *et al.*, 1997; Ikeuchi *et al.*, 1997; Matsumura *et al.*, 1997) are reported to have an earlier age at

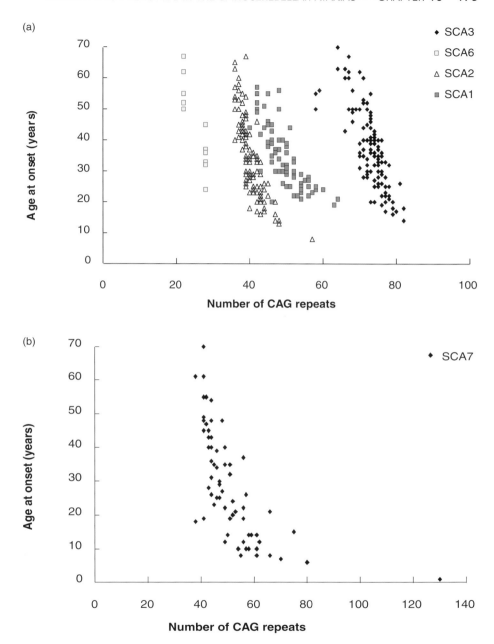

Figure 4. Age at onset vs. number of CAG repeats for the different SCAs. A significant inverse correlation between repeat number and age at onset has been shown for each SCA. Note the different pathological thresholds and different slopes; the SCA1 and SCA3 (a) correlation curves are appreciably steeper than that of SCA7 (b), indicating a greater effect of repeat number on phenotypic expression in SCA7. Data scatter precludes prediction of age at onset from knowledge of repeat number alone. Note the SCA7 individual with 130 repeats whose age at onset was 1 year.

onset, implying a dosage effect.

As discussed above, in all but *SCA6*, the CAG repeat in the pathological allele is unstable and further expansion, which occurs especially during paternal transmission, accounts for anticipation. Although anticipation has been described for *SCA6*, it is much less marked than the other SCAs, and meiotic instability is not generally observed. Hence, 'anticipation' which has been reported in *SCA6* (Ikeuchi *et al.*, 1997; Matsuyama *et al.*, 1997; Schols *et al.*, 1998) may be the result of observer or ascertainment bias, that is, the tendency for investigators to 'look' harder for evidence of the disease in offspring of affected individuals, and thus detect signs of the disease earlier. If anticipation is a real phenomenon in this condition, then another mechanism must be responsible.

3.3 Factors affecting mode of presentation, disease progression and disease severity

In common with age at onset, the rate of disease progression until death, of *SCA1* (Jodice *et al.*, 1994), *SCA3*/MJD (Klockgether *et al.*, 1996) and *SCA7* (David *et al.*, 1998) patients, is negatively correlated with repeat size. The largest *SCA7* expansions are associated with severe and rapidly progressive juvenile forms (Benton *et al.*, 1998, David *et al.*, 1998; Johansson *et al.*, 1998). The frequency of several clinical signs seems also to depend on the repeat number. For example, in *SCA3*/MJD patients, the frequency of pyramidal signs is positively correlated with the size of the expanded repeat whereas the frequency of altered vibration sense is inversely correlated (Durr *et al.*, 1996). Patients with small SCA3/MJD expansions may present with late-onset peripheral neuropathy (areflexia and amyotrophy) (Durr *et al.*, 1996) and/or DOPA-responsive parkinsonism (Tuite *et al.*, 1995). In contrast, *SCA3* patients with large expansions tend to present with earlier onset dystonia. Large *SCA2* expansions tend to be associated with myokymia, myoclonus, dystonia and fasciculations (Cancel *et al.*, 1997) as well as chorea and dementia. In SCA7, the mode of presentation is influenced by the size of the repeat; subjects with ≤49 CAG repeats tend to present with ataxia whereas those with >49 repeats tend to present with maculopathy (Giunti *et al.*, 1999) and the frequency of pyramidal signs, ophthalmoplegia and decreased visual acuity increases with repeat size (David *et al.*, 1998). In SCA6, given the slow progression of the disease, only some of the patients with the largest expansions, associated with earlier onset, develop associated signs (Stevanin *et al.*, 1997).

4. Clinico-pathological correlation

As discussed above, although there are important similarities in the molecular characteristics of the different SCA mutations, there is also considerable variation in the clinical manifestations. In addition to cerebellar ataxia, attributable to dysfunction of the cerebellum or its connections, signs and symptoms which result from degeneration of different parts of the central and/or peripheral nervous system are characteristic of the different SCAs (*Table 1*). However, with the exception of pigmentary retinopathy in SCA7, no single clinical sign is found exclusively in, or is diagnostic of, any one form of SCA. In spite of this important principle, neuropathological studies (Durr *et al.*, 1995, 1996; Gilman *et al.*, 1996; Martin *et al.*, 1994; Robitaille *et al.*, 1995; Tsuchiya *et al.*, 1998) have shown that the clinical phenotype of a given SCA corresponds quite well with the degree of involvement of the relevant part of the nervous system (*Table 2*). For example, in *SCA3*, extrapyramidal signs such as parkinsonism and dystonia occur as a result of basal ganglia involvement, and amyotrophy due to anterior horn cell loss. In *SCA2*, pontine nuclear involvement is manifested by slowed saccadic eye movements. Similarly, visual loss in *SCA7* is usual, and is due to a pigmentary macular dystrophy; pathological examination of the retina shows early degeneration of the photoreceptors, the bipolar and the granular cells, particularly in the foveal and parafoveal regions. Later, the inner retinal layers are affected with patchy loss of epithelial pigment cells and penetration of pigmented cells into the retinal layers (Martin *et al.*, 1994). In the case of *SCA6*, which usually manifests as a relatively 'pure' ataxia, with no other major clinical features, the Purkinje cells of the cerebellum and the inferior olives are the only structures which show significant abnormalities at autopsy.

Neuroimaging studies have also shown a correlation between clinical features and magnetic resonance imaging (MRI) abnormalities (Gomez *et al.*, 1997; Klockgether *et al.*, 1998; Schols *et al.*, 1998; Stevanin *et al.*, 1997; Watanabe *et al.*, 1996). Brainstem and cerebellar atrophy is seen in *SCA1*, *SCA2* and *SCA3*, though this may be most marked in *SCA2*. In contrast, *SCA6* patients usually show cerebellar atrophy alone, with sparing of the brainstem. Atrophy of the basal ganglia, specifically the caudate and putamen, is most marked in *SCA3*, in which extrapyramidal (parkinsonian) features are often observed.

In summary, the expanded trinucleotide repeat results in region-specific neurodegeneration the pattern of which is different for each of the SCAs. The reasons for this region-specificity are unclear (see below).

Table 2. A comparison of the major pathological changes in the different SCAs. Key: − = not affected; +/− = sometimes affected; + = mildly affected; ++ = moderately affected; +++ = severely affected. Ext = globus pallidus, external part; int = internal part.

Areas or neurones affected	SCA 1	SCA 2	SCA 3	SCA 6	SCA 7
Purkinje cells	+	++	+/−	+++	+
Inferior olives	+++	+++	−	+/−	+++
Globus pallidus	+ext	+	++int	−	+
Retina	−	−	−	−	+
Substantia nigra pars compacta	+	++	++	−	++
Nucleus pontis	+	+++	+	−	+
Dentate nucleus	++	−	++	+/−	++
Anterior horn cells	+	+	+	−	+
Axonal neuropathy	+	++	++	++—	
Dorsal columns	+	+++	+	−	+
Spinocerebellar tracts	+	−	++	−	+

5. Pathophysiology of trinucleotide repeat disorders

The CAG repeat in normal *SCA1*, *SCA2*, *SCA3*, *SCA6* and *SCA7* alleles encodes a polyglutamine domain in the specified protein. In abnormal or mutant *SCA* genes with an expanded CAG repeat, this polyglutamine domain is expanded. The normal function of the proteins (referred to as *ataxins*) encoded by *SCA1*, *SCA2*, *SCA3* and *SCA7* is obscure; they have no sequence homology with each other or with other known proteins. However, the function of the protein product of the *SCA6* gene is known. This gene contains 47 exons which encode the α_{1A}-subunit of the voltage-gated P/Q type calcium channel CACNA1A (Zhuchenko *et al.*, 1997). The primary transcript undergoes differential splicing such that the cerebellum preferentially expresses a subunit containing the polyglutamine domain in the C-terminal region (Ishikawa *et al.*, 1999), although, in common with the other ataxins, CACNA1A is expressed throughout the nervous system and all peripheral tissues.

The molecular properties of *SCA8*, the sixth SCA to be cloned, are different to those of the other SCA genes, in that affected individuals have expansions in an untranslated CTA/CTG repeat on chromosome 13q21. As no protein or polyglutamine is translated, it has been suggested that the transcript may exert a pathogenic effect at an RNA level, perhaps as an antisense RNA, but this has not been proved. Similarly, it appears that the expanded CAG repeat which is associated with the disease in *SCA12* is untranslated, and a different pathological mechanism must be involved.

There is now evidence that an expanded polyglutamine domain confers a novel but deleterious, toxic property or 'gain-of-function'. The only resemblance which the mutant protein products of *SCA1*, *SCA2*, *SCA3*, *SCA7* as well as those of the *DRPLA*, *SBMA* and *HD* genes bear to each other is restricted to the polyglutamine domain. This toxic property may only be realized if the polyglutamine undergoes self-aggregation. Transglutaminase-mediated cross-linking of expanded polyglutamine has been shown (Green, 1993) but Perutz *et al.* (1994) suggested another mechanism based on the theoretical ability of the β-pleated strands formed by polyglutamines to aggregate by hydrogen bonds (i.e. non covalently) to form polar zippers; intranuclear inclusions containing ubiquitinated aggregates of mutant protein have been found in the affected brain regions in *SCA1* (Duyckaerts *et al.*, 1999), *SCA2* (Koyano *et al.*, 1999, 2000), *SCA3* (Paulson *et al.*, 1997) and *SCA7* (Holmberg *et al.*, 1998) patients, similar to those found in the brains of DRPLA patients (Hayashi *et al.*, 1998) and of juvenile-onset Huntington's patients and transgenic HD mice (Davies *et al.*, 1997).

There appear to be at least four stages in the pathogenesis of the polyglutamine disorders: (1) cleavage of the polyglutamine tracts from the native protein; (2) their localization to the nucleus; (3) aggregation of the polyglutamine peptides to form intranuclear inclusions; and (4) the triggering of neuronal death. The exact role of each of these steps in disease pathogenesis is unclear.

It is hypothesized that a polyglutamine fragment must be cleaved from the native protein and localized to the nucleus in order to exert its pathological effect. How are the putative polyglutamine containing fragments generated? As a consequence of their role in apoptosis, the caspase family of aspartate-specific cysteine proteases has been the focus of much attention (Martin and Green, 1995).

Evidence that caspase cleavage is important in this process has been shown for huntingtin (Goldberg *et al.*, 1996), for the androgen receptor, atrophin-1 and for ataxin-3 (Wellington *et al.*, 1998). The same group has gone on to show that inhibiting caspase cleavage of huntingtin reduces toxicity and aggregate formation in apoptotically stressed neuronal and non neuronal cells. In addition, caspase-1 is activated in the brains of transgenic mice expressing mutant exon-1 huntingtin. Inhibition of caspase activity delayed the onset of neurological symptoms and the formation of nuclear aggregates by 2 weeks (Ona *et al.*, 1999), even though the transgenic construct lacked the known caspase-1 cleavage site. However, the relationship between caspase cleavage and pathogenesis has yet to be shown *in vivo*.

Truncated fragments of polyglutamine proteins with expanded polyglutamine tracts promote aggregation and toxicity in several *in vitro* models (Hackam *et al.*, 1998; Hollenbach *et al.*, 1999; Lunkes and Mande, 1998; Martindale *et al.*, 1998; Merry *et al.*, 1998; Perez *et al.*, 1998; Scherzinger *et al.*, 1997, 1999). In addition, only antibodies to the polyglutamine-containing part of the protein identify nuclear aggregates of the polyglutamine protein in HD, *SCA3*, SBMA and DRPLA patient tissue (Becher *et al.*, 1998; DiFiglia *et al.*, 1997; Paulson *et al.*, 1997; Skinner *et al.*, 1997). This is good evidence that the uncleaved native protein is not incorporated into the aggregates to a significant degree.

How might aggregates affect nuclear function? The proteasome degradation pathway has been implicated in polyglutamine pathogenesis; numerous studies (Chai *et al.*, 1999; Cummings *et al.*, 1998; Stenoien *et al.*, 1999) have shown the presence of the 20S proteasome in polyglutamine aggregates. It is possible that the normal activity of the proteasome complex is thus disrupted such that normal regulation of protein levels is altered. When proteasome activity is inhibited in *SCA*3-transfected cells, the formation of mutant ataxin-3 aggregates is enhanced (Chai *et al.*, 1999).

Further evidence that protein misfolding is critical to the pathogenesis of the polyglutamine disorders comes from observations that molecular chaperones which mediate the correct folding, assembly and degradation of proteins, localize to aggregates in formed by ataxin-1 (Cummings *et al.*, 1998) and the androgen receptor (Stenoien *et al.*, 1999). The heat shock proteins (HSPs) are a family of chaperones inducible by heat and other stressors, that serve essential functions under stress and non stress conditions. Expression of the Hsp40 chaperone, HJD-2, suppressed aggregation by both proteins in non neuronal cells. Chai *et al.* (1999) have recently shown that Hsp40 and Hsp70 localize to intranuclear aggregates in *SCA3* disease tissue and cells expressing mutant ataxin-3, and that overexpression of Hsp40 chaperones can suppress polyglutamine aggregation and that this suppression correlates with a decrease in neurotoxicity.

It may be that intranuclear inclusions (NI) are toxic to neurones but whether the development of inclusions is either necessary or sufficient to cause cell death remains controversial, as does the question of whether nuclear localization is necessary for the inclusions to exert their toxic effect. NI appear before neuronal loss and before expression of the phenotype in a mouse model of HD (Davies *et al.*, 1997). They may be a correlate of pathogenesis, as their density is correlated to the expansion size (Becher *et al.*, 1998). However, they are also present in epithelial cells of a drosophila model of *SCA3* in which there is no neuronal death or phenotype (Warrick *et al.*, 1998), and have been detected in non affected neuronal tissues in a *SCA7* patient (Holmberg *et al.*, 1998) and in peripheral tissues in an HD mouse model (Sathasivam *et al.*, 1999) indicating that their presence is not sufficient to cause death and/or phenotype. A recent, although controversial (Perutz, 1999), study showed, that visible aggregates are not a prerequisite for pathogenesis in a *SCA1* mouse model (Klement *et al.*, 1998) and that neuronal death is not correlated to NI formation in a cellular model of HD (Saudou *et al.*, 1998). Inclusions have not been found in adult-onset HD brains. Hence, protein aggregation may precede the development of identifiable inclusions, and may be one important step in the process of cell death. Some have proposed that inclusions may represent a protective response to toxicity rather than being the toxic agent themselves (Sisodia, 1998).

It was predicted that intranuclear inclusions would not be found in *SCA6*, as the protein is expressed in the cell membrane. However, a recent study (Ishikawa *et al.*, 1999) has shown that cytoplasmic inclusions containing aggregated calcium-channel subunits are found in this condition. Are these cytoplasmic inclusions toxic to the neurones, or is cell death in *SCA6* the result of altered channel permeability, leading to progressive ionic damage to the neurone? The *SCA6* gene encodes either the P- or the Q-type CACNA1A subunit, which is determined by the presence or absence of an asparagine-proline stretch in domain IV. Toru *et al.* (2000) have shown that when human α_{1A}-subunits carrying various polyglutamine lengths are expressed in human embryonic kidney cells, a negative and a positive shift in the voltage dependence of inactivation was recorded for P-type and Q-type channels, respectively. The authors suggest that this could constitute a mechanism which accounts for the selective degeneration of the cerebellar Purkinje cells, which preferentially express P-type channels.

In a transgenic mouse model of *SCA1* where a Purkinje-cell specific promoter is used to drive the expression of mutant ataxin-1 (Lin *et al.*, 2000), some genes such as prenylcysteine carboxymethyltransferase (PCCMT), type 1 ER inositol triphosphate receptor (IP3R1), and inositol polyphosphate 5-phosphatase (INPP5A), an ER calcium pump (SERCA2), calcium ion channel TRP3 and the glutamate transporter EAAT4 are downregulated. These transcripts encode proteins with effects on cellular systems known to be involved in neuronal degeneration, those of calcium and glutamate metabolism. These changes occurred before the onset of Purkinje cell degeneration. As the mice age, after the development of clinical and pathological signs, another gene alpha-1 antichymotrypsin, an acute phase protein, was up-regulated. These changes only occurred in mice expressing the expanded ataxin-1 in the nucleus; no change was seen in control mice expressing either wild-type ataxin-1 or an expanded ataxin-1 that lacked a nuclear localization domain and was confined to the cytoplasm. The authors confirmed that three of the gene products downregulated in mutant mice (PCCMT, SERCA2 and IP3R1) were also under-expressed in the brain of an early-onset *SCA1* patient. Alpha1-antichymotrypsin was also up-regulated in this *SCA1* brain. The mechanism by which the mutant ataxin-1 produces altered gene expression can only be speculative, but a direct transcriptional effect or an effect on chromatin structure, nuclear architecture or transcription factor stability can be hypothesized. The possibility exists that these changes are merely circumstantial and not directly causative, but perhaps a real effect on cytoplasmic calcium concentration or glutamate neurotransmission is the key step in the pathological process.

6. Conclusions

While much progress has been made in the understanding of the clinical pheno-types of the different SCAs and many of the common features identified, there are many crucial questions which remain unanswered in respect of the molecular properties and pathophysiological consequences of the expanded trinucleotide repeat. Why are the repeats unstable? Why is the degree of meiotic instability disease-specific? Why is the age at onset affected to differing degrees by the length of the repeat in the various SCAs? What are the mechanisms in those conditions such as *SCA8* and *SCA12* where an expanded trinucleotide repeat is untranslated? Are neuronal intranuclear inclusions the real agents of disease or just 'innocent bystanders'? While the answer to any one of these questions may have relevance to the entire group of polyglutamine disorders, important differences between the SCAs suggest that no single model will provide a unifying explanation. The next few years will likely see an explosion in the understanding of these disorders, but many more biochemical and pathological steps may yet lie between the level of our current understanding and a complete mechanism.

Acknowledgments

The authors would like to thank Dr. A. Dürr and Dr. G. Stevanin for contributing data and figures to this work. PFW is a UK Medical Research Council Clinical Training Fellow.

References

Becher, M.W., Kotzuk, J.A., Sharp, A.H., Davies, S.W., Bates, G.P., Price, D.L. and Ross, C.A. (1998) Intranuclear neuronal inclusions in Huntington's disease and dentatorubral and pallidoluysian atrophy: correlation between the density of inclusions and IT15 CAG triplet repeat length. *Neurobiol. Dis.* **4**: 387–397.

Benton, C.S., de Silva, R., Rutledge, S.L., Bohlega, S., Ashizawa, T. and Zoghbi, H.Y. (1998) Molecular and clinical studies in SCA-7 define a broad clinical spectrum and the infantile phenotype. *Neurology* **51**: 1081–1086.

Burk, K., Abele, M., Fetter, M., Dichgans, J., Skalej, M., Laccone, F., Didierjean, O., Brice, A. and Klockgether, T. (1996) Autosomal dominant cerebellar ataxia type I clinical features and MRI in families with SCA1, SCA2 and SCA3. *Brain* **119**: 1497–1505.

Burk, K., Fetter, M., Skalej, M., Laccone, F., Stevanin, G., Dichgans, J. and Klockgether, T. (1997) Saccade velocity in idiopathic and autosomal dominant cerebellar ataxia. *J. Neurol. Neurosurg. Psychiatry* **62**: 662–664.

Cancel, G., Abbas, N., Stevanin, G. *et al.* (1995) Marked phenotypic heterogeneity associated with expansion of a CAG repeat sequence at the spinocerebellar ataxia 3/Machado-Joseph disease locus. *Am. J. Hum. Genet.* **57**: 809–816.

Cancel, G., Durr, A., Didierjean, O. *et al.* (1997) Molecular and clinical correlations in spinocerebellar ataxia 2: a study of 32 families. *Hum. Mol. Genet.* **6**: 709–715.

Cancel, G., Gourfinkel-An, I., Stevanin, G., Didierjean, O., Abbas, N., Hirsch, E., Agid, Y. and Brice, A. (1998) Somatic mosaicism of the CAG repeat expansion in spinocerebellar ataxia type 3/Machado-Joseph disease. *Hum. Mutat.* **11**: 23–27.

Chai, Y., Koppenhafer, S.L., Bonini, N.M. and Paulson, H.L. (1999) Analysis of the role of heat shock protein (Hsp) molecular chaperones in polyglutamine disease. *J. Neurosci.* **19**: 10338–10347.

Chai, Y., Koppenhafer, S.L., Shoesmith, S.J., Perez, M.K. and Paulson, H.L. (1999) Evidence for proteasome involvement in polyglutamine disease: localization to nuclear inclusions in SCA3/MJD and suppression of polyglutamine aggregation in vitro. *Hum. Mol. Genet.* **8**: 673–682.

Chong, S.S., McCall, A.E., Cota, J., Subramony, S.H., Orr, H.T., Hughes, M.R. and Zoghbi, H.Y. (1995) Gametic and somatic tissue-specific heterogeneity of the expanded SCA1 CAG repeat in spinocerebellar ataxia type 1. *Nat. Genet.* **10**: 344–350.

Chung, M.Y., Ranum, L.P., Duvick, L.A., Servadio, A., Zoghbi, H.Y. and Orr, H.T. (1993) Evidence for a mechanism predisposing to intergenerational CAG repeat instability in spinocerebellar ataxia type I. *Nat. Genet.* **5**: 254–258.

Cummings, C.J., Mancini, M.A., Antalffy, B., DeFranco, D.B., Orr, H.T. and Zoghbi, H.Y. (1998) Chaperone suppression of aggregation and altered subcellular proteasome localization imply protein misfolding in SCA1. *Nat. Genet.* **19**: 148–154.

David, G., Abbas, N., Stevanin, G. *et al.* (1997) Cloning of the SCA7 gene reveals a highly unstable CAG repeat expansion. *Nat. Genet.* **17**: 65–70.

David, G., Durr, A., Stevanin, G. *et al.* (1998) Molecular and clinical correlations in autosomal dominant cerebellar ataxia with progressive macular dystrophy (SCA7). *Hum. Mol. Genet.* **7**: 165–170.

Davies, S.W., Turmaine, M., Cozens, B.A. *et al.* (1997) Formation of neuronal intranuclear inclusions underlies the neurological dysfunction in mice transgenic for the HD mutation. *Cell* **90**: 537–548.

Del-Favero, J., Krols, L., Michalik, A. *et al.* (1998) Molecular genetic analysis of autosomal dominant cerebellar ataxia with retinal degeneration (ADCA type II) caused by CAG triplet repeat expansion. *Hum. Mol. Genet.* **7**: 177–186.

Dichgans, M., Schols, L., Herzog, J. *et al.* (1999) Spinocerebellar ataxia type 6: evidence for a strong founder effect among German families. *Neurology* **52**: 849–851.

DiFiglia, M., Sapp, E., Chase, K.O., Davies, S.W., Bates, G.P., Vonsattel, J.P. and Aronin, N. (1997) Aggregation of huntingtin in neuronal intranuclear inclusions and dystrophic neurites in brain. *Science* **277**: 1990–1993.

Durr, A., Chneiweiss, H., Khati, C., Stevanin, G., Cancel, G., Feingold, J., Agid, Y. and Brice, A. (1993) Phenotypic variability in autosomal dominant cerebellar ataxia type I is unrelated to genetic heterogeneity. *Brain* **116**: 1497–1508.

Durr, A., Smadja, D., Cancel, G. *et al.* (1995) Autosomal dominant cerebellar ataxia type I in Martinique (French West Indies). Clinical and neuropathological analysis of 53 patients from three unrelated SCA2 families. *Brain* **118**: 1573–1581.

Durr, A., Stevanin, G., Cancel, G. *et al.* (1996) Spinocerebellar ataxia 3 and Machado-Joseph disease: clinical, molecular, and neuropathological features. *Ann. Neurol.* **39**: 490–499.

Duyckaerts, C., Durr, A., Cancel, G. and Brice, A. (1999) Nuclear inclusions in spinocerebellar ataxia type 1. *Acta Neuropathol. (Berl)* **97**: 201–207.

Eichler, E.E., Holden, J.J., Popovich, B.W., Reiss, A.L., Snow, K., Thibodeau, S.N., Richards, C.S., Ward, P.A. and Nelson, D.L. (1994) Length of uninterrupted CGG repeats determines instability in the FMR1 gene. *Nat. Genet.* **8**: 88–94.

Freudenreich, C.H., Stavenhagen, J.B. and Zakian, V.A. (1997) Stability of a CTG/CAG trinucleotide repeat in yeast is dependent on its orientation in the genome. *Mol. Cell Biol.* **17**: 2090–2098.

Geschwind, D.H., Perlman, S., Figueroa, K.P., Karrim, J., Baloh, R.W. and Pulst, S.M. (1997) Spinocerebellar ataxia type 6. Frequency of the mutation and genotype–phenotype correlations. *Neurology* **49**: 1247–1251.

Gilman, S., Sima, A.A., Junck, L., Kluin, K.J., Koeppe, R.A., Lohman, M.E. and Little, R. (1996) Spinocerebellar ataxia type 1 with multiple system degeneration and glial cytoplasmic inclusions. *Ann. Neurol.* **39**: 241–255.

Giunti, P., Stevanin, G., Worth, P.F., David, G., Brice, A. and Wood, N.W. (1999) Molecular and clinical study of 18 families with ADCA type II: evidence for genetic heterogeneity and de novo mutation. *Am. J. Hum. Genet.* **64**: 1594–1603.

Giunti, P., Sweeney, M.G. and Harding, A.E. (1995) Detection of the Machado–Joseph disease/spinocerebellar ataxia three trinucleotide repeat expansion in families with autosomal dominant motor disorders, including the Drew family of Walworth. *Brain* **118**: 1077–1085.

Goldberg, Y.P., McMurray, C.T., Zeisler, J. *et al.* (1995) Increased instability of intermediate alleles in families with sporadic Huntington disease compared to similar sized intermediate alleles in the general population. *Hum. Mol. Genet.* **4**: 1911–1918.

Goldberg, Y.P., Nicholson, D.W., Rasper, D.M. *et al.* (1996) Cleavage of huntingtin by apopain, a proapoptotic cysteine protease, is modulated by the polyglutamine tract. *Nat. Genet.* **13**: 442–449.

Gomez, C.M., Thompson, R.M., Gammack, J.T., Perlman, S.L., Dobyns, W.B., Truwit, C.L., Zee, D.S., Clark, H.B. and Anderson, J.H. (1997) Spinocerebellar ataxia type 6: gaze-evoked and vertical nystagmus, Purkinje cell degeneration, and variable age of onset. *Ann. Neurol.* **42**: 933–950.

Green, H. (1993) Human genetic diseases due to codon reiteration: relationship to an evolutionary mechanism. *Cell* **74**: 955–956.

Group, T.H.S.D.C. (1993) A novel gene containing a trinucleotide repeat that is expanded and unstable on Huntington's disease chromosomes. *Cell* **72**: 971–983.

Hackam, A.S., Singaraja, R., Wellington, C.L., Metzler, M., McCutcheon, K., Zhang, T., Kalchman, M. and Hayden, M.R. (1998) The influence of huntingtin protein size on nuclear localization and cellular toxicity. *J. Cell. Biol.* **141**: 1097–1105.

Harding, A.E. (1982) The clinical features and classification of the late onset autosomal dominant cerebellar ataxias. A study of 11 families, including descendants of the 'the Drew family of Walworth'. *Brain* **105**: 1–28.

Hayashi, Y., Kakita, A., Yamada, M. *et al.* (1998) Hereditary dentatorubral-pallidoluysian atrophy: detection of widespread ubiquitinated neuronal and glial intranuclear inclusions in the brain. *Acta Neuropathol. (Berl)* **96**: 547–552.

Hirayama, K., Takayanagi, T., Nakamura, R. *et al.* (1994) Spinocerebellar degenerations in Japan: a nationwide epidemiological and clinical study. *Acta Neurol. Scand. Suppl.* **153**: 1–22.

Hollenbach, B., Scherzinger, E., Schweiger, K., Lurz, R., Lehrach, H. and Wanker, E.E. (1999) Aggregation of truncated GST-HD exon 1 fusion proteins containing normal range and expanded glutamine repeats. *Philos. Trans. R. Soc. Lond. B. Biol. Sci.* **354**: 991–994.

Holmberg, M., Duyckaerts, C., Durr, A. *et al.* (1998) Spinocerebellar ataxia type 7 (SCA7): a neurodegenerative disorder with neuronal intranuclear inclusions. *Hum. Mol. Genet.* **7**: 913–918.

Igarashi, S., Takiyama, Y., Cancel, G. *et al.* (1996) Intergenerational instability of the CAG repeat of the gene for Machado- Joseph disease (MJD1) is affected by the genotype of the normal chromosome: implications for the molecular mechanisms of the instability of the CAG repeat. *Hum. Mol. Genet.* **5**: 923–932.

Ikeda, Y., Shizuka, M., Watanabe, M., Okamoto, K. and Shoji, M. (2000) Molecular and clinical analyses of spinocerebellar ataxia type 8 in Japan. *Neurology* **54**: 950–955.

Ikeuchi, T., Takano, H., Koide, R. *et al.* (1997) Spinocerebellar ataxia type 6: CAG repeat expansion in alpha1A voltage- dependent calcium channel gene and clinical variations in Japanese population. *Ann. Neurol.* **42**: 879–884.

Imbert, G., Saudou, F., Yvert, G. *et al.* (1996) Cloning of the gene for spinocerebellar ataxia 2 reveals a locus with high sensitivity to expanded CAG/glutamine repeats. *Nat. Genet.* **14**: 285–291.

Ishikawa, K., Fujigasaki, H., Saegusa, H. *et al.* (1999) Abundant expression and cytoplasmic aggregations of α_{1A} voltage-dependent calcium channel protein associated with neurodegeneration in spinocerebellar ataxia type 6. *Hum. Mol. Genet.* **8**: 1185–1193.

Jodice, C., Malaspina, P., Persichetti, F. *et al.* (1994) Effect of trinucleotide repeat length and parental sex on phenotypic variation in spinocerebellar ataxia I. *Am. J. Hum. Genet.* **54**: 959–965.

Jodice, C., Mantuano, E., Veneziano, L. *et al.* (1997) Episodic ataxia type 2 (EA2) and spinocerebellar ataxia type 6 (SCA6) due to CAG repeat expansion in the CACNA1A gene on chromosome 19p. *Hum. Mol. Genet.* **6**: 1973–1978.

Johansson, J., Forsgren, L., Sandgren, O., Brice, A., Holmgren, G. and Holmberg, M. (1998) Expanded CAG repeats in Swedish spinocerebellar ataxia type 7 (SCA7) patients: effect of CAG repeat length on the clinical manifestation. *Hum. Mol. Genet.* **7**: 171–176.

Kang, S., Jaworski, A., Ohshima, K. and Wells, R.D. (1995) Expansion and deletion of CTG repeats from human disease genes are determined by the direction of replication in E. coli. *Nat. Genet.* **10**: 213–218.

Kawaguchi, Y., Okamoto, T., Taniwaki, M. *et al.* (1994) CAG expansions in a novel gene for Machado-Joseph disease at chromosome 14q32.1. *Nat. Genet.* **8**: 221–228.

Kawakami, H., Maruyama, H., Nakamura, S., Kawaguchi, Y., Kakizuka, A., Doyu, M. and Sobue, G. (1995) Unique features of the CAG repeats in Machado-Joseph disease. *Nat. Genet.* **9**: 344–345.

Klement, I.A., Skinner, P.J., Kaytor, M.D., Yi, H., Hersch, S.M., Clark, H.B., Zoghbi, H.Y. and Orr, H.T. (1998) Ataxin-1 nuclear localization and aggregation: role in polyglutamine-induced disease in SCA1 transgenic mice. *Cell* **95**: 41–53.

Klockgether, T., Kramer, B., Ludtke, R., Schols, L. and Laccone, F. (1996) Repeat length and disease progression in spinocerebellar ataxia type 3. *Lancet* **348**: 830.

Klockgether, T., Skalej, M., Wedekind, D. *et al.* (1998) Autosomal dominant cerebellar ataxia type I. MRI-based volumetry of posterior fossa structures and basal ganglia in spinocerebellar ataxia types 1, 2 and 3. *Brain* **121**: 1687–1693.

Koefoed, P., Hasholt, L., Fenger, K., Nielsen, J.E., Eiberg, H., Buschard, K. and Sorensen, S.A. (1998) Mitotic and meiotic instability of the CAG trinucleotide repeat in spinocerebellar ataxia type 1. *Hum. Genet.* **103**: 564–569.

Koob, M.D., Benzow, K.A., Bird, T.D., Day, J.W., Moseley, M.L. and Ranum, L.P. (1998) Rapid cloning of expanded trinucleotide repeat sequences from genomic DNA. *Nat. Genet.* **18**: 72–75.

Koyano, S., Uchihara, T., Fujigasaki, H., Nakamura, A., Yagishita, S. and Iwabuchi, K. (1999) Neuronal intranuclear inclusions in spinocerebellar ataxia type 2: triple-labeling immunofluorescent study. *Neurosci. Lett.* **273**: 117–120.

Koyano, S., Uchihara, T., Fujigasaki, H., Nakamura, A., Yagishita, S. and Iwabuchi, K. (2000) Neuronal intranuclear inclusions in spinocerebellar ataxia type 2. *Ann. Neurol.* **47**: 550.

La Spada, A.R., Roling, D.B., Harding, A.E., Warner, C.L., Spiegel, R., Hausmanowa-Petrusewicz, I., Yee, W.C. and Fischbeck, K.H. (1992) Meiotic stability and genotype-phenotype correlation of the trinucleotide repeat in X-linked spinal and bulbar muscular atrophy. *Nat. Genet.* **2**: 301–304.

La Spada, A.R., Wilson, E.M., Lubahn, D.B., Harding, A.E. and Fischbeck, K.H. (1991) Androgen receptor gene mutations in X-linked spinal and bulbar muscular atrophy. *Nature* **352**: 77–79.

Lerer, I., Merims, D., Abeliovich, D., Zlotogora, J. and Gadoth, N. (1996) Machado-Joseph disease: correlation between the clinical features, the CAG repeat length and homozygosity for the mutation. *Eur. J. Hum. Genet.* **4**: 3–7.

Lima, L. and Coutinho, P. (1980) Clinical criteria for diagnosis of Machado-Joseph disease: report of a non-Azorena Portuguese family. *Neurology* **30**: 319–322.

Lin, X., Antalffy, B., Kang, D., Orr, H.T. and Zoghbi, H.Y. (2000) Polyglutamine expansion down-regulates specific neuronal genes before pathologic changes in SCA1. *Nat. Neurosci.* **3**: 157–163.

Lopes-Cendes, I., Maciel, P., Kish, S. *et al.* (1996) Somatic mosaicism in the central nervous system in spinocerebellar ataxia type 1 and Machado-Joseph disease. *Ann. Neurol.* **40**: 199–206.

Lunkes, A. and Mandel, J.L. (1998) A cellular model that recapitulates major pathogenic steps of Huntington's disease. *Hum. Mol. Genet.* **7**: 1355–1361.

Maciel, P., Gaspar, C., DeStefano, A.L. *et al.* (1995) Correlation between CAG repeat length and clinical features in Machado-Joseph disease. *Am. J. Hum. Genet.* **57**: 54–61.

Marie, P. (1893) Sur l'heredoataxie cerebelleuse. *Semaines de Medicine, Paris* **13**: 444–447.

Martin, J.J., Van Regemorter, N., Krols, L. *et al.* (1994) On an autosomal dominant form of retinal-cerebellar degeneration: an autopsy study of five patients in one family. *Acta Neuropathol.* **88**: 277–286.

Martin, S.J. and Green, D.R. (1995) Protease activation during apoptosis: death by a thousand cuts? *Cell* **82**: 349–352.

Martindale, D., Hackam, A., Wieczorek, A. *et al.* (1998) Length of huntingtin and its polyglutamine tract influences localization and frequency of intracellular aggregates. *Nat. Genet.* **18**: 150–154.

Maruyama, H., Nakamura, S., Matsuyama, Z. *et al.* (1995) Molecular features of the CAG repeats and clinical manifestation of Machado-Joseph disease. *Hum. Mol. Genet.* **4**: 807–812.

Matilla, T., McCall, A., Subramony, S.H. and Zoghbi, H.Y. (1995) Molecular and clinical correlations in spinocerebellar ataxia type 3 and Machado-Joseph disease. *Ann. Neurol.* **38**: 68–72.

Matsumura, R., Futamura, N., Fujimoto, Y., Yanagimoto, S., Horikawa, H., Suzumura, A. and Takayanagi, T. (1997) Spinocerebellar ataxia type 6. Molecular and clinical features of 35 Japanese patients including one homozygous for the CAG repeat expansion. *Neurology* **49**: 1238–1243.

Matsuyama, Z., Kawakami, H., Maruyama, H. *et al.* (1997) Molecular features of the CAG repeats of spinocerebellar ataxia 6 (SCA6). *Hum. Mol. Genet.* **6**: 1283–1287.

Maurer, D.J., O'Callaghan, B.L. and Livingston, D.M. (1996) Orientation dependence of trinucleotide CAG repeat instability in Saccharomyces cerevisiae. *Mol. Cell. Biol.* **16**: 6617–6622.

Merry, D.E., Kobayashi, Y., Bailey, C.K., Taye, A.A. and Fischbeck, K.H. (1998) Cleavage, aggregation and toxicity of the expanded androgen receptor in spinal and bulbar muscular atrophy. *Hum. Mol. Genet.* **7**: 693–701.

Moseley, M.L., Schut, L.J., Bird, T.D., Koob, M.D., Day, J.W. and Ranum, L.P. (2000) SCA8 CTG repeat: en masse contractions in sperm and intergenerational sequence changes may play a role in reduced penetrance. *Hum. Mol. Genet.* **9**: 2125–2130.

Nagafuchi, S., Yanagisawa, H., Sato, K. *et al.* (1994) Dentatorubral and pallidoluysian atrophy expansion of an unstable CAG trinucleotide on chromosome 12p. *Nat. Genet.* **6**: 14–18.

Ona, V.O., Li, M., Vonsattel, J.P. *et al.* (1999) Inhibition of caspase-1 slows disease progression in a mouse model of Huntington's disease. *Nature* **399**: 263–267.

Orr, H.T., Chung, M.Y., Banfi, S. *et al.* (1993) Expansion of an unstable trinucleotide CAG repeat in spinocerebellar ataxia type 1. *Nat. Genet.* **4**: 221–226.

Paulson, H.L., Perez, M.K., Trottier, Y. *et al.* (1997) Intranuclear inclusions of expanded polyglutamine protein in spinocerebellar ataxia type 3. *Neuron* **19**: 333–344.

Perez, M.K., Paulson, H.L., Pendse, S.J., Saionz, S.J., Bonini, N.M. and Pittman, R.N. (1998) Recruitment and the role of nuclear localization in polyglutamine-mediated aggregation. *J. Cell. Biol.* **143**: 1457–1470.

Perutz, M.F. (1999) Glutamine repeats and neurodegenerative diseases: molecular aspects. *Trends Biochem. Sci.* **24**: 58–63.

Perutz, M.F., Johnson, T., Suzuki, M. and Finch, J.T. (1994) Glutamine repeats as polar zippers: their possible role in inherited neurodegenerative diseases. *Proc. Natl Acad. Sci. USA* **91**: 5355–5358.

Pulst, S.M., Nechiporuk, A., Nechiporuk, T. *et al.* (1996) Moderate expansion of a normally biallelic trinucleotide repeat in spinocerebellar ataxia type 2. *Nat. Genet.* **14**: 269–276.

Quan, F., Janas, J. and Popovich, B.W. (1995) A novel CAG repeat configuration in the SCA1 gene: implications for the molecular diagnostics of spinocerebellar ataxia type 1. *Hum. Mol. Genet.* **4**: 2411–2413.

Ranen, N.G., Stine, O.C., Abbott, M.H. *et al.* (1995) Anticipation and instability of IT-15 (CAG)n repeats in parent-offspring pairs with Huntington disease. *Am. J. Hum. Genet.* **57**: 593–602.

Robitaille, Y., Schut, L. and Kish, S.J. (1995) Structural and immunocytochemical features of olivopontocerebellar atrophy caused by the spinocerebellar ataxia type 1 (SCA-1) mutation define a unique phenotype. *Acta Neuropathol.* **90**: 572–581.

Sanpei, K., Takano, H., Igarashi, S. *et al.* (1996) Identification of the spinocerebellar ataxia type 2 gene using a direct identification of repeat expansion and cloning technique, DIRECT. *Nat. Genet.* **14**: 277–284.

Sathasivam, K., Hobbs, C., Turmaine, M. *et al.* (1999) Formation of polyglutamine inclusions in non-CNS tissue. *Hum. Mol. Genet.* **8**: 813–822.

Saudou, F., Finkbeiner, S., Devys, D. and Greenberg, M.E. (1998) Huntingtin acts in the nucleus to induce apoptosis but death does not correlate with the formation of intranuclear inclusions. *Cell* **95**: 55–66.

Scherzinger, E., Lurz, R., Turmaine, M. *et al.* (1997) Huntingtin-encoded polyglutamine expansions form amyloid-like protein aggregates *in vitro* and *in vivo*. *Cell* **90**: 549–558.

Scherzinger, E., Sittler, A., Schweiger, K., Heiser, V., Lurz, R., Hasenbank, R., Bates, G.P., Lehrach, H. and Wanker, E.E. (1999) Self-assembly of polyglutamine-containing huntingtin fragments into amyloid-like fibrils: implications for Huntington's disease pathology. *Proc. Natl Acad. Sci. USA* **96**: 4604–4609.

Schols, L., Amoiridis, G., Buttner, T., Przuntek, H., Epplen, J.T. and Riess, O. (1997) Autosomal dominant cerebellar ataxia: phenotypic differences in genetically defined subtypes? *Ann. Neurol.* **42**: 924–932.

Schols, L., Amoiridis, G., Epplen, J.T., Langkafel, M., Przuntek, H. and Riess, O. (1996) Relations between genotype and phenotype in German patients with the Machado-Joseph disease mutation. *J. Neurol. Neurosurg. Psychiatry* **61**: 466–470.

Schols, L., Gispert, S., Vorgerd, M. *et al.* (1997) Spinocerebellar ataxia type 2. Genotype and phenotype in German kindreds. *Arch. Neurol.* **54**: 1073–1080.

Schols, L., Kruger, R., Amoiridis, G., Przuntek, H., Epplen, J.T. and Riess, O. (1998) Spinocerebellar ataxia type 6: genotype and phenotype in German kindreds. *J. Neurol. Neurosurg. Psychiatry* **64**: 67–73.

Schols, L., Vieira-Saecker, A.M., Schols, S., Przuntek, H., Epplen, J.T. and Riess, O. (1995) Trinucleotide expansion within the MJD1 gene presents clinically as spinocerebellar ataxia and occurs most frequently in German SCA patients. *Hum. Mol. Genet.* **4**: 1001–1005.

Schweitzer, J.K. and Livingston, D.M. (1998) Expansions of CAG repeat tracts are frequent in a yeast mutant defective in Okazaki fragment maturation. *Hum. Mol. Genet.* **7**: 69–74.

Shizuka, M., Watanabe, M., Ikeda, Y., Mizushima, K., Kanai, M., Tsuda, T., Abe, K., Okamoto, K. and Shoji, M. (1998) Spinocerebellar ataxia type 6: CAG trinucleotide expansion, clinical characteristics and sperm analysis. *Eur. J. Neurol.* **5**: 381–387.

Silveira, I., Alonso, I., Guimaraes, L. *et al.* (2000) High germinal instability of the (CTG)n at the SCA8 locus of both expanded and normal alleles. *Am. J. Hum. Genet.* **66**: 830–840.

Silveira, I., Coutinho, P., Maciel, P. *et al.* (1998) Analysis of SCA1, DRPLA, MJD, SCA2, and SCA6 CAG repeats in 48 Portuguese ataxia families. *Am. J. Med. Genet.* **81**: 134–138.

Sisodia, S.S. (1998) Nuclear inclusions in glutamine repeat disorders: are they pernicious, coincidental, or beneficial? *Cell* **95**: 1–4.

Skinner, P.J., Koshy, B.T., Cummings, C.J., Klement, I.A., Helin, K., Servadio, A., Zoghbi, H.Y. and Orr, H.T. (1997) Ataxin-1 with an expanded glutamine tract alters nuclear matrix-associated structures. *Nature* **389**: 971–974.

Sobue, G., Doyu, M., Nakao, N., Shimada, N., Mitsuma, T., Maruyama, H., Kawakami, S. and Nakamura, S. (1996) Homozygosity for Machado-Joseph disease gene enhances phenotypic severity. *J. Neurol. Neurosurg. Psychiatry* **60**: 354–356.

Stenoien, D.L., Cummings, C.J., Adams, H.P., Mancini, M.G., Patel, K., DeMartino, G.N., Marcelli, M., Weigel, N.L. and Mancini, M.A. (1999) Polyglutamine-expanded androgen receptors form aggregates that sequester heat shock proteins, proteasome components and SRC-1, and are suppressed by the HDJ-2 chaperone. *Hum. Mol. Genet.* **8**: 731–741.

Stevanin, G., Durr, A., David, G. *et al.* (1997) Clinical and molecular features of spinocerebellar ataxia type 6. *Neurology* **49**: 1243–1246.

Stevanin, G., Herman, A., Durr, A., Jodice, C., Frontali, M., Agid, Y. and Brice, A. (2000) Are (CTG)n expansions at the SCA8 locus rare polymorphisms? *Nat. Genet.* **24**: 213; discussion 215.

Stevanin, G., Le Guern, E., Ravise, N. *et al.* (1994) A third locus for autosomal dominant cerebellar ataxia type I maps to chromosome 14q24.3-qter: evidence for the existence of a fourth locus. *Am. J. Hum. Genet.* **54**: 11–20.

Takano, H., Cancel, G., Ikeuchi, T. *et al.* (1998) Close associations between prevalences of dominantly inherited spinocerebellar ataxias with CAG-repeat expansions and frequencies of large normal CAG alleles in Japanese and Caucasian populations. *Am. J. Hum. Genet.* **63**: 1060–1066.

Takano, H., Onodera, O., Takahashi, H. *et al.* (1996) Somatic mosaicism of expanded CAG repeats in brains of patients with dentatorubral-pallidoluysian atrophy: cellular population-dependent dynamics of mitotic instability. *Am. J. Hum. Genet.* **58**: 1212–1222.

Takiyama, Y., Igarashi, S., Rogaeva, E.A. *et al.* (1995) Evidence for inter-generational instability in the CAG repeat in the MJD1 gene and for conserved haplotypes at flanking markers amongst Japanese and Caucasian subjects with Machado-Joseph disease. *Hum. Mol. Genet.* **4**: 1137–1146.

Takiyama, Y., Sakoe, K., Amaike, M., Soutome, M., Ogawa, T., Nakano, I. and Nishizawa, M. (1999) Single sperm analysis of the CAG repeats in the gene for dentatorubral-pallidoluysian atrophy (DRPLA): the instability of the CAG repeats in the DRPLA gene is prominent among the CAG repeat diseases. *Hum. Mol. Genet.* **8**: 453–457.

Toru, S., Murakoshi, T., Ishikawa, K. *et al.* (2000) Spinocerebellar ataxia type 6 mutation alters P-type calcium channel function. *J. Biol. Chem.* **275**: 10893–10898.

Tsuchiya, K., Ishikawa, K., Watabiki, S. *et al.* (1998) A clinical, genetic, neuropathological study in a Japanese family with SCA 6 and a review of Japanese autopsy cases of autosomal dominant cortical cerebellar atrophy. *J. Neurol. Sci.* **160**: 54–59.

Tuite, P.J., Rogaeva, E.A., St George-Hyslop, P.H. and Lang, A.E. (1995) Dopa-responsive parkinsonism phenotype of Machado-Joseph disease: confirmation of 14q CAG expansion. *Ann. Neurol.* **38**: 684–687.

Ueno, S., Kondoh, K., Kotani, Y., Komure, O., Kuno, S., Kawai, J., Hazama, F. and Sano, A. (1995) Somatic mosaicism of CAG repeat in dentatorubral-pallidoluysian atrophy (DRPLA). *Hum. Mol. Genet.* **4**: 663–666.

Vincent, J.B., Neves-Pereira, M.L., Paterson, A.D. *et al.* (2000) An unstable trinucleotide-repeat region on chromosome 13 implicated in spinocerebellar ataxia: a common expansion locus. *Am. J. Hum. Genet.* **66**: 819–829.

Wadia, N., Pang, J., Desai, J., Mankodi, A., Desai, M. and Chamberlain, S. (1998) A clinicogenetic analysis of six Indian spinocerebellar ataxia (SCA2) pedigrees. The significance of slow saccades in diagnosis. *Brain* **121**: 2341–2355.

Warrick, J.M., Paulson, H.L., Gray-Board, G.L., Bui, Q.T., Fischbeck, K.H., Pittman, R.N. and Bonini, N.M. (1998) Expanded polyglutamine protein forms nuclear inclusions and causes neural degeneration in Drosophila. *Cell* **93**: 939–949.

Watanabe, M., Abe, K., Aoki, M. *et al.* (1996) Analysis of CAG trinucleotide expansion associated with Machado-Joseph disease. *J. Neurol. Sci.* **136**: 101–107.

Wellington, C.L., Ellerby, L.M., Hackam, A.S. *et al.* (1998) Caspase cleavage of gene products associated with triplet expansion disorders generates truncated fragments containing the polyglutamine tract. *J. Biol. Chem.* **273**: 9158–9167.

Wells, R.D., Parniewski, P., Pluciennik, A., Bacolla, A., Gellibolian, R. and Jaworski, A. (1998) Small slipped register genetic instabilities in Escherichia coli in triplet repeat sequences associated with hereditary neurological diseases. *J. Biol. Chem.* **273**: 19532–19541.

Worth, P.F., Houlden, H., Giunti, P., Davis, M.B. and Wood, N.W. (2000) Large, expanded repeats in *SCA8* are not confined to patients with cerebellar ataxia. *Nat. Genet.* **24**: 214–215.

Zhuchenko, O., Bailey, J., Bonnen, P. *et al.* (1997) Autosomal dominant cerebellar ataxia (*SCA6*) associated with small polyglutamine expansions in the alpha 1A-voltage-dependent calcium channel. *Nat. Genet.* **15**: 62–69.

11

Disorders of cholesterol biosynthesis

David R. Fitzpatrick

1. Introduction

Elucidating the mechanisms that control the intracellular concentration of choles-terol has been one of the most exciting stories in modern biomedical sciences. The work has led to fundamental biological insights into how cells internalize complex macromolecular structures (Brown and Goldstein, 1986) and how feedback control occurs at a molecular level (Brown and Goldstein, 1999; Brown *et al.*, 2000). In addition to its well-known roles in cellular and metabolic homeostasis in the mature organism it is now clear that many developmental processes require these mechanisms to function normally. This chapter will review the phenotypic and molecular aspects of five genetically determined deficiencies in the biosynthetic pathway that mediated *de novo* synthesis of cholesterol in humans.

2. The function of membrane cholesterol

The majority (~90%) of free cholesterol within the cell is located in plasma membranes (Lange *et al.*, 1989) where it is present in roughly equimolar amounts to the sum of all other lipids. Cholesterol promotes 'packing' of unsaturated acyl chains of phospholipids within the bilayer thus reducing the permeability of the membrane to small molecules. This conformational change also increases the width by ~33% when compared to cholesterol-poor membranes such as those in the ER. This physical property enables the cell to target proteins to compartments within the cell utilizing the length of their hydrophobic transmembrane regions (Bretscher and Munro, 1993).

The distribution of cholesterol within a membrane is non-random. Functional elements known as rafts are detergent-insoluble microdomains that are enriched in cholesterol, sphingolipids and glycosylphosphatidylinositol (see below).

3. Related sterols

Cholesterol is the major sterol in all animals but does not have a monopoly amongst eukaryotic organisms (*Figure 3*). Yeast synthesize a distinct but closely-related

Genotype to Phenotype second edition, edited by S. Malcolm and J. Goodship.
© 2001 BIOS Scientific Publishers Ltd, Oxford.

sterol, ergosterol. In plants the side chain can have either a methyl or ethyl addition to carbon 24 that is absent in cholesterol. Bacteria do not have sterols as a major component of their cell walls.

4. The origins of cellular cholesterol

Jejunal epithelial cells actively absorb dietary cholesterol in the form of bile acid miscelles. Here it is converted to acyl-cholesterol esters that are then assembled with triglycerides and lipoproteins to form chylomicrons (CM). CM are secreted into lymph and undergo peripheral lipolysis in the systemic circulation to form core remnants that enter hepatocytes by receptor-mediated endocytosis (*Figure 1*) (Brown and Goldstein, 1986). It is not clear what mechanisms exist to regulate the intestinal absorption of cholesterol, however, alterations in composition of lumenal bile acids (synthesized in hepatocytes from cholesterol) are likely to be important. Although cholesterol is a constituent of almost all food, the majority of cholesterol in human cells is the product of *de novo* biosynthesis in extrahepatic tissues (Dietschy *et al.*, 1993).

5. Endogenous synthesis of cholesterol

All 27 carbon atoms in cholesterol are derived from acetyl CoA. This biosynthetic pathway utilises >30 separate enzymatic reactions and can be divided into three stages:

(1)   the synthesis of mevalonate – site of feedback control (*Figure 1*);
(2)   the synthesis of squalene – formation of non-sterol end-products (*Figure 1*);
(3)   the synthesis of cholesterol – critical for normal morphogenesis (*Figure 2*).

5.1 The synthesis of mevalonate and feedback control

This pathway begins with the formation of 3-hydroxy-3-methylglutaryl CoA (HMG CoA) catalyzed by acetoacetyl CoA thiolase and HMG CoA synthase. Mevalonate is then formed by the action of HMG CoA reductase (*Figure 1*). These are rate-limiting reactions that are under complex regulatory control. Several independent mechanisms have evolved in order to maintain the intracellular concentrations of free cholesterol within a very narrow range by balancing the contribution of LDL receptor-mediated uptake with *de novo* biosynthesis. An increase in LDL receptor-derived cholesterol potently inhibits the transcription of HMG CoA synthase, HMG CoA reductase and LDL receptor genes. This transcriptional repression is mediated by the SCAP-mediated inhibition of proteolytic cleavage of the sterol response element binding protein 1 (SREBP1) by S1P protease (Brown and Goldstein, 1999). The exact nature of the interaction between SCAP and free cholesterol is not yet clear. This inhibition of cleavage prevents the nuclear localization of the NH-terminal fragment of SREBP1, which is a transcription factor, and thus reduces binding to the sterol response element in the promotor region of these genes. HMG CoA reductase is also under post-transcriptional feedback control by alterations in the translation and degradation rates of the enzyme by exogenous cholesterol and non sterol end-products (Goldstein and Brown, 1990).

Pharmocological inhibition of HMG CoA reductase using the Statin group of drugs is highly effective in lowering serum cholesterol and preventing death from ischemic heart disease (Brown and Goldstein, 1996). These are likely to become amongst the most commonly used drugs in developed countries.

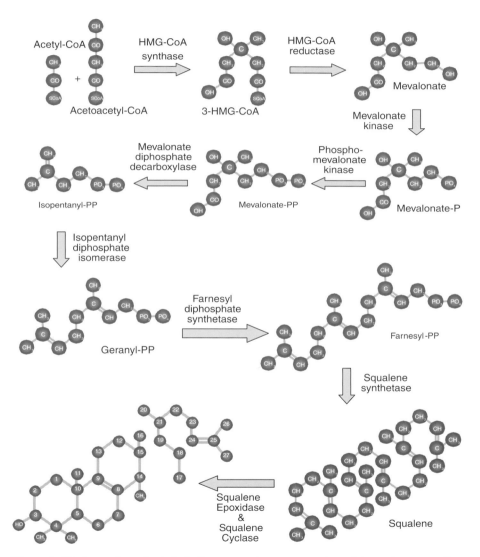

Figure 1. Cholesterol synthesis I. A diagrammatic representation of the early stages of cholesterol biosynthesis catalyzed by enzymes encoded by the following genes: *HMGCS1 EC4.1.3.5* (5p14-p13), *HMGCR EC1.1.1.34* (5q13.3), *MVK EC2.7.1.36* (12q24), *PMVK EC2.7.4.2* (1p13-q23), *MVD EC4.1.1.33* (16q24), *IDI1 EC5.3.3.2* (10p15.1), *FDPS EC2.5.1.10* (?), *FDFT1 EC2.5.1.21* (8p23.1-p22), *SQLE EC1.14.99.7* (8q24.1) *LSS EC5.4.99.7* (21q22.3).

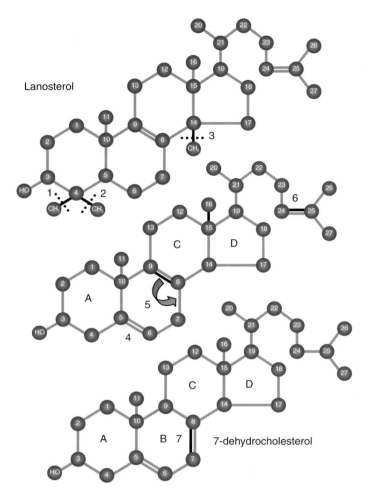

Figure 2. Cholesterol synthesis II. Diagramatic representation of the enzymatic steps in the conversion of Lanosterol to Cholesterol. 1 and 2 = C4 demethylation (ERG26 and ERG27) 3 = C14 demethylation (ERG11) 4 = C5 desaturase (ERG3) 5 = C8-C7 isomerase (ERG2) 6 = C24(C25) reductase 7 = C7(C8) reductase.

5.2 The synthesis of squalene and non-sterol pathway products

Mevalonate pyrophosohate (-PP) is formed by the action of mevalonate kinase. The next three intermediates in the pathway also function as non sterol end-products within the cell. Isopentenyl pyrophosphate (C5) is required for the formation of isopentenyl adenine present in some tRNA. Geranyl pyrophosphate (C10) is used for post-translational modification (isoprenylation) of proteins, as is farnesyl-PP (C15). However, the latter compound is also contributes to the synthesis of dolichols, Heme A and ubiquinone. Squalene (C30) is synthesized by the condensation of two farnesyl-PP molecules (Goldstein and Brown, 1990).

5.3 Isoprenylation

The post-translational modification of proteins by addition of non sterol products of the cholesterol biosynthetic pathway was first described in the mating factors of fungi (Anderegg et al., 1988). It became of interest in mammalian systems when it was discovered that p21^{ras} proteins were farnesylated and that cell growth could be blocked by inhibition of mevalonate synthesis (Casey et al., 1989). Since then many other proteins including ras-related proteins and G-proteins subunits have been shown to be modified both by farnesyl (C15) and geranylgeranyl (C20) covalent additions. These modifications occur mostly at C-terminal cysteine residues of the peptide and are thought to enable membrane association of the modified protein. The specific modification is important for biological activity of proteins and myristylation cannot substitute for isoprenylation although it also allows membrane association. Choroideraemia is caused by mutations in rab geranylgeranyl transferase (Seabra et al., 1992).

5.4 The synthesis of cholesterol

Lanosterol is formed by the cyclisation of squalene to form a recognisably cholesterol-like C30 molecule. The subsequent steps in the synthesis of cholesterol from lanosterol are less well studied than the earlier steps. The isolation of ergosterol auxotropic (ERG) mutants in *Saccharomyces cerevisiae* and cloning the mutated genes has been useful in characterizing several orthologous genes in humans (Chambon et al., 1991; Karst and Lacroute, 1977; Lorenz and Parks, 1987; Nes et al., 1993; Servouse and Karst, 1986). The significant structural differences between these sterols mean that some reactions in the ergosterol pathways have no corresponding step in humans and vice versa (*Table 1*). The major events in the lanosterol-to-cholesterol pathway are: (1) three demethylation events at carbon 4, 4 and 14; (2) 'moving' the double bond from 8(9) to 5(6); 3. reduction of the 24(25) side chain double bond.

It is not clear that these processes form distinct pathways. As almost every conceivable sterol intermediary can be detected *in vivo* the biosynthesis may behave as a matrix rather than a linear pathway. With regard to this point the

Table 1. Genes involved in lanosterol-to-cholesterol/ergosterol biosynthesis in humans and *Saccharomyces cerevisiae*

Enzyme	Yeast gene	Human gene	Location
C8-C7 sterol isomerase	ERG2	EBP	Xp11.23
C5 desaturase	ERG3	SC5DL	11q23.3
C24(28) reductase	ERG4	No equivalent	–
C14 reductase	ERG24	?LBR	1q32
C22 desaturase	ERG5	No equivalent	–
C4 methyl oxidase	ERG25	SC4MOL	4q32–34
C24 methyl transferase	ERG6	No equivalent	–
C14 α-demethylase	ERG11	CYP51	7q21
C3 dehydrogenase	ERG26	NSDHL	Xq28
C7 reductase	No equivalent	DHCR7	11q12–13
C24(25) reductase	No equivalent	?	20
C3 keto-reductase	ERG27	?	?

subcellular localization of the enzymes involved in cholesterol synthesis has been the subject of some debate. It was a long-held belief that the endoplasmic reticulum was the sole site of synthesis. It is clear that at least some and perhaps all of the enzymatic reactions involved in cholesterol biosynthesis also occur in peroxisomes (Biardi and Krisans, 1996; Brunnsgard et al., 1989; Reinhart et al., 1987). Many of these genes have now been cloned and immunolocalization will resolve this.

6. Inborn errors of cholesterol biosynthesis

To date five human diseases have been associated with genetically determined single enzyme deficiency of this pathway; 1. mevalonate kinase deficiency (malonic aciduria and HIDS syndrome); 2. 3-sterol dehydrogenase (C4 decarboxylase) deficiency (CHILD syndrome); 3. 8(7)-sterol isomerase deficiency (Conradi Hunermann syndrome and CHILD syndrome); 4. 7-dehydrocholesterol reductase deficiency (Smith–Lemli–Opitz syndrome); 5. 24-sterol reductase deficiency (desmosterolosis).

7. Mevalonate kinase deficiency

7.1 Summary

Associated phenotypes	Mevalonic aciduria (autosomal recessive); HIDS (autosomal recessive)
Gene	MVK
Map location	12q24
SwissProt/TrEMBL	Q03426 (KIME_HUMAN)
OMIM number	251170
Clinical features (from Poll-The et al., 2000)	Fever (100%); arthralgia (100%); lymphadenopathy (100%); hyper-IgD (85%); hepatomegaly (56%); mental handicap (56%); neurological signs (48%); dysmorphisms (44%); failure to thrive (44%); cataracts (22%)

7.2 Clinical features

Mevalonic aciduria was first described in the 1980s by two groups (Berger et al., 1985; Hoffmann et al., 1986). Berger et al., (1985) described a relatively mild case of a with cerebellar ataxia. Hoffman presented a more severe phenotype of a 2-year-old boy with severe failure to thrive, developmental delay, anemia, hepatosplenomegaly, central cataracts, and dysmorphic facies. Mevalonate kinase deficiency was subsequently confirmed in this family. This condition is frequently lethal in early childhood and is characterized by a relapsing and progressive course. A striking feature on neuroimaging is a progressive cerebellar atrophy (Hoffmann et al., 1993). This feature together with the failure to thrive and anemia may be the result of a general failure of isoprenylation with subsequent failure to maintain normal rates of cell division. An interesting allelic phenotypic variant has been described recently termed hyperimmunoglobulinemia D and periodic fever syndrome (HIDS). This is characterized by intermittent fever associated with abdominal distress, lymphadenopathy and arthralgias (Poll-The *et al.*, 2000).

7.3 Biochemical features

In the classical cases of mevalonate kinase deficiency there is massive excretion of mevalonate in the urine with reported levels from 900–56 000 mmol/nmole creatinine (normal <1.3 mmol/nmole creatinine). The biochemical hallmarks of HIDS are significant mevalonic aciduria (5.3–39.5 mmol/nmole creatinine) and an elevated serum IgD ± elevated IgA. Mevalonate kinase (EC2.7.1.36) catalyses the reaction ATP + mevalonate → ADP + 5-phosphomevalonate and this activity has been localized to the cytoplasm and peroxisomes. The enzyme can be measured in cultured fibroblasts using ^{14}C-mevalonate as a substrate. All affected cases have <4% residual activity compared to control cells.

7.4 Molecular genetics

MVK has a 1188 bp open reading frame with a 396-amino acid predicted protein sequence. This 42 kDa peptide is thought to exist as a homodimer in both the cytoplasmic and peroxisomal compartments. MK belongs to the GHMP kinase ATP-binding domain family of proteins. A well-defined motif [LIVM]-[PK]-×-[GSTA]-×(0,1)-G-L-[GS]-S-S-[GSA]-[GSTAC] in this family of proteins is found between 137 and 148 aa in hMVKp. The TMPRED and TMAP pograms predict two transmembrane domains between 103–125 and 143–162. There are no other striking structural features identified by computer analysis of this protein.

7.5 Molecular pathology and genotype–phenotype correlations

The reported mutations are summarized in *Table 2*. All but one of these mutations are amino acid substitutions rather than truncations of the protein. None alter the

Table 2. MVK mutations

Mutation	Phenotype	Protein stability	Enzyme activity	% MA alleles	%HIDS alleles
−11del92	HIDS	−	−	−	−
H20P (59A→T)	Classical/HIDS	−	0%[a]	7[a]	6[a]
P165L (500C→T)	HIDS	−	−	−	−
T243I (728C→T)	Classical	Normal	1.5%[b]	−	−
L264F (790C→A)	Classical	↓↓	0.3%[b]	−	−
L265P (794T→C)	Classical	↓↓↓	0%[b]	−	−
I268T (803C→T)	Classical/HIDS	−	6.4%[a] 18.7%[b]	25[a]	14[a]
N301T (902A→C)	Classical	↓	1.5%[b]	−	−
V310M (928G→A)	Classical	−	2%[a]	−	−
A334T (1000G→A)	Classical	−	2%[b]	−	−
V377I (1129G→A)	HIDS	−	−	0[a]	52[a]

[a]From Houten *et al.* (2000); [b]From Hinson *et al.* (1999)

motif mentioned above or the previously defined catalytic sites (Potter and Miziorko, 1997; Potter *et al.*, 1997). The V377I and I268T mutation affect highly conserved residues. With the limited data available the best correlation would appear to be between protein stability, enzyme activity and severity of the phenotype. Further characterization of the protein should be directed to the role of the lysine residues at 264 and 265 that appear to be crucial for the maintenance of protein stability.

8. 3-sterol dehydrogenase (3SD) deficiency

8.1 Summary

Associated phenotypes	CHILD syndrome (X-linked dominant)
Gene	NSDHL
Map location	Xq28
SwissProt/TrEMBL	Q15738 NSDL_HUMAN
OMIM number	308050
Clinical features	Unilateral involvement; ichthyosis; 'waxy' erythroderma; limb reduction; hemiatrophy; chondrodysplasia punctata

8.2 Clinical features

CHILD is an acronym for congenital hemidysplasia with ichthyosiform erythro-derma and limb defects. The syndrome is characterized by the congenital or neonatal onset of a unilateral erythema and scaling (Happle *et al.*, 1980). This patchy dermatosis is notable for its blaschoid distribution and the clear demar-cation in the middle of the trunk. The limb defects can vary from mild digital hypoplasia to major reduction deformities and amelia. Similar segmental defects can occur ipsilaterally in any part of the skeleton and brain. Chondrodysplasia punctata has been reported. More than 95% of reported cases are females with a few cases of mother-to-daughter transmission leading CHILD syndrome to be considered an X-linked dominant condition. Several studies have commented on unusual intracellular lamellar or membranous structures that have been seen on electron microscopy (Emami *et al.*, 1992; Hashimoto *et al.*, 1995). Interestingly, these may be a good pathological marker for abnormal cholesterol biosynthesis as they have been noted in related conditions (FitzPatrick *et al.*, 1998).

8.3 Biochemical features

Abnormalities in cholesterol biosynthetic precursors have not been reported in CHILD but, given the molecular defect, there would be predicted to be an accu-mulation of both 4-methyl and 4,4-dimethyl sterol compounds (see below).

8.4 Molecular genetics

Recently mutations in CHILD have been described in the gene *NSDHL* (gbU47105 also termed *H105e3*) that encodes a protein predicted to have *C4* decarboxylase activity (Konig *et al.*, 2000). This gene was originally identified using computerized exon prediction analysis of genomic sequence data from Xq28 between DXS1104 and DXS52 (Levin *et al.*, 1996). It has eight exons, exons 2–8 have an 1119 bp open reading frame encoding a 373 amino acid protein with predicted homology to members of the 3β-hydroxy steroid dehydrogenase family. There is also striking homology to an *E. coli* ORF (c906012–904963). The programs TMPRED, TMAP and Phd-TM all predict a single transmembrane domain between ~295 and 314 aa.

8.5 Molecular pathology and genotype–phenotype correlations

Interest in NSDHL increased when it was discovered that mutations in its ortholog in *Saccharomyces cerevisiae*, open reading frame, YGL001c, caused the ERG26 (3SD deficient) phenotype (Gachotte *et al.*, 1998). The first mammalian mutations were identified in the orthologous gene in the mouse mutants bare patches (Bpa) and striated (Str). Bpa and Str were considered allelic X-dominant disorders. Bpa is characterized by ichthyosis, cataracts, and skeletal defects. Str is a milder phenotype with striations in the coats of affected females that appear ~2 weeks after birth. Both had been localized by linkage analysis to a 600 kb critical region that demonstrates conserved gene order with loci in the NSDHL region of human Xq28. NSDHL mutations have been identified in both 2/4 independent Bpa lines and 3/3 Str lines (*Table 3*; Liu *et al.*, 1999). On subsequent investigation cultured cells from these mice were found to accumulate 4-methyl and 4,4-dimethyl sterol intermediates. Subsequently, human mutations have been identified (Konig *et al.*, 2000). The finding of an affected male infant with CHILD associated with an apparently non mosaic mutation in NSDHL is difficult to explain given the assumption of male lethality and X-inactivation as a basis for the phenotype. This boy has a normal karyotype. More information is required in this very important case particularly regarding enzyme activity measured in normal and abnormal skin in order to develop sensible hypotheses.

Table 3. NSDHL mutations

Mutation	Phenotype	Sex	Side of lesions	No independent alleles	Species
K103X (522A>T)	Bpa	F	Bilateral	1	Mouse
G147DdelTEDLPY (655del18)	Bpa	F	Bilateral	1	Mouse
P98L (508C>T)	Str	F	Bilateral	1	Mouse
insY (1087insTTA)	Str	F	Bilateral	1	Mouse
V109M (540G>A)	Str	F	Bilateral	1	Mouse
R88X (262C>T)	CHILD	M	Right	1	Human
A105V (314C>T)	CHILD	F	Right	2	Human
G205S (613G>A)	CHILD	F	Right	1	Human
Q210X (628C>T)	CHILD	F	Right	1	Human

9.   Sterol isomerase deficiency

9.1 Summary

Associated phenotypes	X-linked dominant chondrodysplasia punctata; CHILD syndrome
Gene	EBP
Map location	Xp11.22
SwissProt/TrEMBL	P70245
OMIM Number	302960
Clinical features	Focal calcification in articular cartilage; cataracts; optic atrophy; follicular atrophoderma; ichthiosis; patchy hyperpigmentation; patchy alopecia; asymmetry; nasal hypoplasia

9.2 Clinical features

Chondrodysplasia punctata (CDP) is a radiological description of focal calcification within cartilage during infancy. In X-linked dominant CDP (CDPX2, also known as Conradi–Hunermann syndrome) the CDP is particularly prominent in the vertebral column and epiphyseal regions of the long bones. The other characteristic features in CDPX2 are sectoral lens opacities, pit-like defects in the skin (folicular atrophoderma), patchy alopecia and body asymmetry (Edidin *et al.*, 1977; Hamaguchi *et al.*, 1995; Happle 1979). In infancy the presentation can be of a 'collodian' baby. Many authors have commented on the extreme phenotypic variability in this condition within families. The extreme rarity of affected males associated with numerous mother-to-daughter transmissions of the disorder mean that the most likely inheritance pattern is X-linked dominant.

9.3 Biochemical features

As yet no direct enzymology is available in patient samples. However, sterols extracted from various tissues including cultured fibroblasts derived have been analyzed in several different individuals with CDPX2. Using gas chromatography/mass spectroscopy an accumulation of many sterol intermediates could be detected with 8(9)-cholesterol and 8-dehydrocholesterol being the most prominent. These findings imply a defect in C8–C7 isomerization. The protein was originally isolated and a homodimeric ER protein from guinea pig liver homogenates with a high affinity binding to emopamil (a phenylalkylamine Ca^{2+} antagonist) (Hanner *et al.*, 1995; Labit-Le Bouteiller *et al.*, 1998; Moebius *et al.*, 1997). The human protein can functionally complement the *Saccharomyces cerevisiae* mutant ERG2 (Moebius *et al.*, 1999).

9.4 Molecular genetics

The *EBP* gene has five exons spanning 7 kb of genomic DNA. The mature mRNA molecule is ubiquitously expressed and contains an ORF predicted to encode a peptide of 230 amino acids. This peptide contains no recognized pfam or

PROSITE motifs. The programs TMPRED, TMAP Phd-TM and DAS all predict five transmembrane domains between ~31–49, 64–82, 117–141, 149–166 and 184–205.

9.5 Molecular pathology and genotype–phenotype correlations

The EBP mutations that have been reported to date in CDPX2, CDP, CHILD syndrome and the tattered mouse are summarized in *Table 4* (Braverman *et al.*, 1999; Derry *et al.*, 1999; Grange *et al.*, 2000). Over half of these mutations are predicted to cause loss of function. No obvious genotype phenotype correlations can be identified as yet. It will be interesting to see if milder mutations will be identified in males with CDP in the future.

Table 4. EBP mutations

Mutation	Phenotype	Sex	Ref	Species
E80K (238G>A)	CDPX2	F	1	Human
C72fs (216–217insT)	CDPX2	F	1	Human
R63X (182C>T)	CDPX2	F	1,2	Human
IVS3+1G>T	CDP	F	1	Human
IVS2–2delA	CDP	F	1	Human
S133R (399C>G)	CDP	F	1	Human
R147H (440G>A)	CDP	F	1,2	Human
W29X (198G>A)	CDPX2	F	2	Human
ΔR62 (del293–295)	CDPX2	F	2	Human
(del345–358)	CDPX2	F	2	Human
W186X (669G>A)	CDPX2	F	2	Human
G107R (454G>A)	Td	F	2	Mouse
	CHILD	F	3	Human

References: 1 = Braverman *et al.* (1999); 2 = Derry *et al.* (1999); 3 = Grange *et al.* (2000)

10. 7-dehydrocholesterol reductase deficiency

10.1 Summary

Inheritance	Autosomal recessive
Gene	DHCR7
Map location	11q13
SwissProt/TrEMBL	Q9UMB7
OMIM number	270400
Clinical features (from (Kelley and Hennekam, 2000))	2/3 syndactyly (97%); mental handicap (95%); microcephaly (84%); postnatal growth failure (82%); anteverted nares (78%); ptosis (70%); genital abnormalities (65%); cardiac defects (54%); postaxial polydactyly (48%); cleft palate (47%); abnormal lung lobulation (45%); renal anomalies (43%)

10.2 Clinical features

SLOS is an autosomal recessive multiple malformation syndrome, which in its most severe form is perinatally lethal. The highest birth incidence (~1 in 20 000) appears to be in Caucasian populations. The clinical features of SLOS have been the subject of several excellent recent reviews (Kelley and Hennekam, 2000; Opitz 1999; Ryan *et al.*, 1998) and will be presented only briefly here (see list in summary above). It is now clear that there is a spectrum of severity in SLOS with the limb bud, craniofacial region, male genital tract and the brain being the most commonly affected developmental processes.

10.3 Biochemical features

Low serum cholesterol in association with an accumulation of 7-dehydrocholes-terol (7DHC) was identified as the biochemical hallmark of SLOS in 1993 (Elias and Irons, 1995; Irons *et al.*, 1993; Tint, 1993; Tint *et al.*, 1994). Confirmation that 7DHC acculmulation was due to deficiency of the enzyme 7DHC reductase followed soon after (Honda *et al.*, 1995).

10.4 Molecular genetics

7-dehydrocholesterol reductase deficiency result from mutations in the gene *DHCR7* (Fitzky *et al.*, 1998; Wassif *et al.*, 1998; Waterham *et al.*, 1998). *DHCR7* maps to 11q13 and consists of at least nine exons spanning ~14 kb. The 475 amino acid open reading frame begins in exon 3 with the stop codon in exon 9. By both protein homology and function it has been asigned to the ERG4_ERG24 family of oxoreductases. The protein has between seven and nine predicted transmembrane regions depending on the program used and is localized to the ER. The N-terminus is thought to be located on the cytoplasmic side of the ER membrane. There has been no detailed functional analysis of specific domains within this protein as yet.

10.5 Molecular pathology and genotype–phenotype correlations

Several reports have suggested a correlation between the severity of the clinical phenotype and the severity of the biochemical disturbance as judged by the level of serum 7DHC and/or cholesterol (Ryan *et al.*, 1998; Witsch-Baumgartner *et al.*, 2000; Yu *et al.*, 2000). Over 50 different causative mutations have been identified. IVS8–1G>C appears to be the most common allele acounting for between 29 and 66% of disease-causing chromosomes (Waterham *et al.*, 2000; Witsch-Baumgartner *et al.*, 2000; Yu *et al.*, 2000). This is a null allele that results in a 134 bp insertion in the mature RNA molecular between exon 8 and 9. Witsch-Baumgartner *et al.*, (2000) presented the most comprehensive analysis of the muta-tional spectrum of SLOS patients. They divided 40 different mutations detected in 84 patients into four groups: null mutations (0), missense mutations affecting a transmembrane domain (TM), missense mutations affecting the 4th cytoplasmic loop (4L) and missense mutations affecting the C-terminus (CT). When these four classes of alleles compared to either the clinical severity score or the 7DHC level the implied order from most severe to least severe was 0>4L>TM>CT. The reason that mutations in the 4th cytoplasmic loop are so deleterious is not clear.

11. 24 sterol reductase deficiency

11.1 Summary

Inheritance	Autosomal recessive (presumed)
Gene	Unknown
Map location	20
OMIM number	602398
Clinical features	Cleft palate; thickening of the alveolus; prominent palatal rugae; gingival nodules; total anomalous pulmonary venous drainage; pulmonary hypoplasia; renal hypoplasia; splenomegaly; unrotated small bowel; hypoplasia of the corpus callosum; dilation of the ventricular system; generalized osteosclerosis; ichthyosis

11.2 Clinical features

To date, only one case of desmosterolosis has been reported (Clayton *et al.*, 1996; FitzPatrick *et al.*, 1998). This female infant was born at 34-weeks-gestation and died at 1 hour-of-age due to extreme respiratory insufficiency. Macrocephaly was apparent with frontal bossing, depressed nasal bridge and low set, posteriorly angulated ears. She had a cleft of the soft palate and thickening of the alveolus and prominent palatal rugae. Multiple gingival nodules were noted. Postmortem examination revealed total anomalous pulmonary venous drainage, pulmonary hypoplasia and renal hypoplasia. The brain showed an immature gyral pattern with poor development of the corpus callosum and gross dilation of the ventricular system. Generalized osteosclerosis was apparent radiologically. A marked phenotypic overlap with Raine syndrome was noted. This is a neonatally lethal condition characterized by generalized osteosclerosis, gingival nodules and severe nasal hypoplasia. However, postmortem material from a case of Raine syndrome showed no accumulation of desmosterol indicating that these are not allelic conditions.

11.3 Biochemical features

GC-MS analysis of all tissues from this infant showed accumulation of desmosterol. Analysis of the parents' plasma sterols showed elevated concentrations of desmosterol consistent with carrier status. On the basis of these biochemical data it was concluded that an autosomal recessive deficiency of Δ^{24}-reductase activity was the likely cause of the developmental anomalies in this case. A cell-line could not be established from this case and thus no enzymology has been possible. A specific inhibitor of this enzyme, triparanol, had been studied and was known to be highly teratogenic in rodents. Interestingly, the phenotype found in embryos exposed to triparinol overlapped with that seen in the human infant. Triparanol was marketed in the 1960s as an anti-hyperlipidaemic drug but withdrawn because the formation of cataracts was a common side effect (Kirby, 1967).

11.4 Molecular genetics

This condition is likely to be caused by mutations in the gene encoding Δ^{24}-sterol reductase (D24SR). The *D24SR* gene has not been cloned in any species but has been localized to human chromosome 20 using somatic cell hybrids (Croce *et al.*, 1974).

12. The mechanism of developmental pathology

Although disturbances in cholesterol metabolism have been firmly associated with developmental diseases it remains unclear how abnormal cholesterol synthesis causes malformations. There are two general mechanisms by which aberrant cholesterol synthesis may result in developmental pathology; a relative deficiency of cholesterol (deficiency) and a relative excess of the sterol precursor (intoxication). These are not mutually exclusive and have proven difficult to separate experimentally. For example, in rodents treating the dam with cholesterol prevents the teratogenic effects of specific inhibitors of late-stage cholesterol biosynthesis (*Table 5*; Batta and Salen, 1998). This has been taken as evidence that cholesterol deficiency is the primary defect. However, such treatment both increases serum cholesterol and reduces the precursor through feedback inhibition at the level of HMG CoA reductase.

As cholesterol is a major component of the plasma membrane either deficiency or intoxication may result in a generally malign effect on this structure. In cultured cells, desmosterol has been shown to significantly increase membrane fluidity (Ashraf *et al.*, 1984; Chalikian and Barchi, 1982; Fiehn and Seiler, 1975; Heiniger and Marshall, 1979; Hitzemann and Johnson, 1983; Schroeder *et al.*, 1982). *In vivo* such a change may alter both the movement of embryonic cells and cell–cell interaction. Given the likely role of cholesterol in determining the width of membranes it is also possible that altering the sterol content may lead to the aberrant functioning or miss-targeting of transmembrane proteins (Bretscher and Munro, 1993). One target for future study are recently identified detergent insoluble microdomains within cell membranes called functional rafts (Bagnat *et al.*, 2000; Jacobson and Dietrich, 1999; Kabouridis *et al.*, 2000; Kurzchalia and Parton, 1999; Nusrat *et al.*, 2000; Rietveld *et al.*, 1999; Rietveld and Simons, 1998; Verkade and Simons, 1997; Zager, 2000). These rafts are enriched in sterols,

Table 5. Specific inhibitors of late stage cholesterol biosynthesis

Name	Enzyme inhibited
Triparanol	D24SR
U18666A	D24SR
BM 15,766	7DHCR
AY9944	C8–C7 isomerase

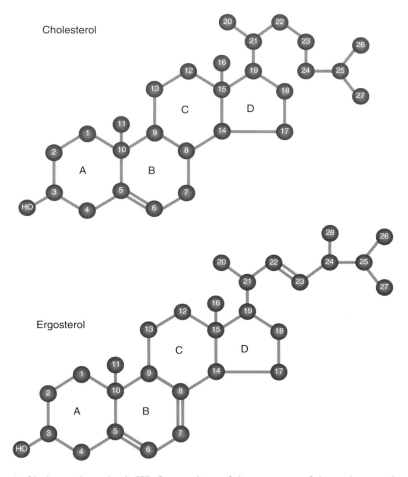

Figure 3. Cholesterol synthesis III. Comparison of the structure of the major vertebrate sterol cholesterol and the major yeast sterol ergosterol.

sphingolipids and glycosylphosphatidylinositol. Rafts associate with sterol- and complex lipid-linked proteins, including hedgehog proteins that are post-translationally modified by the addition of a cholesterol molecule, and they have been postulated as mediators of cell polarization and signal transduction. It will be interesting to determine if the disruption of normal raft function may be the basis of the developmental anomalies seen in both teratogen-mediated and genetic deficiency of late-stage endogenous cholesterol biosynthesis.

13. Future research

Given the number of 'orphan reactions' it is likely that there are other human malformation syndromes will be associated with accumulation of abnormal sterol precursors. Although it is difficult to predict exactly what phenotypic traits to expect the approach pioneered by Richard Kelley is very encouraging. He took an individual phenotypic trait (chondrodysplasia punctata CDP) that was common to the first two late-stage cholesterol biosynthetic defects to be described; SLOS and desmosterolosis. Systematic testing of different forms of chondrodysplasia punctata allowed the biochemical and genetic defect in CDPX2 and CHILD syndrome to be identified. Other traits that may be worth a similar sytematic analysis are thickened alveolar ridges, ichthyosis, holoprosencephaly and osteoscleosis. The stability of sterol compounds and the relative ease of chromatographic separation mean that a low threshold for biochemical testing should be encouraged.

One of the challenges in human genetics is to identify all genes encoding critical enzymatic steps in lanosterol-to-cholesterol biosynthesis. One promising approach is the in silico identification of orthologous human sequences using yeast ERG proteins. Proof that an individual sequence is truly involved in this pathway could utilise homozygous somatic cell knock-outs (Sedivy and Dutriaux, 1999) followed by sterol analysis of the cultured cells. This approach is faciltated by the almost ubiquitous nature of the cholesterol biosynthetic pathway and the ability to culture human cells in the absence of exogenous cholesterol.

References

Anderegg, R.J., Betz, R., Carr, S.A., Crabb, J.W. and Duntze, W. (1988) Structure of Saccharomyces cerevisiae mating hormone a-factor. Identification of S-farnesyl cysteine as a structural component. *J. Biol. Chem.* **263**: 18236–18240.

Ashraf, J., Feix, J.B. and Butterfield, D.A. (1984) Membrane fluidity and myotonia: effects of cholesterol and desmosterol on erythrocyte membrane fluidity in rats with 20,25-diazacholesterol-induced myotonia and on phospholipid liposomes. *Biosci. Rep.* **4**: 115–120.

Bagnat, M., Keranen, S., Shevchenko, A. and Simons, K. (2000) Lipid rafts function in biosynthetic delivery of proteins to the cell surface in yeast. *Proc. Natl Acad. Sci. USA* **97**: 3254–3259.

Batta, A.K. and Salen, G. (1998) Abnormal cholesterol biosynthesis produced by AY 9944 in the rat leads to skeletal deformities similar to the Smith-Lemli-Opitz syndrome [editorial; comment]. *J. Lab. Clin. Med.* **131**: 192–193.

Berger, R., Smit, G.P., Schierbeek, H., Bijsterveld, K. and le Coultre, R. (1985) Mevalonic aciduria: an inborn error of cholesterol biosynthesis? *Clin. Chim. Acta.* **152**: 219–222.

Biardi, L. and Krisans, S.K. (1996) Compartmentalization of cholesterol biosynthesis. Conversion of mevalonate to farnesyl diphosphate occurs in the peroxisomes. *J. Biol. Chem.* **271**: 1784–1788.

Braverman, N., Lin, P., Moebius, F.F. *et al.* (1999) Mutations in the gene encoding 3 beta-hydroxysteroid-delta 8, delta 7- isomerase cause X-linked dominant Conradi-Hunermann syndrome. *Nat. Genet.* **22**: 291–294.

Bretscher, M.S. and Munro, S. (1993) Cholesterol and the golgi apparatus. *Science* **261**: 1280–1281.

Brown, M.S. and Goldstein, J.L. (1986) A receptor-mediated pathway for cholesterol homeostasis. *Science* **232**: 34–47.

Brown, M.S. and Goldstein, J.L. (1996) Heart attacks: gone with the century? [editorial] [see comments]. *Science* **272**: 629.

Brown, M.S. and Goldstein, J.L. (1999) A proteolytic pathway that controls the cholesterol content of membranes, cells, and blood. *Proc. Natl Acad. Sci. USA* **96**: 11041–11048.

Brown, M.S., Ye, J., Rawson, R.B. and Goldstein, J.L. (2000) Regulated intramembrane proteolysis: a control mechanism conserved from bacteria to humans. *Cell* **100**: 391–398.

Brunnsgard, L., Ericsson, J. and Dallner, G. (1989) Intracellular localization of cholesterol biosynthesis. *Acta. Chem. Scand.* **43**: 500–502.

Casey, P.J., Solski, P.A., Der, C.J. and Buss, J.E. (1989) p21ras is modified by a farnesyl isoprenoid. *Proc. Natl Acad. Sci. USA* **86**: 8323–83227.

Chalikian, D.M. and Barchi, R.L. (1982) Membrane desmosterol and the kinetics of the sarcolemmal Na+,K+-ATPase in myotonia induced by 20,25-diazacholesterol. *Exp. Neurol.* **77**: 578–589.

Chambon, C., Ladeveze, V., Servouse, M., Blanchard, L., Javelot, C., Vladescu, B. and Karst, F. (1991) Sterol pathway in yeast. Identification and properties of mutant strains defective in mevalonate diphosphate decarboxylase and farnesyl diphosphate synthetase. *Lipids* **26**: 633–636.

Clayton, P., Mills, K., Keeling, J. and FitzPatrick, D. (1996) Desmosterolosis: a new inborn error of cholesterol biosynthesis [letter; comment]. *Lancet* **348**: 404.

Croce, C.M., Kieba, I., Koprowski, H., Molino, M. and Rothblat, G.H. (1974) Restoration of the conversion of desmosterol to cholesterol in L-cells after hybridization with human fibroblasts. *Proc. Natl Acad. Sci. USA* **71**: 110–113.

Derry, J.M., Gormally, E., Means, G.D. *et al.* (1999) Mutations in a delta 8-delta 7 sterol isomerase in the tattered mouse and X-linked dominant chondrodysplasia punctata. *Nat. Genet.* **22**: 286–290.

Dietschy, J.M., Turley, S.D. and Spady, D.K. (1993) Role of liver in the maintenance of cholesterol and low density lipoprotein homeostasis in different animal species, including humans. *J. Lipid. Res.* **34**: 1637–1659.

Edidin, D.V., Esterly, N.B., Bamzai, A.K. and Fretzin, D.F. (1977) Chondrodysplasia punctata. Conradi-Hunermann syndrome. *Arch. Dermatol.* **113**: 1431–1434.

Elias, E.R. and Irons, M. (1995) Abnormal cholesterol metabolism in Smith-Lemli-Opitz syndrome. *Curr. Opin. Pediatr.* **7**: 710–714.

Emami, S., Rizzo, W.B., Hanley, K.P., Taylor, J.M., Goldyne, M.E. and Williams, M.L. (1992) Peroxisomal abnormality in fibroblasts from involved skin of CHILD syndrome. Case study and review of peroxisomal disorders in relation to skin disease. *Arch. Dermatol.* **128**: 1213–1222.

Fiehn, W. and Seiler, D. (1975) Alteration of erythrocyte (see article) -ATPase by replacement of cholesterol by desmosterol in the membrane. *Experientia* **31**: 773–775.

Fitzky, B.U., Witsch-Baumgartner, M., Erdel, M. *et al.* (1998) Mutations in the Delta7-sterol reductase gene in patients with the Smith-Lemli-Opitz syndrome. *Proc. Natl Acad. Sci. USA* **95**: 8181–8186.

FitzPatrick, D.R., Keeling, J.W., Evans, M.J. *et al.* (1998) Clinical phenotype of desmosterolosis. *Am. J. Med. Genet.* **75**: 145–152.

Gachotte, D., Barbuch, R., Gaylor, J., Nickel, E. and Bard, M. (1998) Characterization of the Saccharomyces cerevisiae ERG26 gene encoding the C-3 sterol dehydrogenase (C-4 decarboxylase) involved in sterol biosynthesis [published erratum appears in *Proc. Natl Acad. Sci. USA* 1999 Feb. 16;**96**(4): 1810]. *Proc. Natl Acad. Sci. USA* **95**: 13794–13799.

Goldstein, J.L. and Brown, M.S. (1990) Regulation of the mevalonate pathway. *Nature* **343**: 425–430.

Grange, D.K., Kratz, L.E., Braverman, N.E. and Kelley, R.I. (2000) CHILD syndrome caused by deficiency of 3beta-hydroxysteroid-delta8, delta7-isomerase [letter] [see comments]. *Am. J. Med. Genet.* **90**: 328–335.

Hamaguchi, T., Bondar, G., Siegfried, E. and Penneys, N.S. (1995) Cutaneous histopathology of Conradi-Hunermann syndrome. *J. Cutan. Pathol.* **22**: 38–41.

Hanner, M., Moebius, F.F., Weber, F., Grabner, M., Striessnig, J. and Glossmann, H. (1995) Phenylalkylamine Ca2+ antagonist binding protein. Molecular cloning, tissue distribution, and heterologous expression. *J. Biol. Chem.* **270**: 7551–7557.

Happle, R. (1979) X-linked dominant chondrodysplasia punctata. Review of literature and report of a case. *Hum. Genet.* **53**: 65–73.

Happle, R., Koch, H. and Lenz, W. (1980) The CHILD syndrome. Congenital hemidysplasia with ichthyosiform erythroderma and limb defects. *Eur. J. Pediatr.* **134**: 27–33.

Hashimoto, K., Topper, S., Sharata, H. and Edwards, M. (1995) CHILD syndrome: analysis of abnormal keratinization and ultrastructure [see comments]. *Pediatr. Dermatol.* **12**: 116–129.

Heiniger, H.J. and Marshall, J.D. (1979) Pinocytosis in L cells: its dependence on membrane sterol and the cytoskeleton. *Cell Biol. Int. Rep.* **3**: 409–420.

Hitzemann, R.J. and Johnson, D.A. (1983) Developmental changes in synaptic membrane lipid composition and fluidity. *Neurochem. Res.* **8**: 121–131.

Hoffmann, G., Gibson, K.M., Brandt, I.K., Bader, P.I., Wappner, R.S. and Sweetman, L. (1986) Mevalonic aciduria–an inborn error of cholesterol and nonsterol isoprene biosynthesis. *N. Engl. J. Med.* **314**: 1610–1614.

Hoffmann, G.F., Charpentier, C., Mayatepek, E. *et al.* (1993) Clinical and biochemical phenotype in 11 patients with mevalonic aciduria. *Pediatrics* **91**: 915–921.

Honda, A., Tint, G.S., Salen, G., Batta, A.K., Chen, T.S. and Shefer, S. (1995) Defective conversion of 7-dehydrocholesterol to cholesterol in cultured skin fibroblasts from Smith-Lemli-Opitz syndrome homozygotes. *J. Lipid. Res.* **36**: 1595–1601.

Irons, M., Elias, E.R., Salen, G., Tint, G.S. and Batta, A.K. (1993) Defective cholesterol biosynthesis in Smith-Lemli-Opitz syndrome [letter] [see comments]. *Lancet* **341**: 1414.

Jacobson, K. and Dietrich, C. (1999) Looking at lipid rafts? [see comments]. *Trends Cell. Biol.* **9**: 87–91.

Kabouridis, P.S., Janzen, J., Magee, A.L. and Ley, S.C. (2000) Cholesterol depletion disrupts lipid rafts and modulates the activity of multiple signaling pathways in T lymphocytes. *Eur. J. Immunol.* **30**: 954–963.

Karst, F. and Lacroute, F. (1977) Ertosterol biosynthesis in Saccharomyces cerevisiae: mutants deficient in the early steps of the pathway. *Mol. Gen. Genet.* **154**: 269–277.

Kelley, R.I. and Hennekam, R.C. (2000) The Smith-Lemli-Opitz syndrome. *J. Med. Genet.* **37**: 321–335.

Kirby, T.J. (1967) Cataracts produced by triparanol. (MER-29). *Trans. Am. Ophthalmol. Soc.* **65**: 494–543.

Konig, A., Happle, R., Bornholdt, D., Engel, H. and Grzeschik, K.H. (2000) Mutations in the NSDHL gene, encoding a 3beta-hydroxysteroid dehydrogenase, cause CHILD syndrome. *Am. J. Med. Genet.* **90**: 339–346.

Kurzchalia, T.V. and Parton, R.G. (1999) Membrane microdomains and caveolae. *Curr. Opin. Cell. Biol.* **11**: 424–431.

Labit-Le Bouteiller, C., Jamme, M.F., David, M. *et al.* (1998) Antiproliferative effects of SR31747A in animal *cell* lines are mediated by inhibition of cholesterol biosynthesis at the sterol isomerase step. *Eur. J. Biochem.* **256**: 342–349.

Lange, Y., Swaisgood, M.H., Ramos, B.V. and Steck, T.L. (1989) Plasma membranes contain half the phospholipid and 90% of the cholesterol and sphingomyelin in cultured human fibroblasts. *J. Biol. Chem.* **264**: 3786–3793.

Levin, M.L., Chatterjee, A., Pragliola, A. *et al.* (1996) A comparative transcription map of the murine bare patches (Bpa) and striated (Str) critical regions and human Xq28. *Genome Res.* **6**: 465–477.

Liu, X.Y., Dangel, A.W., Kelley, R.I. *et al.* (1999) The gene mutated in bare patches and striated mice encodes a novel 3beta-hydroxysteroid dehydrogenase. *Nat. Genet.* **22**: 182–187.

Lorenz, R.T. and Parks, L.W. (1987) Regulation of ergosterol biosynthesis and sterol uptake in a sterol- auxotrophic yeast. *J. Bacteriol.* **169**: 3707–3711.

Moebius, F.F., Reiter, R.J., Hanner, M. and Glossmann, H. (1997) High affinity of sigma 1-binding sites for sterol isomerization inhibitors: evidence for a pharmacological relationship with the yeast sterol C8-C7 isomerase. *Br. J. Pharmacol.* **121**: 1–6.

Moebius, F.F., Soellner, K.E.M., Fiechtner, B., Huck, C.W., Bonn, G. and Glossmann, H. (1999) Histidine77, glutamic acid81, glutamic acid123, threonine126, asparagine194, and tryptophan197 of the human emopamil binding protein are required for in vivo sterol delta 8-delta 7 isomerization. *Biochemistry* **38**: 1119–1127.

Nes, W.D., Janssen, G.G., Crumley, F.G., Kalinowska, M. and Akihisa, T. (1993) The structural requirements of sterols for membrane function in *Saccharomyces cerevisiae.* *Arch. Biochem. Biophys.* **300**: 724–733.

Nusrat, A., Parkos, C.A., Verkade, P. *et al.* (2000) Tight junctions are membrane microdomains. *J. Cell Sci.* **113**: 1771–1781.

Opitz, J.M. (1999) RSH (so-called Smith-Lemli-Opitz) syndrome. *Curr. Opin. Pediatr.* **11**: 353–362.

Poll-The B.T., Frenkel, J., Houten, S.M. *et al.* (2000) Mevalonic aciduria in 12 unrelated patients with hyperimmunoglobulinaemia D and periodic fever syndrome [In Process Citation]. *J. Inherit. Metab. Dis.* **23**: 363–366.

Potter, D. and Miziorko, H.M. (1997) Identification of catalytic residues in human mevalonate kinase. *J. Biol. Chem.* **272**: 25449–25454.

Potter, D., Wojnar, J.M., Narasimhan, C. and Miziorko, H.M. (1997) Identification and functional characterization of an active-site lysine in mevalonate kinase. *J. Biol. Chem.* **272**: 5741–5746.

Reinhart, M.P., Billheimer, J.T., Faust, J.R. and Gaylor, J.L. (1987) Subcellular localization of the enzymes of cholesterol biosynthesis and metabolism in rat liver. *J. Biol. Chem.* **262**: 9649–9655.

Rietveld, A., Neutz, S., Simons, K. and Eaton, S. (1999) Association of sterol- and glycosylphosphatidylinositol-linked proteins with Drosophila raft lipid microdomains. *J. Biol. Chem.* **274**: 12049–12054.

Rietveld, A. and Simons, K. (1998) The differential miscibility of lipids as the basis for the formation of functional membrane rafts. *Biochim. Biophys. Acta.* **1376**: 467–479.

Ryan, A.K., Bartlett, K., Clayton, P. *et al.* (1998) Smith-Lemli-Opitz syndrome: a variable clinical and biochemical phenotype. *J. Med. Genet.* **35**: 558–565.

Schroeder, F., Fontaine, R.N. and Kinden, D.A. (1982) LM fibroblast plasma membrane subfractionation by affinity chromatography on con A-sepharose. *Biochim. Biophys. Acta.* **690**: 231–242.

Seabra, M.C., Brown, M.S., Slaughter, C.A., Sudhof, T.C. and Goldstein, J.L. (1992) Purification of component A of Rab geranylgeranyl transferase: possible identity with the choroideremia gene product. *Cell* **70**: 1049–1057.

Sedivy, J.M. and Dutriaux, A. (1999) Gene targeting and somatic *cell* genetics-a rebirth or a coming of age? *Trends. Genet.* **15**: 88–90.

Servouse, M. and Karst, F. (1986) Regulation of early enzymes of ergosterol biosynthesis in *Saccharomyces cerevisiae*. *Biochem. J.* **240**: 541–547.

Tint, G.S. (1993) Cholesterol defect in Smith-Lemli-Opitz syndrome [letter]. *Am. J. Med. Genet.* **47**: 573–574.

Tint, G.S., Irons, M., Elias, E.R., Batta, A.K., Frieden, R., Chen, T.S. and Salen, G. (1994) Defective cholesterol biosynthesis associated with the Smith-Lemli-Opitz syndrome [see comments]. *N. Engl. J. Med.* **330**: 107–113.

Verkade, P. and Simons, K. (1997) Robert Feulgen Lecture 1997. Lipid microdomains and membrane trafficking in mammalian cells. *Histochem. Cell Biol.* **108**: 211–220.

Wassif, C.A., Maslen, C., Kachilele-Linjewile, S. *et al.* (1998) Mutations in the human sterol delta7-reductase gene at 11q12–13 cause Smith-Lemli-Opitz syndrome. *Am. J. Hum. Genet.* **63**: 55–62.

Waterham, H.R., Oostheim, W., Romeijn, G.J., Wanders, R.J. and Hennekam, R.C. (2000) Incidence and molecular mechanism of aberrant splicing owing to a G→C splice acceptor site mutation causing Smith-Lemli-Opitz syndrome [letter]. *J. Med. Genet.* **37**: 387–389.

Waterham, H.R., Wijburg, F.A., Hennekam, R.C. *et al.* (1998) Smith-Lemli-Opitz syndrome is caused by mutations in the 7- dehydrocholesterol reductase gene [see comments]. *Am. J. Hum. Genet.* **63**: 329–338.

Witsch-Baumgartner, M., Fitzky, B.U., Ogorelkova, M. *et al.* (2000) Mutational spectrum in the Delta7-sterol reductase gene and genotype–phenotype correlation in 84 patients with Smith-Lemli-Opitz syndrome. *Am. J. Hum. Genet.* **66**: 402–412.

Yu, H., Lee, M.H., Starck, L., Elias, E.R. *et al.* (2000) Spectrum of Delta(7)-dehydrocholesterol reductase mutations in patients with the Smith-Lemli-Opitz (RSH) syndrome [In Process Citation]. *Hum. Mol. Genet.* **9**: 1385–1391.

Zager, R.A. (2000) Plasma membrane cholesterol: a critical determinant of cellular energetics and tubular resistance to attack. *Kidney Int.* **58**: 193–205.

12

Mutations in the human *HOX* genes

Frances R. Goodman

1. Introduction

The HOX proteins are a closely related family of transcription factors which play a fundamental role in morphogenesis during embryonic development. First identified in 1978 in the fruit fly *Drosophila* (Lewis, 1978), they were later found to be structurally and functionally conserved throughout the animal kingdom (McGinnis and Krumlauf, 1992). Whereas *Drosophila* has just eight *Hox* genes arranged in a single cluster, humans have 39 *HOX* genes arranged in four clusters named *HOXA* through *HOXD*, each located on a different chromosome (Krumlauf, 1994), as shown in *Figure 1*. These genes are all thought to have evolved from a single prototypic *Hox* gene by a series of tandem duplication events followed by divergence, giving rise first to a single ancestral *Hox* cluster and later to the four separate clusters found in most vertebrates (Carroll, 1995; Kenyon, 1994). Vertebrate *Hox* genes are also classified into 13 different subsets known as 'paralogous groups', on the basis of their sequence similarity and relative position within the clusters. No one cluster contains a representative from all 13 paralogous groups, as different genes were lost from the different clusters early in their separate evolution, but in any given cluster the same subsets have been retained in all vertebrates (Krumlauf, 1994).

HOX proteins help pattern the developing embryo along both the primary (head-to-tail) and secondary (limb and genital bud) body axes. They are therefore important in the development of the central nervous system, axial skeleton, gut and urogenital tract, as well as the limbs and external genitalia (Krumlauf, 1994; Mark *et al.*, 1997; McGinnis and Krumlauf, 1992).

Remarkably, the order of the genes within each cluster corresponds closely to their temporal and spatial expression patterns during development. Thus, genes at the 3' end of each cluster are expressed early, in anterior and proximal regions, whereas genes at the 5' end of each cluster are expressed later, in more posterior and distal regions. This phenomenon, sometimes termed 'temporal and spatial colinearity', implies the existence of cluster-wide regulatory mechanisms for co-ordinating *HOX* gene expression (Kondo and Duboule, 1999). At the molecular

Genotype to Phenotype second edition, edited by S. Malcolm and J. Goodship.

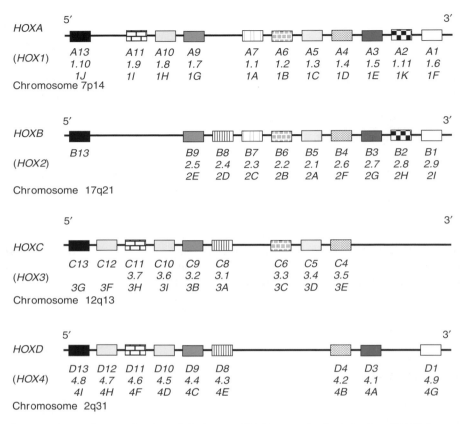

Figure 1. Genomic organization of the human *HOX* genes. Each of the four *HOX* clusters contains 9–11 genes, all orientated in same direction. Current nomenclature (in bold type) and previous nomenclature is given beneath the boxes representing the genes. Different shading is used for each of the 13 paralogous groups.

level, HOX proteins are believed to act by controlling the transcription of specific sets of target genes. Few of these targets have yet been identified, but they are likely to be genes with roles in basic cellular processes, such as proliferation, differentiation, adhesion, migration and death (Graba *et al.*, 1997). All HOX proteins bind specific DNA sequences near their target genes via a highly conserved 60 amino acid DNA-binding domain called the 'homeodomain', which is encoded by a 180-bp sequence element called the 'homeobox'. X-ray crystallography studies of homeodomains complexed with their target DNA show that the homeodomain consists of a flexible N-terminal arm, followed by three alpha-helices, the third of which, the 'recognition helix', sits in the major groove of the DNA and makes base-specific contacts with it (Gehring *et al.*, 1994; Wolberger *et al.*, 1991). Most if not all HOX proteins are also now thought to interact with specific DNA-binding partners, forming multimeric protein complexes which direct transcriptional activation or repression of their targets (Mann and Affolter, 1998; Mann and Chan, 1996).

Until recently, it was widely believed that *HOX* gene mutations would not be found in humans (Sharpe, 1996). Many researchers thought that mutations in such important genes would cause extremely severe developmental defects (as they do in *Drosophila*), resulting in early intrauterine death. Others pointed to the probable functional redundancy among vertebrate *HOX* genes, suggesting that mutations in any one *HOX* gene would cause only minor malformations or no phenotype at all. In fact mice carrying targeted mutations in individual *Hox* genes were subsequently found to have definite but generally non lethal abnormalities (Favier and Dollé, 1997; Krumlauf, 1994), but it was not until 1996 that a human malformation syndrome was first shown to be caused by mutations in a *HOX* gene. This syndrome was synpolydactyly, and the underlying gene was *HOXD13* (Muragaki *et al.*, 1996). In the following year, another human malformation syndrome, hand–foot–genital syndrome (HFGS) was shown to be caused by a mutation in a closely-related gene, *HOXA13* (Mortlock and Innis, 1997). *HOXD13* and *HOXA13* are both located at the extreme 5′ ends of their respective clusters and are important in the development of the distal limbs and lower urogenital tract (Innis, 1997). This chapter first describes the range of mutations identified to date in these two genes and then considers how they may act. It concludes by discussing the likelihood that additional *HOX* gene mutations remain to be found, a possibility strengthened by the very recent discovery of a novel mutation in *HOXA11* in patients with a combination of limb and hematological abnormalities (Thompson and Nguyen, 2000).

2. Synpolydactyly (SPD) and *HOXD13*

SPD is a rare dominantly-inherited congenital limb malformation in which there is a distinctive combination of syndactyly (joined digits) and polydactyly (extra digits). Typically, patients have a soft tissue web between their third and fourth fingers and between their fourth and fifth toes, within which lies an extra finger or toe (*Figure 2a–e*). Incomplete penetrance (a normal phenotype in some obligate mutation carriers) and variable expressivity (differences in the severity of the phenotype in affected individuals) are common. Thus, from one to all four limbs can be involved, and the severity of involvement ranges from partial soft tissue webbing to virtually complete reduplication of a digit.

The clue to the molecular basis of SPD came from linkage studies of a remarkable family from an isolated village in rural Turkey with 182 living affected individuals, in which the condition could be traced back over seven generations, spanning at least 140 years (Sayli *et al.*, 1995). These studies mapped the SPD locus to chromosome 2q31, where the *HOXD* gene cluster is located (Sarfarazi *et al.*, 1995), and mutations were then identified in three affected American families in the most 5′ *HOXD* gene, *HOXD13* (Muragaki *et al.*, 1996). The mutations in question, moreover, turned out to be highly unusual. In normal individuals, exon one of *HOXD13* contains a series of triplet repeats (*Figure 3a*) which encode an N-terminal 15-residue polyalanine tract (*Figure 3b*). The three affected American families carried different-sized expansions of these repeats, resulting in an additional seven, eight and 10 alanine residues respectively (*Figure 3e*). Similar expansions have subsequently been reported in 22 further SPD families (Akarsu *et al.*,

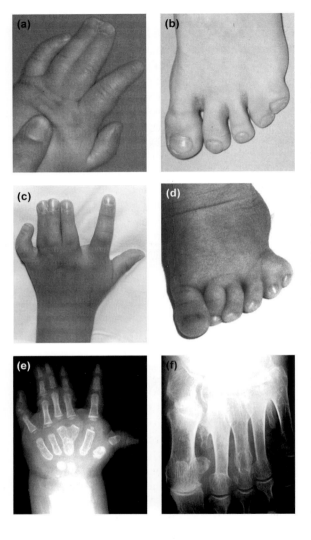

Figure 2. Limb abnormalities caused by mutations in *HOXD13*. (a) Syndactyly of the third and fourth fingers in a child heterozygous for a 14-bp deletion in exon 1 (Δ2 in *Figure 3*). (b) Syndactyly of the fourth and fifth toes in a child heterozygous for a 7-residue polyalanine tract expansion. (c) Duplicated third finger in a child heterozygous for a 14-residue polyalanine tract expansion. (d) Duplicated fifth toe in a child heterozygous for a 9-residue polyalanine tract expansion. (e) Radiograph showing duplicated third finger, broad medially-deviated thumb, incurving fifth finger and small middle phalanges in a child heterozygous for a 14-residue polyalanine tract expansion. (f) Radiograph showing partial duplication of the second and fourth metatarsals and a broad big toe in an adult heterozygous for a 1-bp deletion in exon 2 (Δ1 in *Figure 3*).

1996; Baffico *et al.*, 1997; Goodman *et al.*, 1997), and have been shown to remain stable in size over at least seven generations.

A study of families with different-sized expansions has since found that both penetrance and severity of phenotype correlate with expansion size (Goodman *et al.*, 1997). Thus, the larger the expansion, the greater the number of limbs involved and the more complete the extent of the digit duplication. Affected individuals from a family with a 14-alanine expansion, the largest so far discovered, have the most severe limb phenotype (*Figure 2c, e*), including abnormalities of the thumbs, big toes and wrist bones, which are not usually involved in SPD. Affected males from this family also have hypospadias, a urogenital malformation in which the urethra (the tube that carries urine out of the bladder) opens onto the ventral surface of the penis instead of at its tip. This malformation is thought to occur when the urogenital folds fail to fuse during urethral development, and is

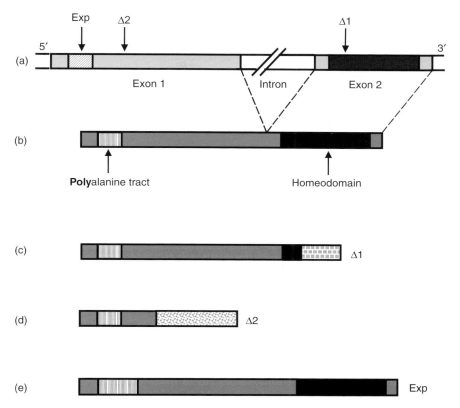

Figure 3. Wild-type and mutant HOXD13. (a) Genomic structure of *HOXD13* showing the two exons, one intron and sites of reported mutations. The diagonally shaded box in exon one represents the imperfect triplet repeat sequence encoding the polyalanine tract, while the solid box in exon two represents the homeobox. Δ = deletion, Exp = expansion. (b) Wild-type HOXD13 protein. (c) and (d) Predicted proteins resulting from a 1-bp deletion in the homeobox (Δ1) and a 14-bp deletion in exon one (Δ2), respectively. The former would contain the first 278 wild-type amino acids followed by 33 novel amino acids, and would lack the last 49 amino acids of the homeodomain. The latter would contain the first 107 wild-type amino acids followed by 115 novel amino acids, and would lack the entire homeodomain. (e) Predicted protein resulting from 7- to 14-residue expansions of the polyalanine tract.

consistent with the observation that *HOXD13* is expressed in the developing urogenital folds in mice (Dollé *et al.*, 1991).

The same genotype–phenotype correlation is also apparent in the very few SPD homozygotes who have been reported. Thus, a patient homozygous for a 7-alanine expansion has very short digits, syndactyly between the third, fourth and fifth fingers, and mildly abnormal wrist bones (Muragaki *et al.*, 1996), but seven patients homozygous for a 9-alanine expansion have a much more severe phenotype, with so-called 'cat's paw' hands containing six to eight rudimentary fingers, complete soft tissue syndactyly and marked malformation of all the wrist bones (Akarsu *et al.*, 1995).

A subtly different form of SPD found in two unrelated families is caused by two different intragenic deletions in *HOXD13* (Goodman *et al.*, 1998). All deletion carriers from both families share a novel foot malformation, whose most prominent feature is partial duplication of the second and sometimes also the fourth rays, with broadening of the big toes (*Figure 2e*). These abnormalities are not found in patients with SPD caused by HOXD13 polyalanine tract expansions. Synpolydactyly of the third and fourth fingers and the fourth and fifth toes also occurs in deletion carriers from these two families, but only at a reduced penetrance. The two deletions affect the homeobox and exon one, respectively (*Figure 3a*), in each case producing a frameshift followed by a long stretch of novel sequence and a premature stop. Both are predicted to result in truncated proteins, one (*Figure 3c*) containing only the first 278 amino acids of the wild-type protein and lacking the last 49 amino acids of the homeodomain, including the entire recognition helix, and the other (*Figure 3d*) containing only the first 107 amino acids of the wild-type protein and lacking the entire homeodomain.

3. Hand–foot–genital syndrome (HFGS) and *HOXA13*

HFGS is another rare dominantly-inherited condition, in which distal limb malformations are accompanied by malformations of the lower urogenital tract. In the hands and feet, the most striking abnormalities are small, proximally-placed thumbs and short, medially-deviated big toes (*Figure 4a–d*). There is also often incurving of the second and fifth fingers, shortening of the second to fifth toes, and delayed ossification, fusion and malformation of the wrist and ankle bones. These abnormalities are fully penetrant, bilateral and symmetrical, with little variation in severity. By contrast, the genital and urinary tract malformations show incomplete penetrance and variable severity. Genital abnormalities include hypospadias in males and double uterus with double cervix in females, which can lead to miscarriage, premature labor and still birth (Donnenfeld *et al.*, 1992; Halal, 1988; Stern *et al.*, 1970). Urinary abnormalities include malformation of the ureters (the tubes connecting the kidneys to the bladder), which can result in chronic infection and kidney failure (Donnenfeld *et al.*, 1992; Halal, 1988; Poznanski *et al.*, 1975).

The clue to the molecular basis of HFGS came from studies of a spontaneous mouse mutant, the *Hypodactyly* mouse (Hummel, 1970). Heterozygous *Hypodactyly* mice (*Hd/+*) have distal limb abnormalities virtually identical to those of HFGS patients, including small first digits, incurving or shortening of the second and fifth digits, and delayed ossification and fusion of many of the wrist and ankle bones. The *Hd* locus was mapped to the vicinity of the mouse *HoxA* gene cluster (Innis *et al.*, 1996), and the underlying mutation was found to be a 50-bp deletion in exon one of *Hoxa13* (Mortlock *et al.*, 1996), as shown in *Figure 5g*. The similarity between the limb abnormalities in *Hd/+* mice and HFGS patients then prompted an examination of the *HOXA13* gene in the family in which HFGS had first been described (Mortlock and Innis, 1997). Affected individuals turned out to be heterozygous for a nonsense mutation in the homeobox (*Figure 5a*), which is predicted to result in a truncated HOXA13 protein lacking

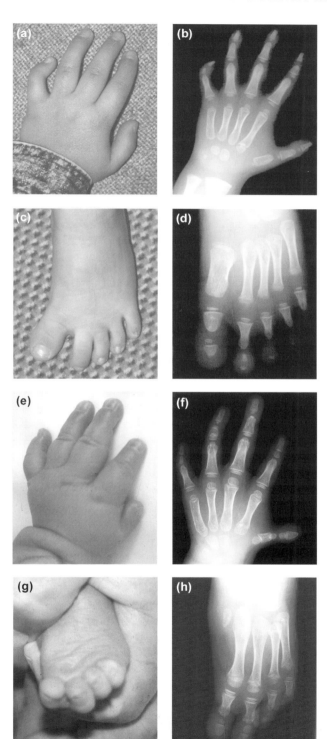

Figure 4. Limb abnormalities caused by mutations in *HOXA13*. (a) and (b) Small proximally-placed thumb, incurving second and fifth fingers with small middle phalanges, and delayed ossification of the wrist bones in a child heterozygous for a nonsense mutation in exon one (X4 in *Figure 5*). (c) and (d) Short medially-deviated big toe and small or absent middle and distal phalanges in toes two to five in the same child. (e) and (f) Extremely small thumb, short second to fifth fingers with small middle phalanges, and delayed ossification of the wrist bones in a child heterozygous for a missense mutation in the homeobox (M in *Figure 5*). (g) and (h) Absent big toe, rudimentary first metatarsal, small or absent middle and distal phalanges in toes two to five, and abnormal ankle bones in the same child.

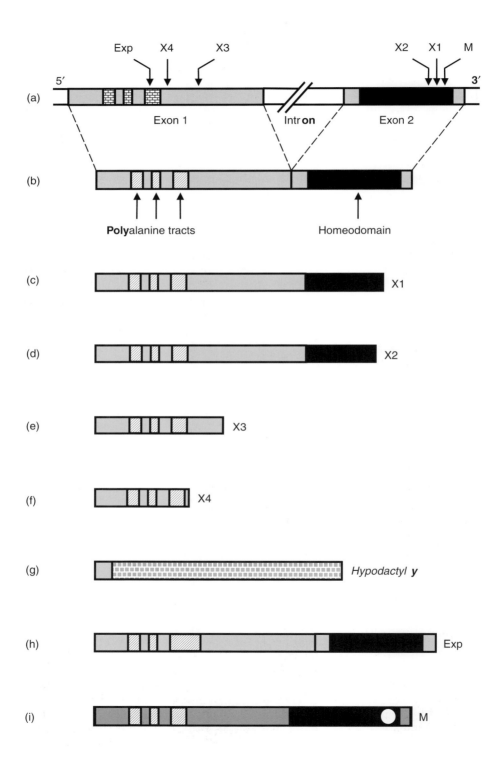

◀ **Figure 5.** Wild-type and mutant *HOXA13*. (a) Genomic structure of *HOXA13*, showing the two exons, one intron and sites of reported mutations. The three diagonally shaded boxes in exon one represent the three successive imperfect triplet repeat sequences encoding the three polyalanine tracts, while the solid box in exon two represents the homeobox. X = nonsense mutation, Exp = expansion, M = missense mutation. (b) Wild-type *HOXA13* protein. (c) to (f) Predicted proteins resulting from the four nonsense mutations. X1 and X2, both in the homeobox, would remove the last 20 and 24 amino acids, respectively. X3 and X4, both in exon one, would remove the last 193 and 253 amino acids, respectively. (g) Mutant Hoxa13 protein in the *Hypodactyly* mouse. The first 25 wild-type amino acids are followed by 275 novel amino acids. (h) Predicted protein resulting from the 8-residue expansion of the third polyalanine tract. (i) Predicted protein resulting from the missense mutation in the homeobox, which substitutes a histidine residue for asparagine 51 of the homeodomain.

---

the last 20 C-terminal amino acids, including three of the four key homeodomain residues responsible for contacting target DNA (*Figure 5c*).

Five additional mutations in *HOXA13* have recently been identified in families with HFGS (Goodman *et al.*, 2000), as shown in *Figure 5a*. Three are also nonsense mutations, one in the homeobox, just upstream of the original mutation, and two in exon one. These mutations are predicted to result in truncated proteins lacking the last 24, 193 and 253 amino acids respectively, including in the former case all four residues that directly contact DNA and in the latter two cases the entire homeodomain (*Figure 5d, e, f*). Interestingly, the fourth new mutation is a polyalanine tract expansion, very similar to those in HOXD13 that cause SPD. Exon one of *HOXA13* contains three successive triplet repeats (*Figure 5a*) encoding three successive N-terminal polyalanine tracts (*Figure 5b*), which in normal individuals are 14, 12 and 18 residues long, respectively. The fourth new mutation identified expands the last of these tracts by an additional eight alanine residues (*Figure 5h*) and has remained stable in size over at least three generations.

All the *HOXA13* mutations mentioned so far cause a typical HFGS phenotype, but the fifth new mutation results in limb abnormalities more severe than any previously reported, including extremely small thumbs, absent big toes and marked shortening of all the other digits (*Figure 4e–h*). This is a missense mutation in the homeobox (*Figure 5i*) which is predicted to replace asparagine 51 of the homeodomain by a histidine.

4. Hemizygosity for the *HOXA* and *HOXD* gene clusters

An interesting recent report describes a patient with a cytogenetically visible *de novo* interstitial deletion on chromosome 7p14, which encompasses the entire *HOXA* cluster (Devriendt *et al.*, 1999). This patient has distal limb and genital tract abnormalities typical of HFGS, which are most probably due to haploinsufficiency for *HOXA13*, together with malformations of the heart and soft palate similar to those found in mice homozygous for a targeted deletion in *Hoxa3* (Chisaka and Capecchi, 1991), which may well be due to haploinsufficiency for *HOXA3*.

The effects of hemizygosity for the *HOXD* cluster are less clear. At least 30 patients have been reported with cytogenetically visible deletions that involve chromosome 2q31, most of whom have multiple malformations, including in some cases limb and genital abnormalities (Boles *et al.*, 1995; Collins, 1996; Del Campo *et al.*, 1999; Nixon *et al.*, 1997; Slavotinek *et al.*, 1999). Six of these deletions have now been shown to remove the entire *HOXD* cluster (Collins, 1996; Del Campo *et al.*, 1999; Nixon *et al.*, 1997; Slavotinek *et al.*, 1999), but one, found in a patient with bilateral split foot, appears not to involve the cluster, suggesting that a gene underlying split hand and split foot may lie just outside the cluster (Boles *et al.*, 1995). In a family with a submicroscopic 2q31 deletion which removes the five most 5′ *HOXD* genes (*HOXD9-HOXD13*) along with the adjacent *EVX2* gene, but does not include any other gene in the region, affected individuals have a limb phenotype similar to SPD (Goodman *et al.*, 1999). This might reflect disturbed regulation of the remaining 3′ *HOXD* genes, but most likely results from haploinsufficiency for the six deleted genes, uncomplicated by the effects of haploinsufficiency for any neighboring genes.

5. Functional effects of the *HOXD13* and *HOXA13* mutations

The mutations so far discovered in *HOXD13* and *HOXA13* fall into three distinct classes: truncation mutations, polyalanine tract expansions and an amino acid substitution in the homeodomain, each of which is likely to act by a different mechanism.

5.1 Truncation mutations

The two intragenic deletions in *HOXD13* (*Figure 3c* and *d*) and the four nonsense mutations in *HOXA13* (*Figure 5c–f*) are predicted to result in proteins lacking part or all of the homeodomain and might therefore be expected to cause a loss of function. Similar mutations in other homeodomain proteins, such as PAX6 (Prosser and van Heyningen, 1998), EMX2 (Brunelli *et al.*, 1996), PITX2 (formerly RIEG) (Semina *et al.*, 1996) and HLXB9 (Ross *et al.*, 1998; Hagan *et al.*, 2000), are thought to act as null alleles, whether because the mutant protein is unstable or cannot bind target DNA. Moreover, as mentioned above, haploinsufficiency for *HOXA13* appears to cause typical HFGS (Devriendt *et al.*, 1999). The abnormalities in patients with the *HOXD13* and *HOXA13* truncation mutations differ, however, from those in the corresponding knock-out mice.

Mice with targeted disruptions of the homeobox of *HOXD13* have been generated by two different groups, with virtually identical results (Davis and Capecchi, 1996; Dollé *et al.*, 1993). Only one third to one half of heterozygotes have any abnormalities, and these consist of a rudimentary extra post-axial digit, usually in the forelimbs, sometimes accompanied by minor wrist bone defects. Homozygotes have more marked limb abnormalities, especially in the forelimbs, including delayed chondrification and ossification of almost all the distal limb bones, shortening of the digits, especially digits two and five, and, in about half, an extra post-axial digit. Homozygotes also have abnormalities of the lowest sacral vertebra and the internal anal sphincter, and males are infertile, with malformation of the penian bone and accessory sex glands (Dollé *et al.*, 1993; Kondo *et al.*, 1996; Podlasek *et al.*, 1997). Neither heterozygotes nor homozygotes, however, have the distinctive foot phenotype or the central synpolydactyly seen in humans with *HOXD13* truncation mutations.

Mice with a targeted disruption of the homeobox of *Hoxa13* and a targeted deletion that removes the entire *Hoxa13* gene have also been generated, and share a common phenotype (Fromental-Ramain *et al.*, 1996; Warot *et al.*, 1997). Heterozygotes have far milder limb abnormalities than humans with *HOXA13* truncation mutations. In the forelimbs there is merely occasional fusion of the bones of the first digit, while in the hindlimbs there is often malformation of the claw and distal bone of the first digit, with occasional soft tissue syndactyly between digits two and three. Homozygotes die *in utero* with severe urinary and genital tract malformations, including displaced ureters, absent bladder and abnormalities of the developing uterus and vagina, as well as premature stenosis of the umbilical arteries, which may explain the mutations' early lethality. In addition, they have absent first digits, with hypoplasia and webbing of digits two to five, and delayed chondrification of most distal limb bones. Interestingly, the limb phenotype in these mice is also less severe than that in *Hypodactyly* mice. *Hd/+* mice have limb abnormalities virtually identical to human HFGS patients, as described earlier. Most *Hd/Hd* mice die *in utero*, for reasons that are as yet unknown. Those that are born usually survive to adulthood but are infertile, due to penian bone defects in males and vaginal and cervical hypoplasia in females (Post and Innis, 1999b); they have only a single incompletely-formed digit on each paw (Mortlock *et al.*, 1996).

The phenotypic differences between humans with *HOXD13* and *HOXA13* truncation mutations and the corresponding knock-out mice can be explained in several ways. First, the truncated human proteins, although unable to bind DNA specifically, might nevertheless exert a deleterious functional effect through their remaining N-terminal portions. The *Drosophila* homeodomain protein *fushi tarazu*, for example, if expressed at very high levels, can regulate target genes even when its homeodomain has been partially deleted, because it retains the capacity to interact with DNA-binding partners through its N-terminal region (Copeland *et al.*, 1996). If this is also true of HOXD13 and HOXA13, however, the truncated proteins must be stably expressed and localize to the nucleus, and their effects must be mediated by the first 107 wild-type amino-acids of HOXD13 and the first 135 wild-type amino acids of HOXA13, respectively, as these are the only regions shared by the different truncated forms of the two proteins.

A second possibility is that the consequences of haploinsufficiency for *HOXD13* or *HOXA13* are rather different in humans and in mice, perhaps because of human/mouse differences in sensitivity to reduced gene dosage or in the role normally played by the two genes. The latter is especially likely in the developing female reproductive tract, which is anatomically different in mice and humans: instead of a single uterine cavity, normal female mice have two separate uterine horns which only join together at the level of the cervix. Differences of this kind now seem to be the explanation for the different phenotypes seen in humans and mice carrying loss-of-function mutations in another homeobox gene,

HLXB9 (Hagan *et al.*, 2000). The limb abnormalities in *Hypodactyly* mice are very similar to those in humans with HFGS, however, and are markedly more severe than those in mice with targeted *Hoxa13* mutations, even when each mouse mutation is crossed onto the genetic background of the other (Post and Innis, 1999a). The *Hypodactyly* mutant protein retains only the first 25 amino acids of Hoxa13, followed by 275 amino acids with no wild-type counterpart (*Figure 5g*). This protein is stable, localizes to the nucleus, and has been found to cause limb reduction defects when strongly overexpressed, but only in three out of 15 transgenic mice (Post *et al.*, 2000). A gain-of-function effect exerted by this protein may underlie the *Hypodactyly* phenotype, but if so, the consequences in mice appear to be very similar to those of haploinsufficiency for *HOXA13* in humans.

The final and perhaps the most likely alternative is that the human truncation mutations and the *Hypodactyly* mutation all act as null alleles, and that the targeted mouse mutations have not produced a straightforward loss of function. The insertion of a selectable marker cassette during the construction of targeted mouse mutations is known to be able to disrupt the expression of neighboring genes, especially if the genes in question are closely linked or clustered and share regulatory elements, as the *Hox* genes do (Olson *et al.*, 1996). This possibility could be explored by studying the site, timing and level of expression of the remaining *Hox* genes in mice with the targeted *Hoxd13* and *Hoxa13* mutations, and, in the longer term, by generating mice carrying targeted *Hoxd13* and *Hoxa13* mutations unaccompanied by additional regulatory sequences, using methods that eliminate the selection cassette, such as the Cre/*lox*P site-specific recombinase system.

5.2 Polyalanine tract expansions

As well as being the first human *HOX* gene mutations to be discovered, the polyalanine tract expansions in HOXD13 that cause SPD proved to be the first instance of a novel class of mutation. Similar pathological expansions were subsequently identified in HOXA13, as described above, as well as in two non homeodomain transcription factors, RUNX2 (formerly CBFA1) (Mundlos *et al.*, 1997) and ZIC2 (Brown *et al.*, 1998), which result in the malformation syndromes cleidocranial dysplasia and holoprosencephaly, respectively. In all four transcription factors, the normal polyalanine tracts are short (15–18 residues) and show little or no polymorphism in length, while the expansions are also short (7–15 extra residues) and meiotically stable. They thus differ sharply from the polyglutamine tract expansions found in neurodegenerative disorders such as Huntington's disease, as well as from the expansions of non coding triplet repeats found in fragile X syndrome, myotonic dystrophy and Friedrich's ataxia. In these cases, the normal triplet repeat sequences are long and show considerable polymorphism in length. Moreover, the expansions are many times longer than the normal repeat sequence and exhibit striking meiotic instability, usually increasing in length when they are transmitted and thus causing increasingly severe disease in successive generations of the affected family (Ashley and Warren, 1995).

The different properties of the two types of disease-causing expansions are due to important differences in the nature of the underlying triplet repeats. In the case of the 'dynamic' expansions, there are long stretches of perfect triplet repeats which predispose to strand slippage on replication. The repeats that encode the polyalanine tracts, however, are imperfect, due to the use of four alternative alanine codons (GCA, GCC, GCG and GCT). Thus, only 10 of the 15 residues of the HOXD13 polyalanine tract are encoded by GCG; residues 5, 12 and 14 are encoded by GCA, residue 8 is encoded by GCT, and residue 15 is encoded by GCC, as shown in *Figure 6*. Such cryptic interruptions are thought to prevent strand slippage. Instead, expansions of the polyalanine tracts are probably caused by unequal crossing-over (non reciprocal recombination) between two normal alleles that have become misaligned during replication (Warren, 1997).

Polyalanine tracts are a common motif in both homeodomain and non homeodomain transcription factors, but their normal function is not understood. One possibility is that they act as flexible spacer elements between other functional domains (Karlin and Burge, 1996); another is that they bind other proteins with which the transcription factors interact, such as DNA binding partners or transcriptional co-factors (Han and Manley, 1993; Licht *et al.*, 1994). It is also unclear what the effect of expanding such tracts would be. In the case of HOXD13 at least, expansions do not appear to perturb the mutant protein's stability or ability to enter the nucleus (Goodman, unpublished data). Moreover, as described above, the penetrance and severity of the associated phenotype increase with increasing expansion size, suggesting a progressive gain of function. This interpretation is supported by the observation that the phenotypes caused by expansions in HOXD13 and RUNX2 are subtly different from those caused by probable loss-of-function mutations in the same genes (Muragaki *et al.*, 1996; Goodman *et al.*, 1997, 1998; Mundlos *et al.*, 1997), although this does not appear to be the case for HOXA13 (Goodman *et al.*, 2000) or ZIC2 (Brown *et al.*, 1998).

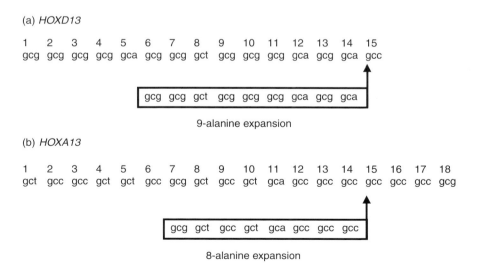

Figure 6. Polyalanine tract expansions. (a) The normal imperfect triplet repeat sequence encoding the 15-residue tract in HOXD13 and the probable insertion site of a 9-residue expansion. (b) The normal imperfect triplet repeat sequence encoding the third 18-residue tract in HOXA13 and the probable insertion site of an 8-residue expansion.

How might an expansion cause a gain-of-function effect? Although it should not affect the mutant protein's ability to bind DNA, it could well disrupt protein–protein interactions that normally occur, and might even allow novel protein–protein interactions to take place. The mutant protein is thus likely to exert a dominant negative effect over the remaining wild-type protein, as it will occupy the same DNA binding site, but prevent or alter the regulation of target genes.

Distal limb malformations similar to those found in SPD occur in mice with a targeted deletion at the 5′ end of the *HoxD* cluster, which removes *Hoxd13*, *Hoxd12* and *Hoxd11* without leaving behind a neomycin resistance cassette (Zákány and Duboule, 1996). In mice heterozygous for this triple deletion, the only abnormality is mild shortening of digits two and five. Homozygotes have a number of digital abnormalities in common with SPD patients, however, including shortening, fusion, webbing and polydactyly. These mice lack both copies of *Hoxd13*, *Hoxd12* and *Hoxd11*, so their phenotype supports the hypothesis that mutant HOXD13 protein carrying an expanded polyalanine tract exerts a dominant negative effect over HOXD11 and HOXD12 as well as over wild-type HOXD13, acting as a 'super' dominant negative.

5.3 Substitution in the homeodomain

The only *HOX* gene missense mutation so far identified is the amino acid substitution in the homeodomain of HOXA13 (*Figure 5i*). This alters a key residue in the recognition helix, which is an asparagine in every known homeodomain protein and directly contacts target DNA (Gehring *et al.*, 1994; Passner *et al.*, 1999; Wolberger *et al.*, 1991). Although the resultant protein's synthesis and stability would probably not be affected, its interactions with DNA would almost certainly be perturbed. The side chain of the substituted amino acid, histidine, is bulky but partially positive at neutral pH, so it might permit DNA binding. The mutant protein might therefore recognize new targets or fail to recognize some of its normal ones, while nevertheless retaining its capacity to regulate transcription once bound. This could produce a novel gain of function, thus explaining the exceptional severity of the resultant limb phenotype (*Figure 4e–h*).

6.   What further HOX gene mutations remain to be found?

The mutations already discovered in *HOXD13* and *HOXA13* suggest that mutations in other *HOX* genes may also cause human malformation syndromes. Most such mutations are unlikely to be simple null alleles, however, as the phenotype caused by hemizygosity for the entire *HOXA* gene cluster can be attributed to haploinsufficiency for just *HOXA3* and *HOXA13* (Devriendt *et al.*, 1999), and hemizygosity for the five most 5′ *HOXD* genes and *EVX2* produces a phenotype very similar to that caused by mutations in *HOXD13* alone (Goodman *et al.*, 1999). Gain-of-function mutations, however, may be expected to produce different phenotypes. Such mutations could include further polyalanine tract expansions and amino acid substitutions in the homeodomain, as well as polyalanine tract contractions, amino acid substitutions in domains that interact with DNA-binding partners or transcriptional co-factors, mutations affecting the regulatory regions of single or multiple genes, and chromosomal re-arrangements perturbing the regulation of an entire cluster.

The gene underlying Kantaputra-type mesomelic dysplasia, a dominantly-inherited limb malformation characterized by marked shortening and bowing of the forearms and lower legs, has recently been mapped to the vicinity of the *HOXD* gene cluster (Fujimoto *et al.*, 1998). Similar limb abnormalities, accompanied by malformations of the cervical and lumbosacral spine, are associated with a balanced translocation involving chromosome 2q31 (Ventruto *et al.*, 1983). Interestingly, these limb abnormalities also closely resemble those found in a radiation-induced semi-dominant mouse mutant, *Ulnaless* (Davisson and Cattanach, 1990). The *Ulnaless* locus is tightly linked to the mouse *HoxD* gene cluster (Peichel *et al.*, 1996), and the expression patterns of *Hoxd11–Hoxd13* in *Ulnaless* limb buds are abnormal, but an underlying mutation in any one *Hoxd* gene has been excluded, as has a large deletion or chromosomal re-arrangement in the region (Hérault *et al.*, 1997; Peichel *et al.*, 1997). In all three cases, the causative mutation is likely to lie in a hitherto unidentified *cis*-acting regulatory element that controls the cluster as a whole.

In addition to their role in development, *HOX* genes, particularly members of the *HOXA* and *HOXB* clusters, also play an important part in normal hematopoiesis. Germline mutations in specific *HOX* genes may therefore cause inherited hematological disorders, while somatic mutations may contribute to hematological malignancies (Look, 1997; van Oostveen *et al.*, 1999). In two unrelated families with an unusual dominantly-inherited combination of amegakaryocytic thrombocytopenia and proximal radio-ulnar synostosis, affected individuals have very recently been shown to carry a single base deletion in the homeobox of *HOXA11*, which is predicted to result in a truncated protein lacking two of the four key amino acids responsible for contacting target DNA (Thompson and Nguyen, 2000). In patients with acute myeloid leukemia, chronic myeloid leukemia and myelodysplastic syndrome, recurrent chromosomal translocations have been identified which give rise to NUP98/HOXA9 and NUP98/HOXD13 fusion proteins (Borrow *et al.*, 1996; Hatano *et al.*, 1999; Nakamura *et al.*, 1996a; Raza-Egilmez *et al.*, 1998; Wong *et al.*, 1999). These are thought to cause leukemic transformation by misregulating genes that are normal *HOX* targets in myeloid cells (Kasper *et al.*, 1999). Activation of *Hoxa7* or *Hoxa9* by proviral integration can lead to myeloid leukemia in mice (Nakamura *et al.*, 1996b), as can overexpression of *Hoxa9* and *Hoxa10* in mouse hematopoietic cells (Kroon *et al.*, 1998; Thorsteinsdottir *et al.*, 1997). Several known DNA-binding partners of HOX proteins, including members of the PBX and MEIS protein families, are also implicated in leukemogenesis in their own right (Look, 1997; van Oostveen *et al.*, 1999), and co-activation of *MEIS1* and *HOXA9* appears to be a common event in myeloid leukemia (Lawrence *et al.*, 1999).

Finally, many *HOX* genes continue to be expressed during adult life in specific tissues, including the gut, kidneys, genital tract and skin, and there is increasing evidence that they are often misexpressed in solid tumors (Cillo *et al.*, 1999). This suggests that somatic *HOX* gene mutations, particularly regulatory mutations, may have a wider role in oncogenesis. HOXA5, for example, has recently been shown to activate transcription of the key tumor suppressor gene *p53*, and in many human breast cancer cell lines and tumors loss of *p53* expression turns out to be secondary to loss of *HOXA5* expression (Raman *et al.*, 2000). This may be due to a somatic mutation in *HOXA5*, but is more commonly associated with methylation of the gene's promoter region, resulting in transcriptional silencing. *HOXA5* itself may thus be an important tumor suppressor gene, and other *HOX* genes may play similar roles in different tumor types. In cancer as in developmental disorders, the search for mutations in the human *HOX* genes may be just beginning.

References

Akarsu, A.N., Akhan, O., Sayli, B.S., Sayli, U., Baskaya, G. and Sarfarazi, M. (1995) A large Turkish kindred with syndactyly type II (synpolydactyly). 2. Homozygous phenotype? *J. Med. Genet.* 32: 435–441.

Akarsu, A.N., Stoilov, I., Yilmaz, E., Sayli, B.S. and Sarfarazi, M. (1996) Genomic structure of *HOXD13* gene: a nine polyalanine duplication causes synpolydactyly in two unrelated families. *Hum. Mol. Genet.* 5: 945–952.

Ashley, C.T. and Warren, S.T. (1995) Trinucleotide repeat expansion and human disease. *Annu. Rev. Genet.* 29: 703–728.

Baffico, M., Baldi, M., Cassan, P.D., Costa, M., Mantero, R., Garani, P. and Camera, G. (1997) Synpolydactyly: clinical and molecular studies on four Italian families. *Eur. J. Hum. Genet.* 5(Suppl. 1): A142.

Boles, R.G., Pober, B.R., Gibson, L.H., Willis, C.R., McGrath, J., Roberts, D.J. and Yang-Feng, T.L. (1995) Deletion of chromosome 2q24-q31 causes characteristic digital anomalies: case report and review. *Am. J. Med. Genet.* 55: 155–160.

Borrow, J., Shearman, A.M., Stanton, V.P. et al. (1996) The t(7;11)(p15;p15) translocation in acute myeloid leukemia fuses the genes for nucleoporin NUP98 and class I homeoprotein HOXA9. *Nat. Genet.* 12: 159–167.

Brown, S.A., Warburton, D., Brown, L.Y., Yu, C., Roeder, E.R., Stengel-Rutkowski, S., Hennekam, R.C.M. and Muenke, M. (1998) Holoprosencephaly due to mutations in *ZIC2*, a homologue of Drosophila odd-paired. *Nat. Genet.* 20: 180–183.

Brunelli, S., Faiella, A., Capra, V., Nigro, V., Simeone, A., Cama, A. and Boncinelli, E. (1996) Germline mutations in the homeobox gene *EMX2* in patients with severe schizencephaly. *Nat. Genet.* 12: 94–96.

Carroll, S.B. (1995) Homeotic genes and the evolution of arthropods and chordates. *Nature* 376: 479–485.

Chisaka, O. and Capecchi, M.R. (1991) Regionally restricted developmental defects resulting from targeted disruption of the mouse homeobox gene *Hox-1.5*. *Nature* 350: 473–479.

Cillo, C., Faiella, A., Cantile, M. and Boncinelli, E. (1999) Homeobox genes and cancer. *Exp. Cell Res.* 248: 1–9.

Collins, A.L. (1996) 2q31.3 is important in distal limb morphogenesis. In: *Proceedings of the Eighth Manchester Birth Defects Conference*, p. 90.

Copeland, J.W.R., Nasiadka, A., Dietrich, B.H. and Krause, H.M. (1996) Patterning of the Drosophila embryo by a homeodomain-deleted Ftz polypeptide. *Nature* 379: 162–165.

Davis, A.P. and Capecchi, M.R. (1996) A mutational analysis of the 5' *HoxD* genes: dissection of genetic interactions during limb development in the mouse. *Development* 122: 1175–1185.

Davisson, M.T. and Cattanach, B.M. (1990) The mouse mutation *Ulnaless* on chromosome 2. *J. Hered.* 81: 151–153.

Del Campo, M., Jones, M.C., Veraksa, A.N., Curry, C.J., Jones, K.L., Mascarello, J.T., Ali-Kahn-Catts, Z., Drumheller, T. and McGinnis, W. (1999) Monodactylous limbs and abnormal genitalia are associated with hemizygosity for the human 2q31 region that includes the *HOXD* cluster. *Am. J. Hum. Genet.* 65: 104–110.

Devriendt, K., Jaeken, J., Matthijs, G., Van Esch, H., Debeer, P., Gewillig, M. and Fryns, J.P. (1999) Haploinsufficiency of the *HOXA* gene cluster, in a patient with hand–foot–genital syndrome, velopharyngeal insufficiency and persistent patent ductus Botalli. *Am. J. Hum. Genet.* 65: 249–251.

Dollé, P., Izpisúa-Belmonte, J. C., Brown, J. M., Tickle, C. and Duboule, D. (1991). *HOX-4* genes and the morphogenesis of mammalian genitalia. *Genes Dev.* 5: 1767–1776.

Dollé, P., Dierich, A., LeMeur, M., Schimmang, T., Schuhbaur, B., Chambon, P. and Duboule, D. (1993) Disruption of the *Hoxd-13* gene induces localized heterochrony leading to mice with neotenic limbs. *Cell* 75: 431–441.

Donnenfeld, A.E., Schrager, D.S. and Corson, S.L. (1992) Update on a family with hand–foot–genital syndrome: hypospadias and urinary tract abnormalities in two boys from the fourth generation. *Am. J. Med. Genet.* 44: 482–484.

Favier, B. and Dollé, P. (1997) Developmental functions of mammalian *Hox* genes. *Mol. Hum. Reprod.* 3: 115–131.

Fromental-Ramain, C., Warot, X., Messadecq, N., LeMeur, M., Dollé, P. and Chambon, P. (1996) *Hoxa-13* and *Hoxd-13* play a crucial role in the patterning of the limb autopod. *Development* 122: 2997–3011.

Fujimoto, M., Kantaputra, P.N., Ikegawa, S. *et al.* (1998) The gene for mesomelic dysplasia Kantaputra type is mapped to chromosome 2q24-q32. *J. Hum. Genet.* 43: 32–36.

Gehring, W.J., Qian, Y.Q., Billeter, M., Furukubo-Tokunaga, K., Schier, A.F., Resendez-Perez, D., Affolter, M., Otting, G. and Wuthrich, K. (1994) Homeodomain-DNA recognition. *Cell* 78: 211–223.

Goodman, F.R., Mundlos, S., Muragaki, Y. *et al.* (1997) Synpolydactyly phenotypes correlate with size of expansions in HOXD13 polyalanine tract. *Proc. Natl Acad. Sci. USA* 94: 7458–7463.

Goodman, F.R., Giovannucci-Uzielli, M.L., Hall, C., Reardon, W., Winter, R. and Scambler, P. (1998) Deletions in *HOXD13* segregate with an identical, novel foot malformation in two unrelated families. *Am. J. Hum. Genet.* 63: 992–1000.

Goodman, F., Majewski, F., Winter, R. and Scambler, P. (1999) Haploinsufficiency for *HOXD8-HOXD13* and *EVX2* causes atypical synpolydactyly. *Am. J. Hum. Genet.* 65(Suppl.): A298.

Goodman, F.R., Bacchelli, C., Brady, A.F. *et al.* (2000) Novel *HOXA13* Mutations and the Phenotypic Spectrum of Hand-Foot-Genital Syndrome. *Am. J. Hum. Genet.* 67: 197–202.

Graba, Y., Aragnol, D. and Pradel, J. (1997) *Drosophila Hox* complex downstream targets and the function of homeotic genes. *BioEssays* 19: 379–388.

Hagan, D.M., Ross, A.J., Strachan, T. *et al.* (2000) Mutation analysis and embryonic expression of the *HLXB9* Currarino syndrome gene. *Am. J. Hum. Genet.* 66: 1504–1515.

Halal, F. (1988) The hand–foot–genital (hand–foot–uterus) syndrome: family report and update. *Am. J. Med. Genet.* 30: 793–803.

Han, K. and Manley, J.L. (1993) Transcriptional repression by the *Drosophila* Even-skipped protein: definition of a minimal repression domain. *Genes Dev.* 7: 491–503.

Hatano, Y., Miura, I., Nakamura, T., Yamazaki, Y., Takahashi, N. and Miura, A.B. (1999) Molecular heterogeneity of the NUP98/HOXA9 fusion transcript in myelodysplastic syndromes associated with t(7;11)(p15;p15). *Br. J. Haematol.* 107: 600–604.

Hérault, Y., Fraudeau, N., Zákány, J. and Duboule, D. (1997) *Ulnaless (Ul)*, a regulatory mutation inducing both loss-of-function and gain-of-function of posterior *Hoxd* genes. *Development* 124: 3493–3500.

Hummel, K. (1970) *Hypodactyly*, a semi-dominant lethal mutation in mice. *J. Hered.* 61: 219–220.

Innis, J.W., Darling, S.M., Kazen-Gillespie, K., Post, L.C., Mortlock, D.P. and Yang, T. (1996) Orientation of the *Hoxa* complex and placement of the *Hd* locus distal to *Hoxa2* on mouse chromosome 6. *Mamm. Genome* 7: 216–217.

Innis, J.W. (1997) Role of *HOX* genes in human development. *Curr. Opin. Pediatr.* 9: 617–622.

Karlin, S. and Burge, C. (1996) Trinucleotide repeats and long homopeptides in genes and proteins associated with nervous system disease and development. *Proc. Natl Acad. Sci. USA* 93: 1560–1565.

Kasper, L.H., Brindle, P.K., Schnabel, C.A., Pritchard, C.E., Cleary, M.L. and van Deursen, J.M. (1999) CREB binding protein interacts with nucleoporin-specific FG repeats that activate transcription and mediate NUP98-HOXA9 oncogenicity. *Mol. Cell. Biol.* 19: 764–776.

Kenyon, C. (1994) If birds can fly, why can't we? Homeotic genes and evolution. *Cell* 78: 175–180.

Kondo, T., Dollé, P., Zákány, J. and Duboule, D. (1996) Function of posterior *HoxD* genes in the morphogenesis of the anal sphincter. *Development* 122: 2651–2659.

Kondo, T. and Duboule, D. (1999) Breaking colinearity in the mouse *HoxD* complex. *Cell* 97: 407–417.

Kroon, E., Krosl, J., Thorsteinsdottir, U., Baban, S., Buchberg, A.M. and Sauvageau, G. (1998) Hoxa9 transforms primary bone marrow cells through specific collaboration with Meis1a but not Pbx1b. *EMBO J.* 17: 3714–3725.

Krumlauf, R. (1994) *Hox* genes in vertebrate development. *Cell* **78**: 191–201.

Lawrence, H.J., Rozenfeld, S., Cruz, C., Matsukuma, K., Kwong, A., Komuves, L., Buchberg, A.M. and Largman, C. (1999) Frequent co-expression of the *HOXA9* and *MEIS1* homeobox genes in human myeloid leukemias. *Leukemia* **13**: 1993–1999.

Lewis, E.B. (1978) A gene complex controlling segmentation in *Drosophila*. *Nature* **276**: 565–570.

Licht, J.D., Hanna-Rose, W., Reddy, J.C., English, M.A., Ro, M., Grossel, M., Shaknovich, R. and Hansen, U. (1994) Mapping and mutagenesis of the amino-terminal transcriptional repression domain of the *Drosophila* Kruppel protein. *Mol. Cell. Biol.* **14**: 4057–4066.

Look, A.T. (1997) Oncogenic transcription factors in the human acute leukemias. *Science* **278**: 1059–1064.

Mann, R.S. and Chan, S.K. (1996) Extra specificity from extradenticle: the partnership between HOX and PBX/EXD homeodomain proteins. *Trends Genet.* **12**: 258–262.

Mann, R.S. and Affolter, M. (1998) Hox proteins meet more partners. *Curr. Opin. Genet. Dev.* **8**: 423–429.

Mark, M., Rijli, F.M. and Chambon, P. (1997) Homeobox genes in embryogenesis and pathogenesis. *Pediatr. Res.* **42**: 421–429.

McGinnis, W. and Krumlauf, R. (1992) Homeobox genes and axial patterning. *Cell* **68**: 283–302.

Mortlock, D.P., Post, L.C. and Innis, J.W. (1996) The molecular basis of *Hypodactyly* (*Hd*): a deletion in *Hoxa13* leads to arrest of digital arch formation. *Nat. Genet.* **13**: 284–289.

Mortlock, D.P. and Innis, J.W. (1997) Mutation of *HOXA13* in hand–foot–genital syndrome. *Nat. Genet.* **15**: 179–180.

Mundlos, S., Otto, F., Mundlos, C. *et al.* (1997) Mutations involving the transcription factor *CBFA1* cause cleidocranial dysplasia. *Cell* **89**: 773–779.

Muragaki, Y., Mundlos, S., Upton, J. and Olsen, B.R. (1996) Altered growth and branching patterns in synpolydactyly caused by mutations in *HOXD13*. *Science* **272**: 548–551.

Nakamura, T., Largaespada, D.A., Lee, M.P. *et al.* (1996a) Fusion of the nucleoporin gene *NUP98* to *HOXA9* by the chromosome translocation t(7;11)(p15;p15) in human myeloid leukemia. *Nat. Genet.* **12**: 154–158.

Nakamura, T., Largaespada, D.A., Shaughnessy, J.D., Jr., Jenkins, N.A. and Copeland, N.G. (1996b) Cooperative activation of *Hoxa* and *Pbx1*-related genes in murine myeloid leukaemias. *Nat. Genet.* **12**: 149–153.

Nixon, J., Oldridge, M., Wilkie, A.O. and Smith, K. (1997) Interstitial deletion of 2q associated with craniosynostosis, ocular coloboma, and limb abnormalities: cytogenetic and molecular investigation. *Am. J. Med. Genet.* **70**: 324–327.

Olson, E.N., Arnold, H.H., Rigby, P.W. and Wold, B.J. (1996) Know your neighbors: three phenotypes in null mutants of the myogenic bHLH gene *MRF4*. *Cell* **85**: 1–4.

Passner, J.M., Ryoo, H.D., Shen, L., Mann, R.S. and Aggarwal, A.K. (1999) Structure of a DNA-bound Ultrabithorax-Extradenticle homeodomain complex. *Nature* **397**: 714–719.

Peichel, C.L., Abbott, C.M. and Vogt, T.F. (1996) Genetic and physical mapping of the mouse *Ulnaless* locus. *Genetics* **144**: 1757–1767.

Peichel, C.L., Prabhakaran, B. and Vogt, T.F. (1997) The mouse *Ulnaless* mutation deregulates posterior *HoxD* gene expression and alters appendicular patterning. *Development* **124**: 3481–3492.

Podlasek, C.A., Duboule, D. and Bushman, W. (1997) Male accessory sex organ morphogenesis is altered by loss of function of *Hoxd-13*. *Dev. Dyn.* **208**: 454–465.

Post, L.C. and Innis, J.W. (1999a) Altered *Hox* expression and increased cell death distinguish *Hypodactyly* from *Hoxa13* null mice. *Int. J. Dev. Biol.* **43**: 287–294.

Post, L.C. and Innis, J.W. (1999b) Infertility in adult *Hypodactyly* mice is associated with hypoplasia of distal reproductive structures. *Biol. Reprod.* **61**: 1402–1408.

Post, L.C., Margulies, E.H., Kuo, A. and Innis, J.W. (2000) Severe limb defects in *Hypodactyly* mice result from the expression of a novel, mutant HOXA13 protein. *Dev. Biol.* **217**: 290–300.

Poznanski, A.K., Kuhns, L.R., Lapides, J. and Stern, A.M. (1975) A new family with the hand–foot–genital syndrome – a wider spectrum of the hand–foot–uterus syndrome. *Birth Defects Original Article Series* **11**: 127–135.

Prosser, J. and van Heyningen, V. (1998) *PAX6* mutations reviewed. *Hum. Mutat.* **11**: 93–108.

Raman, V., Martensen, S.A., Reisman, D., Evron, E., Odenwald, W.F., Jaffee, E., Marks, J. and Sukumar, S. (2000) Compromised *HOXA5* function can limit *p53* expression in human breast tumours. *Nature* **405**: 974–978.

Raza-Egilmez, S.Z., Jani-Sait, S.N., Grossi, M., Higgins, M.J., Shows, T.B. and Aplan, P.D. (1998) *NUP98-HOXD13* gene fusion in therapy-related acute myelogenous leukemia. *Cancer Res.* **58**: 4269–4273.

Ross, A.J., Ruiz-Perez, V., Wang, Y. *et al.* (1998) A homeobox gene, *HLXB9*, is the major locus for dominantly inherited sacral agenesis. *Nat. Genet.* **20**: 358–361.

Sarfarazi, M., Akarsu, A.N. and Sayli, B.S. (1995) Localization of the syndactyly type II (synpolydactyly) locus to 2q31 region and identification of tight linkage to *HOXD8* intragenic marker. *Hum. Mol. Genet.* **4**: 1453–1458.

Sayli, B.S., Akarsu, A.N., Sayli, U., Akhan, O., Ceylaner, S. and Sarfarazi, M. (1995) A large Turkish kindred with syndactyly type II (synpolydactyly). 1. Field investigation, clinical and pedigree data. *J. Med. Genet.* **32**: 421–434.

Semina, E.V., Reiter, R., Leysens, N.J. *et al.* (1996) Cloning and characterization of a novel bicoid-related homeobox transcription factor gene, *RIEG*, involved in Rieger syndrome. *Nat. Genet.* **14**: 392–399.

Sharpe, P. (1996) *HOX* gene mutations – the wait is over. *Nat. Med.* **2**: 748–749.

Slavotinek, A., Schwarz, C., Getty, J.F., Stecko, O., Goodman, F. and Kingston, H. (1999) Two cases with interstitial deletions of chromosome 2 and sex reversal in one. *Am. J. Med. Genet.* **86**: 75–81.

Stern, A.M., Gall, J.C.J., Perry, B.L., Stimson, C.W., Weitkamp, L.R. and Poznanski, A.K. (1970) The hand–foot–uterus syndrome. *J. Pediatr.* **77**: 109–116.

Thompson, A.A. and Nguyen, L.T. (2000) Amegakaryocytic thrombocytopenia and radio-ulnar synostosis are associated with *HOXA11* mutation. *Nat. Genet.* **26**: 397–398.

Thorsteinsdottir, U., Sauvageau, G., Hough, M.R., Dragowska, W., Lansdorp, P.M., Lawrence, H.J., Largman, C. and Humphries, R.K. (1997) Overexpression of *HOXA10* in murine hematopoietic cells perturbs both myeloid and lymphoid differentiation and leads to acute myeloid leukemia. *Mol. Cell. Biol.* **17**: 495–505.

van Oostveen, J., Bijl, J., Raaphorst, F., Walboomers, J. and Meijer, C. (1999) The role of homeobox genes in normal hematopoiesis and hematological malignancies. *Leukemia* **13**: 1675–1690.

Ventruto, V., Pisciotta, R., Renda, S., Festa, B., Rinaldi, M.M., Stabile, M., Cavaliere, M.L. and Esposito, M. (1983) Multiple skeletal familial abnormalities associated with balanced reciprocal translocation 2;8(q32;p13). *Am. J. Med. Genet.* **16**: 589–594.

Warot, X., Fromental-Ramain, C., Fraulob, V., Chambon, P. and Dollé, P. (1997) Gene dosage-dependent effects of the *Hoxa-13* and *Hoxd-13* mutations on morphogenesis of the terminal parts of the digestive and urogenital tracts. *Development* **124**: 4781–4791.

Warren, S.T. (1997) Polyalanine expansion in synpolydactyly might result from unequal crossing-over of *HOXD13*. *Science* **275**: 408–409.

Wolberger, C., Vershon, A.K., Liu, B., Johnson, A.D. and Pabo, C.O. (1991) Crystal structure of a MATa2 homeodomain-operator complex suggests a general model for homeodomain-DNA interactions. *Cell* **67**: 517–528.

Wong, K.F., So, C.C. and Kwong, Y.L. (1999) Chronic myelomonocytic leukemia with t(7;11)(p15;p15) and NUP98/HOXA9 fusion. *Cancer Genet. Cytogenet.* **115**: 70–72.

Zákány, J. and Duboule, D. (1996) Synpolydactyly in mice with a targeted deficiency in the *HoxD* complex. *Nature* **384**: 69–71.

13

PITX2 gene in development

Jeffrey C. Murray and Elena V. Semina

1. Introduction

Organogenesis involves a dynamic balance of the mechanisms regulating cell division, differentiation and death. Transcription factors play an active role in organogenesis and may as well possess a later maintenance function. We and others have recently discovered a homeobox-containing transcription factor gene *PITX2* that is involved in the development of the eye, heart, lung, pituitary, tooth and abdominal region in different species (Semina *et al.*, 1996; Gage and Camper, 1997; Mucchielli *et al.*, 1996; Arakawa *et al.*, 1998; Ryan *et al.*, 1998). Humans that have one affected copy of *PITX2* gene develop Axenfeld–Rieger syndrome (Semina *et al.*, 1996). The *Pitx2*-deficient mice die *in utero* due to multiple defects in the formation of the above mentioned systems. Misexpression of the *Pitx2* in frog and chick embryos results in laterality defects of the heart and gut. Therefore Pitx2 is essential to normal development and represents an important regulator of yet to be defined downstream genes involved in the formation of these systems.

2. Roles of *PITX2* in development

2.1 PITX2 *is responsible for Axenfeld–Rieger syndrome*

Axenfeld–Rieger syndrome is an autosomal-dominant disorder with specific ocular, dental and umbilical anomalies and is one of the phenotypes of Axenfeld–Rieger spectrum (*Figure 1*) (Axenfeld 1920; Rieger 1934, 1935; Berg 1932; Alward 2000). Axenfeld–Rieger syndrome is characterized by complete penetrance but variable expressivity is reported in families. The ocular anomalies can include a prominent annular white line near the limbus at the level of Descement membrane (posterior embryotoxon), hypoplastic iris, irido-corneal adhesions and glaucoma. Dental manifestations vary from small teeth to complete anodontia, missing lateral mandibular incisors are described as the most common feature. Other common craniofacial features of Rieger syndrome include maxillary hypoplasia and dysplastic ears. Umbilical anomalies range from isolated redundant skin at the site of umbilicus to severe hernias or omphalocele. Some Rieger syndrome pedigrees also manifest pituitary and cardiac defects, hearing loss, hydrocephalus and hypospadius (Tsai and Grajewski 1994; Mammi *et al.*, 1998).

Axenfeld–Rieger syndrome can be caused by alterations in several different genes. This was originally proposed because of identification of different chromosomal anomalies in patients with Axenfeld–Rieger syndrome and was later confirmed by linkage studies. The majority of reported cytogenetic abnormalities

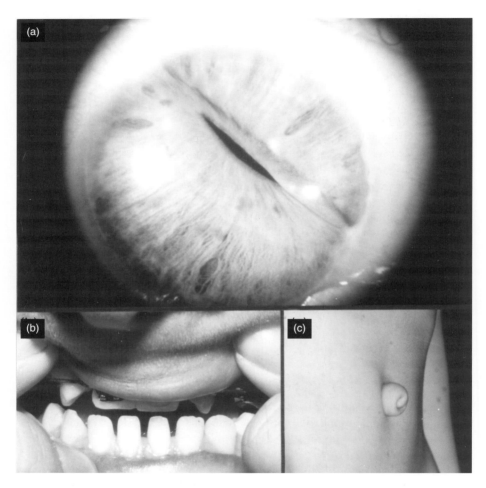

Figure 1. Affected features of Axenfeld–Rieger syndrome. (a) The hypoplastic iris allows the pupillary sphincter muscle to be seen as a band around the pupil. The pupil is distorted (corectopia). The distinct white line in the periphery of the cornea is termed 'posterior embryotoxon' and represents a prominent and anteriorly displaced Schwalbe's line of the trabecular meshwork which is not normally visible. (b) Notice oligodontia and microdontia as evident by the missing lateral incisors (canines are in the lateral incisor position) and small mandibular central incisors. (c) The abdominal area of Axenfeld–Rieger syndrome patient showing caharacteristic umbilical abnormalities due to failure of the periumbilical skin to involute. Reproduced with permission from *Nat. Genetics* **14**: 392—399.

include deletions/translocations affecting 4q, 13q and 6p and trisomy/monosomy 9, 16, 18, 20 and 21, where anomalies of 4, 13 and 6 chromosomes were associated with syndromic features while the other cases were mostly characterized with only ocular findings (reviewed in Rogers, 1988). There are four genetic loci and two causative genes for Axenfeld–Rieger syndrome identified to date: 4q25 (Murray *et al.*, 1992; Semina *et al.*, 1996), 13q14 (Phillips *et al.*, 1996), 6p25 (Mirzayans *et al.*, 2000) and 16q24 (Nishimura *et al.*, 2000).

Positional cloning of the gene responsible for 4q25-linked cases of Rieger syndrome resulted in an identification of the human *PITX2* gene and its mouse homologue *Pitx2* (originally called *RIEG/Rieg*) in 1996 (Semina *et al.*, 1996). The mouse *Pitx2* gene was also identified independently by other researchers in a hunt for homeobox genes expressed in pituitary (originally called *Ptx2*; Gage and Camper, 1997) and brain (originally called *Otlx2* and *Brx1*, respectively; Mucchielli *et al.*, 1996; Kitamura *et al.*, 1997). The human *PITX2* was also isolated as a target of *All1*, the human homolog of *Drosophila trithorax* (Arakawa *et al.*, 1998). The *All1* gene is frequently altered in human acute leukemias resulting in a loss of All1 function that leads to a missing expression of *Pitx2*. These results suggest a role for *PITX2* in tumorigenesis that ought to be further explored.

The *PITX2* gene encodes a homeodomain-containing transcription factor of the *paired-like* group, *bicoid* subgroup (Semina *et al.*, 1996; Amendt *et al.*, 1998). Comparison of the PITX2 homeodomain with others identified the greatest homology with the murine Ptx1/P-OTX, *C. Elegans unc-30*, *Drosophila Otd* and murine proteins of Otx family. All of these proteins share a lysine residue at position 9 of the third helix of the homeodomain, which is characteristic of *bicoid*-related proteins and has been shown to determine the specificity of binding to 3xCC dinucleotide following the TAAT core (Gehring, 1993; Gehring *et al.*, 1994).

The *PITX2* gene was shown to lie in close proximity to chromosomal break-points that occurred in two independent patients with the disorder and was found to be mutated in several unrelated families with Rieger syndrome (Semina *et al.*, 1996). We suggested that haploinsufficiency is a likely etiologic mechanism of Rieger syndrome supported by reports of large deletions in some Rieger patients that were now shown to include the complete gene (Schinzel *et al.*, 1997; Flomen *et al.*, 1998). Several new translocation breakpoints have been recently reported with one being mapped as far as ~90 kb upstream of the *PITX2* gene (Flomen *et al.*, 1998), suggesting an existence of the complex regulatory system involved in execution of normal *PITX2* expression. The phenomenon, when disease cases are associated with chromosomal rearrangement outside the transcription and promoter regions, is recognized as 'position effect' and it was described for a number of transcription factor genes (Kleinjan and van Heyningen, 1998). There are a number of mechanisms that can explain this etiology as, for example, removal of a locus-specific enhancer elements or a chromatin-organizing region from the gene that results in reduction in its expression.

The mouse homolog, *Pitx2*, was isolated and showed remarkable homology with the human gene at both the nucleotide and protein levels. By *in situ* hybridization on whole mount embryos and sections, the *Pitx2* mRNA was detected in the mesenchyme around the developing eye, in the epithelium and mesenchyme of the maxilla and mandible, dental lamina, in the umbilical region, pituitary, midbrain region and limbs (*Figure 2b*) (Semina *et al.*, 1996). The ocular, dental and umbilical sites of *Pitx2* expression are in a good agreement with the affected phenotype of Rieger syndrome patients.

The *PITX2* gene was recently shown to be involved in two other disorders, whose differential diagnosis from Rieger syndrome was mostly based on specific ocular features- iris hypoplasia (Alward *et al.*, 1998), iridogoniodysgenesis syndrome (Kulak *et al.*, 1998) and Peter's anomaly (Doward *et al.*, 1999). The collective term Axenfeld–Rieger syndrome was recently suggested for these conditions due to significant phenotypic and genotypic overlap (Alward, 2000). Our studies show that approximately 34% of Axenfeld–Rieger syndrome and 1% of isolated anterior chamber anomalies patients demonstrate mutations in the *PITX2* gene. Most mutations affect the homeobox region of the gene encoding homeodomain, which plays a major role in a target DNA-motif recognition and binding. No correlation has been detected between the position of the mutation in the gene and the severity of the phenotype. Moreover, the same mutation has been found to have different phenotypic expression in two unrelated families: full Axenfeld–Rieger syndrome in one family and the ocular anomaly without any other systemic defects in a second family.

Other genes involved in pathogenesis of disorders similar to Axenfeld–Rieger syndrome include *PAX6* (aniridia, Peter's anomaly, iris hypoplasia; Prosser and van Heyningen, 1998), *FREAC3/FKHL7* (Axenfeld–Rieger anomaly, iridogoniodysgenesis, iris hypoplasia; Nishimura *et al.*, 1998; Kume *et al.*, 1998; Mears *et al.*, 1998), *PITX3* (anterior segment mesenchymal dysgenesis; Semina *et al.*, 1997, 1998), *JAG1* (Alagille syndrome; Krantz *et al.*, 1998) and *EYA1* (anterior segment dysgenesis; Azuma *et al.*, 2000), *LIMX1B* (nail-patella syndrome; Chen *et al.*, 1998; Vollrath *et al.*, 1998), *FGFR2* (craniosynostosis syndrome; Okajima *et al.*, 1999) and *FOXE3* (Semina *et al.*, 2001). These genes may be involved in the same pathway(s) during development that now can be elucidated.

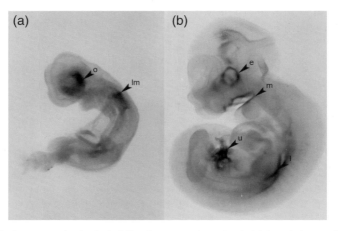

Figure 2. Whole-mount *in situ* hybridization on mouse day 8.5 (a) and day 11 (b) embryos with anti-sense digoxeginin-labeled riboprobe derived from the 3′UTR of the mouse *Pitx2* cDNA. The staining is seen in the areas of the optic eminence (o) and asymmetrically in the left lateral plate mesoderm (lm) in day 8.5-pc embryos and in the eye (e), maxilla and mandible (m), umbilical cord (u) and limb (l) in day 11-pc embryos.

2.2 Pitx2 *and left–right patterning*

An exciting development in expanding the function of the *Pitx2* gene was the discovery that it plays a key role in the development of asymmetric structures in vertebrates (Yoshioka *et al.*, 1998; Logan *et al.*, 1998; Piedra *et al.*, 1998; Ryan *et al.*, 1998; St Amand *et al.*, 1998). The early expression of *Pitx2* was shown to be asymmetric in mice, chicken, *Xenopus* and even some invertebrates, such as lancelet (*Figure 2a*) (Yasui *et al.*, 2000). Misexpression of *Pitx2* changed the relative position of visceral organs and the direction of body rotation in chick and *Xenopus* embryos (Ryan *et al.*, 1998). Moreover, *Pitx2* expression was shown to be regulated sequentially by Shh and Nodal and to be disturbed in mouse mutants with laterality defects (Yoshioka *et al.*, 1998; Logan *et al.*, 1998; Piedra *et al.*, 1998; Ryan *et al.*, 1998). Further development of this theme was provided by studies of *Pitx2*-deficient mice. The *Pitx2–/–* homozygous mice demonstrate some asymmetry defects (see below) in the development of the heart and lung.

Different isoforms of *Pitx2* have been identified and their roles are beginning to be elucidated. To date, three *Pitx2* isoforms are detectable in human, mouse, *Xenopus*, chicken and zebrafish (*Figure 3*) (Gage and Camper, 1997; Arakawa *et al.*, 1998; Essner *et al.*, 2000; Schweickert *et al.*, 2000). All of the isoforms share the homeobox and the C-terminal region with *a* and *b* isoforms being different by an alternatively spliced small internal exon and isoform *c* being expressed from a different promoter. It was shown that isoforms demonstrate different expression patterns and distinct upstream and downstream genetic pathways. With respect to left-right patterning, only *Pitx2c* isoform was shown to have asymmetric pattern in *Xenopus* and chicken embryos (Yu *et al.*, 2001). Misexpression of different isoforms can alter left-right development, but *Pitx2c* has a slightly stronger effect on the laterality of the heart in *Xenopus* and zebrafish (Schweickert *et al.*, 2000; Essner *et al.*, 2000). In agreement with this, elimination of the *Pitx2c* and not of the *Pitx2a* expression from the left lateral plate mesoderm by antisense oligonucleotide resulted in randomization of heart looping (Yu *et al.*, 2001). It is notable that all of the *PITX2*-Rieger syndrome/anomaly mutations that are identified to date, are located in the last two *PITX2* exons encoding homeodomain and C-terminal region, which are common for all of the isoforms. The isoform-specific mutations may be responsible for a different phenotype(s) that is yet to be discovered.

2.3 Mouse models

Generation of mice with germ-line gene mutations through gene targeting in embryonic stem cells has provided an opportunity to study *Pitx2* function by allowing experimental studies in an organismal context. *Pitx2*-deficient mice have been generated in several laboratories mostly through alteration of the exons encoding homeobox region of the gene (Lu *et al.*, 1999; Lin *et al.*, 1999; Kitamura *et al.*, 1999; Gage *et al.*, 1999b). *Pitx2*-null mice died *in utero* around embryonic day (E) 9.25–15. The *Pitx2–/–* embryos exhibited a failure to close the ventral body wall and thorax with thoracic and abdominal organs being positioned externally, defects in the cardiovascular system including heart positioning on the right rather than the midline, common atrium and hypoplasia of ventricles, right pulmonary isomerism, pituitary gland and dental hypoplasia and enophthalmos.

Genomic structure of *PITX2*

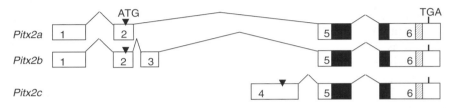

Figure 3. Genomic organization and schematics of the *PITX2* isoforms: intron sizes (in kb) are shown on the top and exons sizes at the bottom, exons are numbered, homeobox region is shown in black, positions of the initiation ATG codons are shown by arrows.

Rieger syndrome patients with mutations in the *PITX2* gene do not feature any visible heart, lung, brain or pituitary anomalies, which may be explained by the dominant nature of this disorder as one normal copy of the *PITX2* gene is present in these patients. The *PITX2* mutations that affect both copies of the gene may be responsible for more severe phenotypes or result in embryonic death.

2.4 Biochemical studies of Pitx2 protein

The Pitx2 protein is a homeodomain-containing transcription factor that belongs to the *paired*-like class by overall similarity and *bicoid*-like subclass because of the defining lysine at residue 50 of the homeodomain. This residue is located at position 9 within the recognition helix of the homeodomain and is a major determinant of DNA binding specificity (Gehring, 1993; Gehring *et al.*, 1994). The lysine residue at this position is predicted to selectively recognize the 3xCC dinucleotide adjacent to the TAAT core. Consistent with this prediction, it has been demonstrated that Pitx2 can bind the DNA sequence 5'-TAATCC-3' (Amendt *et al.*, 1998).

Axenfeld–Rieger syndrome mutations most frequently affect the homeodomain region and result in a loss of binding to the *bicoid* site or transactivation activity, or in generation of unstable protein (Amendt *et al.*, 1998, 2000). Analysis of the five mutant PITX2 proteins demonstrated that the mutant protein forms associated with milder anterior segment phenotype retained most of the protein activities as examined by the DNA binding and transactivation studies (Kozlowski and Walter, 2000). In this study, the mutant proteins associated with iris hypoplasia and iridogoniodysgenesis syndrome showed different degrees of reduced activity, whereas the Axenfeld–Rieger syndrome mutant PITX2 proteins proved to be non-functional. In agreement with this study, Saadi *et al.* (2001) describes a mutation that changes the lysine at position 50 of the homeodomain to glutamic acid. The mutation was shown to have a dominant-negative effect and to result in the severe ocular condition resembling partial aniridia in this Rieger syndrome patient.

The C-terminal end of Pitx2 has been shown to be involved in protein-protein interactions with, for example, the pituitary transcription factor, Pit-1. The addition of Pit-1 to Pitx2 results in increased binding and synergistic activation of the prolactin promoter (Amendt *et al.*, 1998). Deletion analysis of *Pitx2* revealed that the C-terminal 39 amino acids represses DNA binding activity and is required for protein–protein interactions with Pit-1. Interaction of Pit-1 with this C39 domain masks the inhibitory effect and results in increased activity of the Pitx2 protein (Amendt *et al.*, 1999). The C39 region contains a 14-amino-acid motif that was originally identified as being conserved between several *paired*-like homeodomain proteins (Semina *et al.*, 1996).

The Pitx2 protein isoforms were detected in mouse embryonic day-12 extracts as 32 and 37 kD polypeptides (Hjalt et al. 2000). Pitx2 was found to be expressed during the development of the eye, tooth, umbilicus, pituitary, gut, heart and limb, all consistent with sites of its mRNA expression (Semina et al. 1996; Gage and Camper 1997; Hjalt et al. 2000).

Identification of downstream targets of PITX2 proteins becomes increasingly important to learn about developmental processes that are governed by this gene. Several genes were described that are regulated by PITX2 in the pituitary (Amendt *et al.*, 1998; Tremblay *et al.*, 2000). Recently the mouse procollagen lysyl hydroxylase (*Plod-2*) gene was identified as a likely downstream target of Pitx2 by chromatin precipitation approach (Hjalt *et al.*, 2001). Interestingly, mutations in the other family member, PLOD-1, are associated with Ehlers–Danlos syndrome, kyphoscoliosis type (type VI, EDVI), that shares some of the same affected sites with Axenfeld–Rieger syndrome.

3. The Pitx family

The *Pitx2* gene belongs to the Pitx family currently consisting of three genes, *Pitx1–3*, that were identified in many different species (reviewed in Gage *et al.*, 1999a). The three genes show a significant overlap in expression patterns and may therefore compensate for each other at some sites as has been shown for some other homeobox-containing genes. The *Pitx1* gene was shown to be important for limb, mandible and pituitary development as concluded from the mouse knockout studies and misexpression experiments in chicken (Lanctot *et al.*, 1999; Szeto *et al.*, 1999; Logan and Tabin, 1999), while the *PITX3/Pitx3* gene was found to be involved in the anterior segment dysgenesis ocular phenotype in humans and *aphakia* in mice by linkage and mutation studies (Semina *et al.*, 1997, 1998, 2000; Rieger *et al.*, 2001).

4. Conclusion

It is now well established that mutations within the homeodomain and C-terminal regions of the *PITX2* gene can result in Axenfeld–Rieger syndrome and related ocular anomalies in humans. There is a clear variability of expression of this phenotype that is likely due to gene–gene or possible gene–environment interactions. No isoform-specific mutations have yet been identified that might suggest that alternate phenotypes are possible. Clear evidence that the gene plays a role in left-right axis determination also suggest that related anomalies should be searched for in humans. The gene itself contains likely regulatory elements both in introns and in the upstream regions that should also be candidates for future mutation searches and gene–phenotype correlation. In summary, this gene plays a critical role in ocular, dental, pituitary, umbilical and paired organ development and is a candidate for a wide range of human genetic defects.

References

Alward, W.L.M. (2000) Axenfeld-Rieger syndrome in the age of molecular genetics. *Am. J. Ophthalmol.* **130**: 107–115.

Alward, W.L.M., Semina, E.V., Kalenak, J.W., Heon, E., Sheth, B.P., Stone, E.M. and Murray, J.C. (1998) Autosomal dominant iris hypoplasia is caused by a mutation in the Rieger syndrome (RIEG/PITX2) gene. *Am. J. Ophthalmol.* **125**: 98–100.

Amendt, B.A., Semina, E.V. and Alward, W.L. (2000) Rieger syndrome: a clinical, molecular, and biochemical analysis. *Cell. Mol. Life Sci.* **57**: 1652–1666.

Amendt, B.A., Sutherland, L.B. and Russo, A.F. (1999) Multifunctional role of the Pitx2 homeodomain protein C-terminal tail. *Mol. Cell. Biol.* **19**: 7001–7010.

Amendt, B.A., Sutherland, L.B., Semina, E.V. and Russo, A.F. (1998) The molecular basis of Rieger syndrome. Analysis of Pitx2 homeodomain protein activities. *J. Biol. Chem.* **273**: 20066–20072.

Arakawa, H., Nakamura, T., Zhadanov, A.B., *et al.* (1998) Identification and characterization of the *ARP1* gene, a target for the human acute leukemia *ALL1* gene. *Proc. Natl. Acad. Sci. USA* **95**: 4573–4578.

Axenfeld, T. (1920) Embryotoxon corneae posteris. *Ber Dtsch Ophthalmol Ges* **42**: 381–382.

Azuma, N., Hirakiyama, A., Inoue, T., Asaka, A. and Yamada, M. (2000) Mutations of a human homologue of the *Drosophila* eyes absent gene (*EYA1*) detected in patients with congenital cataracts and ocular anterior segment anomalies. *Hum. Mol. Genet.* **9**: 363–366.

Berg, F. (1932) Erbliches jugendliches glaukom. *Acta Ophthalmol.* **10**: 568–587.

Chen, H., Lun, Y., Ovchinnikov, D., Kokubo, H., Oberg, K.C., Pepicelli, C.V., Gan, L., Lee, B. and Johnson, R.L. (1998) Limb and kidney defects in Lmx1b mutant mice suggest an involvement of LMX1B in human nail patella syndrome. *Nat. Genet.* **19**: 51–55.

Doward, W., Perveen, R., Lloyd, I.C., Ridgway, A.E., Wilson, L. and Black, G.C. (1999) A mutation in the *RIEG1* gene associated with Peters' anomaly. *J. Med. Genet.* **36**: 152–155.

Essner, J.J., Branford, W.W., Zhang, J. and Yost, H.J. (2000) Mesendoderm and left-right brain, heart and gut development are differentially regulated by pitx2 isoforms. *Development* **127**: 1081–1093.

Flomen, R.H., Vatcheva, R., Gorman, P.A., Baptista, P.R., Groet, J., Barisic, I., Ligutic, I. and Nizetic, D. (1998) Construction and analysis of a sequence-ready map in 4q25: Rieger syndrome can be caused by haploinsufficiency of RIEG, but also by chromosome breaks approximately 90 kb upstream of this gene. *Genomics* **47**: 409–413.

Gage, P.J. and Camper, S.A. (1997) Pituitary homeobox 2, a novel member of the bicoid-related family of homeobox genes, is a potential regulator of anterior structure formation. *Hum. Mol. Genet.* **6**: 457–464.

Gage, P.J., Suh, H. and Camper, S.A. (1999a) The bicoid-related *Pitx* gene family in development. *Mamm. Genome* **10**: 197–200.

Gage, P.J., Suh, H. and Camper, S.A. (1999b) Dosage requirement of Pitx2 for development of multiple organs. *Development* **126**: 4643–4651.

Gehring, W.J. (1993) Exploring the homeobox. *Gene* **135**: 215–221.

Gehring, W.J., Qian, Y.Q., Billeter, M., *et al.* (1994) Homeodomain-DNA recognition. *Cell* **78**: 211–223.

Hjalt, T.A., Amendt, B.A. and Murray, J.C. (2001) PITX2 regulates Procollagen Lysyl Hydroxylase (*PLOD*) gene expression. implications for the pathology of Rieger syndrome. *J. Cell. Biol.* **152**: 545–552.

Hjalt, T.A., Semina, E.V., Amendt, B.A. and Murray, J.C. (2000) The Pitx2 protein in mouse development. *Dev. Dyn.* **218**: 195–200.

Kitamura, K., Miura, H., Yanazawa, M., Miyashita, T. and Kato, K. (1997) Expression patterns of Brx1 (Rieg gene), Sonic hedgehog, Nkx2.2, Dlx1 and Arx during zona limitans intrathalamica and embryonic ventral lateral geniculate nuclear formation. *Mech. Dev.* **67**: 83–96.

Kitamura, K., Miura, H., Miyagawa-Tomita, S., *et al.* (1999) Mouse Pitx2 deficiency leads to anomalies of the ventral body wall, heart, extra- and periocular mesoderm and right pulmonary isomerism. *Development* **126**: 5749–5758.

Kozlowski, K. and Walter, M.A. (2000) Variation in residual PITX2 activity underlies the phenotypic spectrum of anterior segment developmental disorders. *Hum. Mol. Genet.* **9**: 2131–2139.

Krantz, I.D., Colliton, R.P., Genin, A., Rand, E.B., Li, L., Piccoli, D.A. and Spinner, N.B. (1998) Spectrum and frequency of jagged1 (*JAG1*) mutations in Alagille syndrome patients and their families. *Am. J. Hum. Genet.* **62**: 1361–1369.

Kleinjan, D.J. and van Heyningen, V. (1998) Position effect in human genetic disease. *Hum. Mol. Genet.* **7**: 1611–1618.

Kulak, S.C., Kozlowski, K., Semina, E.V., Pearce, W.G. and Walter, M.A. (1998) Mutation in the *RIEG1* gene in patients with iridogoniodysgenesis syndrome. *Hum. Mol. Genet.* **7**: 1113–1117.

Kume, T., Deng, K.Y., Winfrey, V., Gould, D.B., Walter, M.A. and Hogan, B.L. (1998) The forkhead/winged helix gene *Mf1* is disrupted in the pleiotropic mouse mutation congenital hydrocephalus. *Cell* **93**: 985–996.

Lanctot, C., Moreau, A., Chamberland, M., Tremblay, M.L. and Drouin, J. (1999) Hindlimb patterning and mandible development require the *Ptx1* gene. *Development* **126**: 1805–1810.

Lin, C.R., Kioussi, C., O'Connell, S., Briata, P., Szeto, D., Liu, F., Izpisua-Belmonte, J.C. and Rosenfeld, M.G. (1999) Pitx2 regulates lung asymmetry, cardiac positioning and pituitary and tooth morphogenesis. *Nature* **401**: 279–282.

Logan, M. and Tabin, C.J. (1999) Role of Pitx1 upstream of *Tbx4* in specification of hindlimb identity. *Science* **283**: 1736–1739.

Logan, M., Pagan-Westphal, S.M., Smith, D.M., Paganessi, L. and Tabin, C.J. (1998) The transcription factor Pitx2 mediates situs-specific morphogenesis in response to left-right asymmetric signals. *Cell* **94**: 307–317.

Lu, M.F., Pressman, C., Dyer, R., Johnson, R.L. and Martin, J.F. (1999) Function of Rieger syndrome gene in left-right asymmetry and craniofacial development. *Nature* **401**: 276–278.

Mammi, I., De Giorgio, P., Clementi, M. and Tenconi, R. (1998) Cardiovascular anomaly in Rieger syndrome: heterogeneity or contiguity? *Acta Ophthalmol. Scand.* **76**: 509–512.

Mears, A.J., Jordan, T., Mirzayans, F., *et al.* (1998) Mutations of the Forkhead/Winged-helix gene, *FKHL7*, in patients with Axenfeld–Rieger anomaly. *Am. J. Hum. Genet.* **63**: 1316–1328.

Mirzayans, F., Gould, D.B., Heon, E., Billingsley, G.D., Cheung, J.C., Mears, A.J. and Walter, M.A. (2000) Axenfeld–Rieger syndrome resulting from mutation of the *FKHL7* gene on chromosome 6p25. *Eur. J. Hum. Genet.* **8**: 71–74.

Mucchielli, M.L., Martinez, S., Pattyn, A., Goridis, C. and Brunet, J.F. (1996) *Otlx2*, an *Otx*-related homeobox gene expressed in the pituitary gland and in a restricted pattern in the forebrain. *Mol. Cell. Neurosci.* **8**: 258–271.

Murray, J.C., Bennett, S.R., Kwitek, A.E., *et al.* (1992) Linkage of Rieger syndrome to the region of the epidermal growth factor gene on chromosome 4. *Nat. Genetics* **2**: 46–49.

Nishimura, D.Y., Swiderski, R.E., Alward, W.L., *et al.* (1998) The forkhead transcription factor gene FKHL7 is responsible for glaucoma phenotypes which map to 6p25. *Nat. Genetics* **19**: 140–147.

Nishimura, D.Y., Searby, C.C., Borges, A.S., *et al.* (2000) Identification of a fourth Rieger syndrome locus at 16q24. *Am. J. Hum. Genet.* **67**: 2146.

Okajima, K., Robinson, L.K., Hart, M.A., *et al.* (1999) Ocular anterior chamber dysgenesis in craniosynostosis syndromes with a fibroblast growth factor receptor 2 mutation. *Am. J. Med. Genet.* **85**: 160–170.

Phillips, J.C., del Bono, E.A., Haines, J.L., Pralea, A.M., Cohen, J.S., Greff, L.J. and Wiggs, J.L. (1996) A second locus for Rieger syndrome maps to chromosome 13q14. *Am. J. Hum. Genet.* **59**: 613–619.

Piedra, M.E., Icardo, J.M., Albajar, M., Rodriguez-Rey, J.C. and Ros, M.A. (1998) Pitx2 participates in the late phase of the pathway controlling left-right asymmetry. *Cell* **94**: 319–324.

Prosser, J. and van Heyningen, V. (1998) PAX6 mutations reviewed. *Hum. Mutat.* 11: 93–108.

Rieger, H. (1934) Verlagerung und schitzform der pupille mit hypopasie des irisvordblattes. *Z. Augenheilk.* 84: 98–99.

Rieger, H. (1935) Beitraege zur kenntnis seltener und entrundung der pupille. *Albrecht von Graefes Arch. Klin. Exp. Ophthalmol.* **133**: 602–635.

Rieger, D.K., Reichenberger, E., McLean, W., Sidow, A. and Olsen, B.R. (2001) A double-deletion mutation in the *Pitx3* gene causes arrested lens development in aphakia mice. *Genomics* 72: 61–72.

Rogers, R.C. (1988) Rieger syndrome. *Proc. Greenwood. Genet. Center* 7: 9–13.

Ryan, A.K., Blumberg, B., Rodriguez-Esteban, C., *et al.* (1998) Pitx2 determines left-right asymmetry of internal organs in vertebrates. *Nature* **394**: 545–551.

Saadi, I., Semina, E.V., Harris, D.J., Murphy, K.P., Murray, J.C. and Russo, A.F. (2001) Identification of a dominant negative homeodomain mutation in Rieger syndrome. *J. Biol. Chem.* **276**: 23034–23041.

Schinzel, A., Brecevic, L., Dutly, F., Baumer, A., Binkert, F. and Largo, R.H. (1997) Multiple congenital anomalies including the Rieger eye malformation in a boy with interstitial deletion of (4) (q25–q27) secondary to a balanced insertion in his normal father: evidence for haplotype insufficiency causing the Rieger malformation. *J. Med. Genet.* **34**: 1012–1014.

Schweickert, A., Campione, M., Steinbeisser, H. and Blum, M. (2000) Pitx2 isoforms: involvement of Pitx2c but not Pitx2a or Pitx2b in vertebrate left-right asymmetry. *Mech. Dev.* **90**: 41–51.

Semina, E.V., Brownell, I., Mintz-Hittner, H.A., Murray, J.C. and Jamrich, M. (2001) Mutations in the human forkhead transcription factor FOXE3 with anterior segment ocular dysgenesis and cataracts. *Hum. Mol. Genet.* **10**: 231–236.

Semina, E.V., Ferrell, R.E., Mintz-Hittner, H.A., *et al.* (1998) A novel homeobox gene *PITX3* is mutated in families with autosomal-dominant cataracts and ASMD. *Nat. Genetics* **19**: 167–170.

Semina, E.V., Murray, J.C., Reiter, R., Hrstka, R.F. and Graw, J. (2000) Deletion in the promoter region and altered expression of *Pitx3* homeobox gene in aphakia mice. *Hum. Mol. Genet.* **9**: 1575–1585.

Semina, E.V., Reiter, R., Leysens, N.J., *et al.* (1996) Cloning and characterization of a novel bicoid-related homeobox transcription factor gene, RIEG, involved in Rieger syndrome. *Nat. Genetics* 14: 392–399.

Semina, E.V., Reiter, R.S. and Murray, J.C. (1997) Isolation of a new homeobox gene belonging to the *Pitx/Rieg* family: expression during lens development and mapping to the aphakia region on mouse chromosome 19. *Hum. Mol. Genet.* **6**: 2109–2116.

St Amand, T.R., Ra, J., Zhang, Y., Hu, Y., Baber, S.I., Qiu, M. and Chen, Y. (1998) Cloning and expression pattern of chicken *Pitx2*: a new component in the SHH signaling pathway controlling embryonic heart looping. *Biochem. Biophys. Res. Commun.* **247**: 100–105.

Szeto, D.P., Rodriguez-Esteban, C., Ryan, A.K., *et al.* (1999) Role of the Bicoid-related homeodomain factor Pitx1 in specifying hindlimb morphogenesis and pituitary development. *Genes Dev.* **13**: 484–494.

Tremblay, J.J., Goodyer, C.G. and Drouin, J. (2000) Transcriptional properties of Ptx1 and Ptx2 isoforms. *Neuroendocrinology* **71**: 277–286.

Tsai, J.C. and Grajewski, A.L. (1994) Cardiac valvular disease and Axenfeld-Rieger syndrome. *Am. J. Ophthalmol.* 118: 255–256.

Vollrath, D., Jaramillo-Babb, V.L., Clough, M.V., McIntosh, I., Scott, K.M., Lichter, P.R. and Richards, J.E. (1998) Loss-of-function mutations in the LIM-homeodomain gene, **LMX1B**, in nail-patella syndrome. *Hum. Mol. Genet.* **7**: 1091–1098.

Yasui, K., Zhang, S., Uemura, M. and Saiga, H. (2000) Left–right asymmetric expression of *BbPtx*, a *Ptx*-related gene, in a lancelet species and the developmental left-sidedness in deuterostomes. *Development* **127**: 187–195.

Yoshioka, H., Meno, C., Koshiba, K., *et al.* (1998) *Pitx2*, a bicoid-type homeobox gene, is involved in a lefty-signaling pathway in determination of left-right asymmetry. *Cell* **94**: 299–305.

Yu, X., St Amand, T.R., Wang, S., Li, G., Zhang, Y., Hu, Y., Nguyen, L., Qiu, M., Chen, Y. (2001) Differential expression and functional analysis of Pitx2 isoforms in regulation of heart looping in the chick. *Development* **128**: 1005–1013.

14

The hedgehog pathway and developmental disorders

Allen E. Bale

1. Introduction

Hedgehog is a secreted molecule that influences the differentiation of a variety of tissues during development. Hedgehog, its receptor, 'patched', and many downstream members of the hedgehog signal transduction pathway were originally discovered by developmental biologists studying embryogenesis in the fruit fly, *Drosophila melanogaster* (Nusslein Volhard and Wieschaus, 1980). The whimsical name, 'hedgehog', is derived from the appearance of the mutant fruit fly embryo, which is covered with bristles and in this sense resembles the small mammals known as hedgehogs. *Drosophila* hedgehog works in concert with other molecules to lay down the basic framework of the embryo, determining anterior–posterior relationships in developing structures (Peifer and Bejsovec, 1992). Genes involved in this type of embryonic patterning are known as 'segment polarity genes'. The general flavor of a segment polarity defect in *Drosophila* is loss of anterior–posterior orientation and may include both mirror-image duplication of structures that should have distinct anterior and posterior sides and fusion of paired structures where one is normally anterior and the other posterior. The bristle-covered appearance of the *hedgehog* mutant embryo is caused by development of bristles in both the anterior and posterior half of body segments which normally would have bristles only in a stripe in the anterior half.

Vertebrate homologs of many segment polarity genes have been identified, and in most cases a single gene in *Drosophila* corresponds to a family of related homologs in vertebrates. The vertebrate hedgehog pathway plays a fundamental role in development of the central and peripheral nervous systems, skeleton, limbs, lungs, skin, hair, and germ cells (Wicking *et al.*, 1999). Disruption of hedgehog signalling in developing vertebrate embryos can lead to defects analogous to segment polarity abnormalities in *Drosophila*. Mouse models as well as several human disease states have helped determine the normal role of this pathway in

Genotype to Phenotype second edition, edited by S. Malcolm and J. Goodship.

mammalian development. Surprisingly, mutations of some of the members of this pathway are also associated with human cancer as well as birth defects. For example, germline mutations of *patched* cause Gorlin syndrome, an autosomal dominant disorder characterized by multiple skin cancers and other tumors, as well as congenital anomalies of the brain, bones, and teeth (Hahn *et al.*, 1996).

2. Biochemistry of the hedgehog pathway (*Figure 1*)

Hedgehog encodes a 45 kDa protein that undergoes autocatalytic cleavage and modification to give a 20 kDa, active N-terminal fragment covalently bound to cholesterol (Porter *et al.*, 1996). The role of cholesterol in hedgehog signalling is not known, but it may be important in limiting diffusion of the hedgehog molecule and the spatial distribution of its effects. The cholesterol moiety potentiates, but is not absolutely essential for hedgehog signalling. Three vertebrate homologs of the *Drosophila hedgehog* gene have been identified including *sonic hedgehog* (SHH), desert hedgehog *(DHH)* and Indian hedgehog *(IHH)*. *SHH* is the most broadly expressed member of this family and is probably responsible for the major effects on development of the brain, spinal cord, axial skeleton, and limbs (Chiang *et al.*, 1996). *IHH* has been implicated in regulation of cartilage differentiation in the growth of long bones (Lanske *et al.*, 1996; St-Jacques *et al.*, 1999; Vortkamp *et al.*, 1996), and *DHH* exerts its effect mainly in the developing germline and in Schwann cells of the peripheral nervous system (Bitgood *et al.*, 1996; Mirsky *et al.*, 1999).

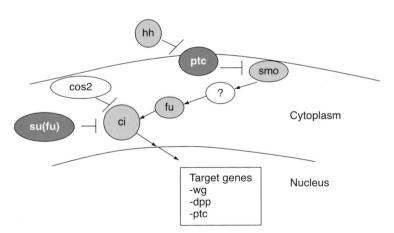

Figure 1. Elements of the *Drosophila* hedgehog pathway. Members of the hedgehog (hh) family bind to patched (ptc) releasing smoothened (smo) to transduce a signal. A downstream complex composed of fused (fu), suppressor of fused (su(fu)), costal 2 (cos2), and cubitus interruptus (ci) dissociates, and an active form of ci translocates to the nucleus where it switches on transcription of the target genes, wingless (wg), decapentaplegic (dpp), and patched (ptc). Vertebrate homologs of all of these genes have been identified, and in most cases a single gene in *Drosophila* corresponds to a family of related homologs in vertebrates.

The hedgehog signal is received and transduced at the membrane via a receptor complex (Marigo *et al.*, 1996; Stone *et al.*, 1996) consisting of patched and smoothened. Patched is the component that specifically binds hedgehog. The cDNA sequence of *patched* predicts a 1500 amino acid glycoprotein with 12 membrane-spanning domains (Hooper *et al.*, 1989; Nakano *et al.*, 1989) and two large extracellular loops that are required for hedgehog binding. There are several patched homologs in humans. *Patched1 (PTCH)* is probably the major receptor molecule for all three forms of human hedgehog, and mutations in this gene are associated with the a wide variety of birth defects (Hahn *et al.*, 1996; Goodrich *et al.*, 1997). *Patched2 (PTCH2)* is a close homolog of *PTCH* . Its normal function is not known, although it is mutated in rare medulloblastomas and basal cell carcinomas (Motoyama *et al.*, 1998; Smyth *et al.*, 1999). *TRC8* has homology to the region of *PTCH* encoding the membrane spanning domains and second extracellular loop. It was found by cloning a translocation breakpoint in a renal cell carcinoma family and is mutated in some sporadic renal cell tumors (Gemmill *et al.*, 1998). The transmembrane domains of patched show an intriguing homology to the 'cholesterol sensing' motifs of the Niemann–Pick disease protein (NPC1) and 3-hydroxy-3-methylglutaryl coenzyme A (HMG-CoA) reductase (Carstea *et al.*, 1997; Loftus *et al.*, 1997). The functional significance of this homology is not clear as there is no evidence that patched participates in cholesterol homeostasis, but this motif may have a broader role in intracellular trafficking of receptors and their ligands. The interaction between patched and the hedgehog may in some way involve the cholesterol sensing domain.

Smoothened is a 115 kDa protein with structural similarity to serpentine, G-protein coupled receptors (Alcedo *et al.*, 1996; van den Heuvel and Ingham, 1996). There appears to be only one human smoothened gene (*SMO*), though it is homologous to members of the 'frizzled' (WNT receptor) family. In the absence of hedgehog, smoothened and patched form an inactive complex. When hedgehog binds to patched, the complex is altered and smoothened and is then free to transduce the signal. Hedgehog binding does not physically release smoothened from patched, and presumably the release of inhibition reflects a modification or conformational change in smoothened. Smoothened is presumed to transduce the signal to downstream members of the pathway through a G-protein, but no such interacting protein has been identified in *Drosophila* or vertebrates.

Based on epistatic interactions in *Drosophila*, fused, suppressor of fused, costal 2 and cubitus interruptus (ci) lie in the pathway downstream from smoothened (Ingham, 1998). The ultimate member of the pathway in *Drosophila* is ci, a 155 kDa zinc finger transcription factor homologous to the GLI family in vertebrates. In cells not exposed to hedgehog (hh), ci forms a complex with costal-2, fused and supressor of fused at the microtubules. In this form, Ci can be cleaved to a 75 kDa N terminal fragment that retains the zinc finger domain and can apparently translocate to the nucleus and repress downstream target genes (Ohlmeyer and Kalderon, 1998). In the presence of hh, the complex dissociates and full-length ci is thought to mature into a short-lived transcriptional activator which translocates to the nucleus and transcriptionally activates target genes. Targets of the *Drosophila* hedgehog pathway include *wingless* (WNT family in vertebrates), and *decapentaplegic (TGF beta* family in vertebrates), and *PTCH*, itself. The wingless and decapentaplegic proteins may be the main mediators of hedgehog effect both by autocrine effects in the cells responding to the hh signal, and paracrine effects on surrounding tissues. The upregulation of PTCH expression results in the presentation of large amounts of PTCH protein at the cell membrane, which sequesters hedgehog and limits its spread beyond the cells in which it is produced (Chen and Struhl, 1996).

3. What human disease and mouse models tell us about the function of the hedgehog pathway

3.1 Switching off the pathway

Holoprosencephaly. Holoprosencephaly is a developmental disorder characterized by incomplete cleavage of the forebrain during embryogenesis leading to variable defects of the brain and face. Examples of developmental defects include a single ventricle instead of two lateral ventricles, absence of corpus callosum, absent or abnormal pituitary, lack of olfactory lobes, lack of optic nerves, cyclopia and other eye defects, hypotelorism, and midline clefting of lip, palate, nose.

Estimates of the incidence of this disorder range from 4:100 000 to eight per 100 000 live births (Croen *et al.*, 1996; Olsen *et al.*, 1997). The disease is genetically heterogeneous. Twenty five to 40% of cases are cytogenetically abnormal (mostly trisomy 13, some trisomy 18, all other abnormalities rare). Among cytogenetically normal cases, a substantial minority have a family history, and the majority of recognized familial cases are autosomal dominant with variable expressivity. In several autosomal dominant families, the disease gene was mapped to chromosome 7q36, a region containing the *SHH* gene. Molecular analysis showed that mutations of *SHH* underlie this disease (Belloni *et al.*, 1996; Roessler *et al.*, 1996) in both the familial forms mapping to chromsome 7q and in a small percentage of sporadic patients. The pathogenesis of this disorder is believed to reflect haploinsufficiency for SHH protein, and the marked variable expressivity is presumed to be due to random stochastic effects and possibly the effects of modifying genes. That loss of SHH function leads to holoprosencephaly indicates that vertebrate hedgehog is involved in early differentiation of ventral brain structures. SHH, itself, probably does not have a direct role in right–left segmentation of CNS structures but rather a broader role in differentiation of ventral CNS precursors and induction of other genes directly involved in bilateral subdivision of the brain.

The phenotype of human holoprosencephaly due to SHH mutation is almost entirely limited to the brain and face with no apparent dysmorphology of other structures in which SHH is believed to function in embryogenesis (Gianotti *et al.*, 1999; Odent *et al.*, 1999). The lack of limb and skeletal defects probably reflects less sensitivity to haploinsufficiency in these developing structures. In contrast to the human disease, the mouse model homozygous for targetted disruption of SHH shows both a holoprosencephaly-like phenotype and abnormal development of the spine, ribs, and distal limbs (Chiang *et al.*, 1999). The limb abnormalities include absence of digits and fusion of anterior and posterior distal limb bones. The haploinsufficiency state in the mouse is not associated with a phenotype.

Grieg cephalopolysyndactyly and Pallister–Hall syndrome. Three related autosomal dominant disorders are caused by mutation in human *GLI3*, a zinc finger gene related to Drosophila *cubitus interruptus* (Biesecker, 1997). Grieg cephalopolysyndactyly is characterized by syndactyly, polydactyly, broad or bifid thumbs, unusual skull shape, and hip dislocations. The mutations that cause this phenotype range from large deletions to premature stops resulting in truncation of the protein upstream from the zinc finger domain (Vortkamp *et al.*, 1991). Pallister–Hall syndrome has a wide variety of features including virtually all of the findings in Grieg cephalopolysyndactyly plus hypothalamic hamartoblastomas, hypopituitarism, abnormalities of the palate, tongue, and jaw, short limbs, nail dysplasia, imperforate anus, laryngeal cleft, renal dysplasia, and congenital heart disease. The mutations found in patients with this syndrome lead to truncation of the protein just downstream of the zinc finger domain and could be producing a product similar to the 75 kDa, repressor forms of the ci/GLI protein family (Kang *et al.*, 1997). Hence, these mutations may function as dominant negatives. A third syndrome related to mutations in GLI3 is isolated postaxial polydactyly type 3 (Radhakrishna *et al.*, 1997). The mutations in this disorder trucate the protein farther downstream than those in Pallister–Hall and are assumed to produce a partially functional protein.

Smith–Lemli–Opitz syndrome (SLOS). SLOS is an autosomal recessive malformation syndrome associated with hypospadius, cryptorchidism, syndactyly and polydactyly, ptosis, broad nasal tip, microcephaly and a variety of other malformations. Some cases have facial clefting and holoprosencephaly. The molecular basis for this syndrome is a defect of cholesterol biosynthesis at the level of 3 beta-hydroxy-steroid-delta7-reductase (Shefer *et al.*, 1995). The SHH protein normally undergoes post-translational modifications including addition of cholesterol (Incardona and Eaton, 2000). The holoprosencephaly and possibly polydactyly associated with this syndrome may be due to failure to modify SHH protein correctly (Kelley *et al.*, 1996). Hence, SML leads to a secondary defect in the SHH protein.

3.2 Switching on the pathway

Gorlin syndrome and patched. Gorlin syndrome (also known as the nevoid basal cell carcinoma syndrome and basal cell nevus syndrome) is an autosomal dominant disorder characterized by predisposition to basal cell carcinomas (*BCCs*) of the skin, medulloblastomas, and ovarian fibromas. Other neoplasms that occur at an increased incidence in NBCCS include fibrosarcomas, meningiomas, rhabdomyosarcomas, and cardiac fibromas. In addition to benign and malignant tumors, malformations are a striking component of NBCCS. The syndrome is associated with pits of the palms and soles, keratocysts of the jaw and other dental malformations, cleft palate, characteristic coarse facies, strabismus, dysgenesis of the corpus callosum, calcification of the falx cerebri, spina bifida occulta and other spine anomalies, bifid ribs and other rib anomalies, polydactyly, ectopic calcification, mesenteric cysts, macrocephaly, and generalized overgrowth (Bale *et al.*, 1991; Gorlin, 1995; Kimonis *et al.*, 1997). Although not all patients with this syndrome are tall, some patients may reach gigantic proportions and exhibit features reminiscent of acromegaly.

The prevalence of Gorlin syndrome has been estimated at one per 56 000, and 1–2% of medulloblastomas (Evans *et al.*, 1991) and approximately 0.1% of BCCs are attributable to the syndrome (Springate, 1986). Two thirds of cases have a positive family history with the remainder being sporadic. Males and females are affected equally, and there is no evidence for imprinting. There are very few cases of non penetrance, but the extent of expression of the many features of the syndrome is variable (Anderson *et al.*, 1967). Unilateral Gorlin syndrome is not uncommon and no doubt reflects somatic mutation in one cell of an early embryo.

Since the syndrome was delineated in the late 1950s and 1960s (Gorlin and Goltz, 1962; Howell and Caro, 1959), numerous laboratory investigations have been undertaken to identify the underlying molecular basis. The prominence of developmental defects makes this syndrome unusual among autosomal dominant cancer predisposition syndromes. Nevertheless, NBCCS shares with other disorders the multiplicity, random distribution, and early age of onset for neoplasms. Statistical analysis of the distribution of BCCs in affected individuals suggested that tumors in the syndrome arise through a two-hit mechanism (Strong, 1977) and that the underlying defect might be mutation in a tumor suppressor gene (Cavenee *et al.*, 1983). This theory was strongly supported by the mapping of the NBCCS gene to chromosome 9q22–31 and the demonstration that the exact same region was deleted in a high percentage of BCCs and other tumors related to the disorder (Gailani *et al.*, 1992).

Positional cloning identified the human homolog of a *Drosophila patched* as a candidate gene (Hahn *et al.*, 1996; Johnson *et al.*, 1996). Vertebrate patched was known to function in the developing neural tube, pharyngeal pouches, somites and limb buds (Goodrich *et al.*, 1996). Many of the clinical features of NBCCS, including abnormalities of the brain, craniofacial structures, ribs, vertebrae and limbs correlate well with the apparent role of *patched* in the development of these structures, making *patched* a good candidate gene for this syndrome. Furthermore, bifid ribs and polydactyly were 'segment polarity-like' features. Screening of the *patched* coding region revealed a wide spectrum of mutations in NBCCS patients, with the majority predicted to result in premature protein truncation (Chidambaran *et al.*, 1996; Wicking *et al.*, 1997). Mutations are spread throughout the entire gene with no apparent clustering. The extensive phenotypic variability does not correlate with the nature of or location of mutations in *patched*. Different kindreds with identical mutations differ dramatically in the extent of clinical features, suggesting that genetic background or environmental factors may have an important role in modifying the spectrum of both developmental and neoplastic traits.

The pathophysiology of generalized and symmetric developmental defects in Gorlin syndrome probably relates to haploinsufficiency for the protein. Generalized overgrowth, for example, suggests that loss of just one copy of the *patched* exerts an effect on growth and differentiation. A possible mechanism relating overgrowth to deficiency of patched involves cartilage differentiation (*Figure 2*). Growth of long bones during childhood occurs via proliferation of chondrocytes at the epiphyses. Following puberty, the last of the cartilage cells differentiate, the epiphyses close, and linear growth stops. Indian hedgehog is involved in control of bone growth via regulation of parathormone-related

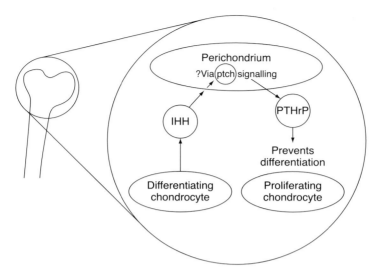

Figure 2. Regulation of chondrocyte differentiation by Indian hedgehog (IHH) and parathormone-related peptide (PTHrP). IHH is secreted by prehypertrophic (early differentiating) chondrocytes. Perichondrial cells receiving an IHH signal secrete parathormone-related peptide (PTHrP). PTHrP prevents differentiation of additional chondrocytes. Patched (PTCH) expression is induced in perichondrial cells by IHH and is presumed to be a mediator of IHH effect on PTHrP expression. Because PTCH normally opposes IHH effects, the reduced gene dosage of PTCH in Gorlin syndrome patients should lead to increased IHH effects. Increased secretion of PTHrP would shift the balance toward greater chondrocyte proliferation and increased growth of long bones.

peptide (PTHrP) (Lanske *et al.*, 1996; St-Jacques *et al.*, 1999; Vortkamp *et al.*, 1996). This substance, which is secreted by perichondrial cells of the epiphyses in response to hedgehog signaling, prevents differentiation of chondrocytes and thereby permits continued linear bone growth. By analogy to the relationship between hedgehog and patched in *Drosophila*, it seems likely that the action of hedgehog is mediated through inhibition of patched function. Presumably patched activity constitutively represses expression of PTHrP. This model predicts that in the absence of patched, PTHrP expression would be high, resulting in a shift of the balance toward proliferation rather than differentiation of chondrocytes.

The scattered, non symmetric developmental defects in Gorlin syndrome probably occur through a two-hit mechanism involving loss of patched function in embryonic or fetal cells (*Figure 3*). Like the neoplasms in the syndrome many of the anomalies are multiple and appear in a random pattern (e.g. spina bifida occulta, bifid ribs, and keratocysts of the jaw), but isolated defects of the same type are seen occasionally in the general population. The jaw cysts in patients with NBCCS are lined with keratinizing epithelium, which is similar to the lining of the oral cavity. This epithelium probably derives from tooth precursors which either migrated abnormally or failed to regress. Molecular studies of this epithelial tissue show loss of the normal allele resulting in hemizygosity for the mutation

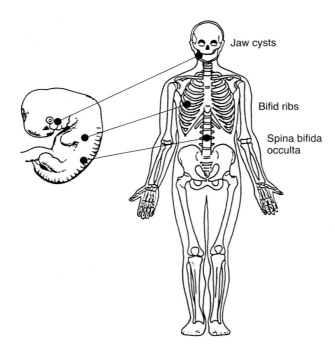

Jaw cysts

Bifid ribs

Spina bifida
occulta

Figure 3. A two-hit model
for birth defects in Gorlin
syndrome. Cells in a
Gorlin syndrome fetus
suffering a 'second hit' at
the patched locus develop
into abnormal clones that
give rise to developmental
defects. Loss of the
normal allele has been
demonstrated in jaw cysts.
This model could also
explain bifid ribs and
spinal bifida occulta.

(Levanat *et al.*, 1996). Sporadic keratocysts have also been shown to have *PTCH* mutations. These data are consistent with a two-hit model in which inactivation of both copies of the gene in a tooth precursor cell leads to a dysregulated clone manifested later as a developmental abnormality. The mouse models for this syndrome (heterozygous patched knockouts) (Goodrich *et al.*, 1997; Hahn *et al.*, 1998) recapitulate the features of the human disease almost precisely and support a two-hit model for birth defects. Ionizing radiation *in utero* increases the frequency of the scattered, random types of birth defects consistent with developmental defects arising from embryonic and fetal cells that have suffered a second, radiation-induced hit.

Neoplasia in the syndrome can be explained by activation of downstream members of the hedgehog pathway in cells that have lost both normal copies of *PTCH*. For example, the human GLI genes are functionally activated by loss of patched, and GLI1 has previously been shown to act as an oncogene in brain tumors including medulloblastomas (Kinzler *et al.*, 1987). Furthermore, mouse models over expressing GLI2 in epidermis develop skin tumors that resemble BCCs (Grachtchouk *et al.*, 2000). WNT1, a vertebrate homolog of *Drosophila* wingless, is known to cause mammary tumors in mice when activated (Nanni *et al.*, 1994). Switching on the WNT pathway in humans results in a variety of tumors including medulloblastomas (APC mutations in Turcot syndrome) (Hamilton *et al.*, 1995) and skin tumors (Chan *et al.*, 1999). Decapentaplegic is a member of the transforming growth factor-beta (TGFB) superfamily, with closest homology to the vertebrate bone-morphogenetic-protein (BMP) subfamily. Members of the TGF-beta family have complex roles in cell growth and differentiation. The mad (mothers against dpp) gene is a component of the signal transduction pathway of dpp in Drosophila, and two human homologs of the mad gene, DPC4 and MADR2, have been shown to act as tumor suppressors (Eppert *et al.*, 1996; Hahn *et al.*, 1996; Thiagalingam *et al.*, 1996).

Sporadic BCCs. In addition to germline mutations in Gorlin syndrome, *patched* has been found to be mutated in sporadic tumors associated with the syndrome (Gailani *et al.*, 1996; Unden *et al.*, 1996). The majority of BCCs, both sporadic and hereditary, show allelic loss for chromosome 9q22. Minute BCCs are as likely as large tumors to have chromosome 9 allelic loss, and all histological subtypes, whether primary or recurrent, have a high frequency of allelic loss on chromosome 9. Tumors with allelic loss on chromosome 9 sometimes show additional areas of loss on other chromosomes, but no tumors have loss on other chromosomes without involvement of chromosome 9 (Gailani *et al.*, 1992; van der Riet *et al.*, 1994). Identification of *patched* has made possible direct mutation screening in sporadic tumors. Meticulous sequence analysis has revealed inactivating mutations in both homologs of the gene in the great majority of BCCs.

The mutational spectrum of *patched* in BCCs suggests that environmental factors other than ultraviolet B (UVB), the predominant carcinogenic component of sunlight, may play a role in tumorigenesis. UVB typically causes formation of photodimers that result in G.C–A.T transitions opposite dipyrimidine sites (Ananthaswamy *et al.*, 1990). Mutations in p53, which occur in 40–56% of BCCs, are almost always UVB-related (Ziegler *et al.*, 1993) and mutations in the ras family of proto-oncogenes are often of the type caused by UVB (van der Schroeff *et al.*, 1990). However, fewer than 50% of the *patched* mutations in sporadic BCCs have the typical UVB signature (Gailani *et al.*, 1996). Other carcinogens, possibly ultraviolet A or cosmic rays, may play a synergistic role with UVB in BCC etiology.

Given that constitutive activation of SMO is likely to upregulate transcription of hedgehog target genes in the same way as PTCH inactivation, it is not surprising that activating mutations in the *SMO* gene have been detected in 10–20% of sporadic BCCs (Xie *et al.*, 1998). One common mutation (Trp535Leu) in the 7th transmembrane domain of Smo has been detected in most BCCs lacking PTCH mutations. In contrast to wild-type smoothened, the mutant form has been shown to result in constitutive smoothened signalling in an *in vitro* focus forming assay and in transgenic mice expressing this mutant gene under control of an epidermis promoter.

As hedgehog itself is primarily responsible for activation of this pathway, it is feasible that it also may be mutated in associated tumors, a premise supported by the finding that transgenic mice overexpressing *Sonic hedgehog* develop BCC-like skin lesions (Oro *et al.*, 1997). Accordingly, a single recurrent mutation in *Sonic hedgehog* was initially reported in a range of tumor types including BCC, but failure of other workers to detect this mutation suggests that it is extremely rare.

Taken together, inactivation of *PTCH* or oncogenic activation of *SMO* occurs in a large proportion of BCCs, suggesting that dysregulation of hedgehog signalling is a requirement for BCC formation. The term gatekeeper was coined by Kinzler and Vogelstein (1996) to describe genes which must be inactivated or activated to give rise to a particular type of tumor. Although it has previously been suggested that *patched* is the gatekeeper gene for BCC formation (Sidransky, 1996), it may be more accurate to regard the receptor complex consisting of patched and smoothened as the BCC gatekeeper.

Sporadic medulloblastoma and other tumor types. Mutations in genes of the hedgehog pathway have also been described in a range of tumors other than BCCs. In keeping with the predisposition of NBBCS patients to develop the childhood brain tumor medulloblastoma, mutations of both *PTCH* and *SMO* have been detected in sporadic medulloblastomas and other primitive neuroectodermal tumors (Pietsch *et al.*, 1997; Raffel *et al.*, 1997; Wolter *et al.*, 1997). NBCCS patients develop the desmoplastic histological subtype of medulloblastoma, and while one study reported *PTCH* mutations exclusively in this subtype (Pietsch *et al.*, 1997), mutations in classic medulloblastomas have also been described (Wolter *et al.*, 1997). In all cases, the incidence of mutations in these tumors was significantly less than that observed in BCCs, indicating it is likely that only a subset of these neoplasias are caused by mutations in *PTCH* or in members of the hedgehog signalling pathway. This notion is also supported by the frequency of loss of heterozygosity of the *PTCH* locus in medulloblastomas which is reduced when compared to BCCs (Pietsch *et al.*, 1997).

Other tumors which carry *PTCH* mutations include tricoepitheliomas, esophageal squamous cell carcinomas, and transitional cell carcinomas of the bladder (Maesawa *et al.*, 1998; McGarvey *et al.*, 1998). Although mutations in *PTCH* are yet to be demonstrated in sporadic rhabdomyosarcomas, mice heterozygous for a null allele of *patched* develop this soft tissue tumor of the muscle at a high frequency (Hahn *et al.*, 1998). As this tumor is also found at an increased incidence in NBCCS individuals, it is likely that this pathway plays a role in its pathogenesis.

4. Conclusions

Although the role of the hedgehog pathway in embryonic development was discovered more than 20 years ago, many of the mediators of hedgehog signal are yet to be identified. The biochemical interactions that are known to occur as a result of exposure of cells to hedgehog do not suggest a purely linear pathway, and the word 'network' would better describe the series of interactions that mediate and modulate the effects of hedgehog on nuclear targets genes. Even in *Drosophila*, there are several lines of evidence that indicate more complexity than would be expected for a linear pathway. For example, mutation of hedgehog and mutation of smoothened do not produce identical phenotypes. Multiple vertebrate homologs of many members of *Drosophila* pathway allow for more complexity due to branch points at almost every level of signalling.

Human hereditary diseases associated with mutations in hedgehog-related genes reflect the complexity of hedgehog signalling. In a purely linear pathway, it might be expected that any mutation inhibiting the signal would result in the same phenotype. That holoprosencephaly and Grieg cephalopolysyndactyly have few phenotypic features in common is probably a result of branching of the pathway downstream from hedgehog as well as different sensitivities to haploinsufficiency for different gene products. Similarly, naturally occurring mutations in patched are restricted almost entirely to BCCs and medulloblastomas. Gli1 however, is mutated almost entirely in brain tumors. If switching on the pathway at any level had the same effect, one might expect the tissue specificity of patched and Gli mutations to be the same. It also appears that certain common features can be caused by dysregulation that either enhances or suppresses the hedgehog signal. For example, dysgenesis of the corpus callosum is common in both holoprosencephaly and Gorlin syndrome. Likewise, polydactyly is a feature of both Grieg cephalopolysyndactyly and Gorlin syndrome.

Despite difficulties predicting exactly which phenotypes might be associated with mutations in particular genes, there is a general 'flavor' to human hereditary diseases associated with this pathway including midline brain defects, polydactyly, syndactyly, mirror-image duplications and other defects of segmental structures of the axial skeleton, and in some cases cancer predisposition. Vertebrate homologs of *suppressor of fused, fused,* and *costal 2* have recently been reported. Mouse models or insightful mutation analysis of human hereditary diseases will provide insight into the normal function of these segment polarity genes in vertebrate development.

References

Alcedo, J., Ayzenzon, M., Von Ohlen, T., Noll, M. and Hooper, J.E. (1996) The Drosophila smoothened gene encodes a seven-pass membrane protein, a putative receptor for the hedgehog signal. *Cell* **86**: 221–232.

Ananthaswamy, H.N. and Pierceall, W.E. (1990) Molecular mechanisms of ultraviolet radiation carcinogenesis. *Photochem. Photobiol.* **52**: 1119–1136.

Anderson, D.E., Taylor, W.B., Falls, H.F. and Davidson, R.T. (1967) The nevoid basal cell carcinoma syndrome. *Am. J. Hum. Genet.* **19**: 12–22.

Bale, S.J., Amos, C.I., Parry, D.M. and Bale, A.E. (1991) The relationship between head circumference and height in normal adults and in the nevoid basal cell carcinoma syndrome and neurofibromatosis type 1. *Am. J. Med. Genet.* **40**: 206–210.

Belloni, E., Muenke, M., Roessler, E. *et al.* (1996) Identification of sonic hedgehog as a candidate gene responsible for holoprosencephaly. *Nat. Genet.* **14**: 353–356.

Biesecker, L.G. (1997) Strike three for GLI3. *Nature Genet.* **17**: 259–260.

Bitgood, M.J., Shen, L. and McMahon, A.P. (1996) Sertoli cell signaling by Desert hedgehog regulates the male germline. *Curr. Biol.* **6**: 298–230.

Carstea, E.D., Morris, J.A., Coleman, K.G. *et al.* (1997) Neimann-Pick C1 disease gene: Homology to mediators of cholesterol homeostasis. *Science* **277**: 228–231.

Cavenee, W.K., Dryja, T.P., Phillips, R.A. *et al.* (1983) Expression of recessive alleles by chromosomal mechanisms in retinoblastoma. *Nature* **305**: 779–784.

Chan, E.F., Gat, U., McNiff, J.M. and Fuchs E. (1999) A common human skin tumour is caused by activating mutations in beta-catenin. *Nat. Genet.* **21**: 410–413.

Chen, Y. and Struhl, G. (1996) Dual roles for patched in sequestering and transducing Hedgehog. *Cell* **87**: 553–563.

Chiang, C., Litingtung, Y., Lee, E., Young, K.E., Corden, J.L., Westphal, H. and Beachy, P.A. (1996) Cyclopia and defective axial patterning in mice lacking sonic hedgehog gene function. *Nature* **383**: 407–413.

Chidambaram, A., Goldstein, A.M., Gailani, M.R. *et al.* (1996) Mutations in the human homologue of the Drosophila patched gene in Caucasian and African-American nevoid basal cell carcinoma syndrome patients. *Cancer Res.* **56**: 4599–4601.

Croen, L.A., Shaw, G.M. and Lammer, E.J. (1996) Holoprosencephaly: epidemiologic and clinical characteristics of a California population. *Am. J. Med. Genet.* **64**: 465–472.

Eppert, K., Scherer, S.W., Ozcelik, H. *et al.* (1996) MADR2 maps to 18q21 and encodes a TGFB-regulated MAD-related protein that is functionally mutated in colorectal carcinoma. *Cell* **86**: 543–552.

Evans, D.G.R., Farndon, P.A., Burnell, L.D., Gattamaneni, H.R. and Birch, J.M. (1991) The incidence of Gorlin syndrome in 173 consecutive cases of medulloblastoma. *Br. J. Cancer* **64**: 959–961.

Gailani, M., Stahle-Backdahl, M., Leffell, D. *et al.* (1996) The role of the human homologue of Drosophila patched in sporadic basal cell carcinomas. *Nat. Genet.* **14**: 78–81.

Gailani, M.R., Bale, S.J., Leffell, D.J. *et al.* (1992) Developmental defects in Gorlin syndrome related to a putative tumor suppressor gene on chromosome 9. *Cell* **69**: 111–117.

Gemmill, R.M., West, J.D., Boldog, F., Tanaka, N., Robinson, L.J., Smith, D.I., Li, F. and Drabkin, H.A. (1998) The hereditary renal cell carcinoma 3;8 translocation fuses FHIT to a patched-related gene, TRC8. *Proc. Nat. Acad. Sci.* **95**: 9572–9577.

Giannotti, A., Imaizumi, K., Jones, K.L. *et al.* (1999) SHH mutations cause a significant proportion of autosomal dominant holoprosencephaly. *Hum. Mol. Gen.* **8**: 2479–2488.

Goodrich, L.V., Johnson, R.L., Milenkovic, L., McMahon, J.A. and Scott, M.P. (1996) Conservation of the hedgehog/patched signaling pathway from flies to mice: induction of a mouse patched gene by hedgehog. *Genes Devel.* **10**: 301–312.

Goodrich, L.V., Milenkovic, L., Higgins, K.M. and Scott, M.P. (1997) Altered neural cell fates and medulloblastoma in mouse patched mutants. *Science* **277**: 1109–1113.

Gorlin, R.J. (1995) Nevoid basal cell carcinoma syndrome. *Dermatol. Clin.* **13**: 113–125.

Gorlin, R.J. and Goltz, R.W. (1962) Multiple nevoid basal-cell epithelioma, jaw cysts and bifid rib. A syndrome. *N. Engl. J. Med.* **262**: 908–912.

Grachtchouk, M., Mo, R., Yu, S., Zhang, X., Sasaki, H., Hui, C.C. and Dlugosz, A.A. (2000) Basal cell carcinomas in mice overexpressing Gli2 in skin. *Nat. Genet.* **24**: 216–217.

Hahn, S.A., Schutte, M., Hoque, A.T. *et al.* (1996) DPC4, a candidate tumor suppressor gene at human chromosome 18q21.1. *Science* **271**: 350–353.

Hahn, H., Wicking, C., Zaphiropoulos, P.G. *et al.* (1996) Mutations of the human homolog of Drosophila patched in the nevoid basal cell carcinoma syndrome. *Cell* **85**: 841–851.

Hahn, H., Wojnowski, L., Zimmer, A.M., Hall, J., Miller, G. and Zimmer, A. (1998) Rhabdomyosarcomas and radiation hypersensitivity in a mouse model of Gorlin syndrome. *Nature Med.* **4**: 619–622.

Hamilton, S.R., Liu, B., Parsons, R.E. *et al.* (1995) The molecular basis of Turcot's syndrome. *New Eng. J. Med.* **332**: 839–847.

Hooper, J.E. and Scott, M.P. (1989) The Drosophila patched gene encodes a putative membrane protein required for segmental patterning. *Cell* **59**: 751–765.

Howell, B. and Caro, M.R. (1959) The basal cell nevus: Its relationship to multiple cutaneous cancers and associated anomalies of development. *Arch. Dermatol.* **79**: 67–80.

Incardona, J.P. and Eaton, S. (2000) Cholesterol in signal transduction. *Cell Biol.* **12**: 193–203.

Ingham, P.W. (1998) Transducing hedgehog: the story so far. *EMBO* **17**: 3505–3511.

Johnson, R.L., Rothman, A.L., Xie, J. *et al.* (1996) Human homolog of patched, a candidate gene for the basal cell nevus syndrome. *Science* **272**: 1668–1671.

Kang, S., Graham, J.M. Jr., Haskins-Olney, A. and Biesecker, L.G. (1997) GLI3 frameshift mutations cause autosomal dominant Pallister-Hall syndrome. *Nat. Genet.* **15**: 266–268.

Kelley, R.L., Roessler, E., Hennekam, R.C., Feldman, G.L., Kosaki, K., Jones, M.C., Palumbos, J.C. and Muenke, M. (1996) Holoprosencephaly in RSH/Smith-Lemli-Opitz syndrome: does abnormal cholesterol metabolism affect the function of Sonic Hedgehog? *Am. J. Med. Gen.* **66**: 478–484.

Kimonis, V.E., Goldstein, A.M., Pastakia, B., Yang, M.L., Kase, R., DiGiovanna, J.J., Bale, A.E. and Bale, S.J. (1997) Clinical features in 105 persons with nevoid basal cell carcinoma syndrome. *Am. J. Med. Genet.* **69**: 299–308.

Kinzler, K., Bigner, S., Bigner, D. *et al.* (1987) Identification of an amplified, highly expressed gene in a human glioma. *Science* **236**: 70–73.

Kinzler, K.W. and Vogelstein, B. (1997). Gatekeepers and caretakers. *Nature* **386**: 761–763.

Lanske, B., Karaplis, A.C., Lee, K. *et al.* (1996) PTH/PTHrP receptor in early development and Indian hedgehog-regulated bone growth. *Science* **273**: 663–666.

Levanat, S.L., Gorlin, R.J., Fallet, S., Johnson, D.R., Fantasia, J.E. and Bale, A.E. (1996) A two-hit model for developmental defects in Gorlin syndrome. *Nat. Genet.* **12**: 85–87.

Loftus, S.K., Morris, J.A., Carstea, E.D. *et al.* (1997) Murine Model of Meimann-Pick C Disease: Mutation in a Cholesterol Homeostasis Gene. *Science* 277: 232–235.

Marigo, V., Davey, R.A., Zuo, Y., Cunningham, J.M. and Tabin, C.J. (1996) Biochemical evidence that patched is the Hedgehog receptor. *Nature* 384: 176–179.

Maesawa, C., Tamura, G., Iwaya, T. *et al.* (1998) Mutations in the human homologue of the Drosophila patched gene in esophageal squamous cell carcinoma. *Genes, Chromosomes Cancer* 21: 276–279.

McGarvey, T.W., Maruta, Y., Tomaszewski, J.E., Linnenbach, A.J. and Malkowicz, S.B. (1998). PTCH gene mutations in invasive transitional cell carcinoma of the bladder. *Oncogene* 17: 1167–1172.

Mirsky, R., Parmantier, E., McMahon, A.P. and Jessen, K.R. (1999) Schwann cell-derived desert hedgehog signals nerve sheath formation. *Ann. NY Acad. Sci.* 883: 196–202.

Motoyama, J., Takabatake, T., Takeshima, K. and Hui, C. (1998). Ptch2, a second mouse Patched gene is co-expressed with Sonic hedgehog. *Nat. Genet.* 18: 104–106.

Nakano, Y., Guerrero, I., Hidalgo, A., Taylor, A., Whittle, J.R. and Ingham, P.W. (1989) A protein with several possible membrane-spanning domains encoded by the Drosophila segment polarity gene patched. *Nature* 341: 508–513.

Nanni, L., Ming, J.E., Bocian, M., Steinhaus, K., Bianchi, D.W., Die-Smulders, C. and Nusse, R. (1994) The Wnt family in tumorigenesis and in normal development. *J. Steroid Biochem. Molec. Biol.* 43: 9–12.

Nusslein-Volhard, C. and Wieschaus, E. (1980) Mutations affecting segment number and polarity in Drosophila. *Nature* 287: 795–801.

Odent, S., Atti -Bitach, T., Blayau, M. *et al.* (1999) Expression of the Sonic hedgehog (SHH) gene during early human development and phenotypic expression of new mutations causing holoprosencephaly. *Hum. Mol. Genet.* 8: 1683–1689.

Ohlmeyer, J.T. and Kalderon, D. (1998) Hedgehog stimulates maturation of Cubitus interruptus into a labile transcriptional activator. *Nature* 396: 749–753.

Olsen, C.L. Hughes, J.P. Youngblood, L.G. and Sharpe-Stimac, M. (1997) Epidemiology of holoprosencephaly and phenotypic characteristics of affected children: New York State, 1984–1989. *Am. J. Med. Genet.* 73: 217–226.

Oro, A.E., Higgins, K.M., Hy, S.L., Bonifas, J.M., Epstein, E.H. and Scott, M.P. (1997) Basal cell carcinomas in mice overexpressing sonic hedgehog. *Science* 276: 817–821.

Peifer, M. and Bejsovec, A. (1992) Knowing your neighbors: cell interaction determine intrasegmental patterning in drosophila. *Trends. Genet.* 8: 243–248.

Pietsch, T., Waha, A., Koch, A. *et al.* (1997) Medulloblastomas of the desmoplastic variant carry mutations of the human homologue of *Drosophila patched*. *Cancer Res.* 57: 2085–2088.

Porter, J.A., Young, K.E. and Beachy, P.A. (1996) Cholesterol modification of hedgehog signaling proteins in animal development. *Science* 274: 255–259.

Radhakrishna, U., Wild, A., Grzeschik, K.H. and Antonarakis, S.E. (1997) Mutation in GLI3 in postaxial polydactyly type A . *Nat. Genet.* 17(3): 269–271.

Raffel, C., Jenkins, R.B., Frederick, L., Hebrink, D., Alderete, B., Fults, D.W. and James, C.D. (1997) Sporadic medulloblastomas contain *PTCH* mutations. *Cancer Res.* 57: 842–845.

Roessler, E., Belloni, E., Gaudenz, K. *et al.* (1996) Mutations in the human sonic sedgehog gene cause holoprosencephaly. *Nat. Genet.* 14: 357–360.

Shefer, S., Salen, G., Batta, A.K., Honda, A., Tint, G.S., Irons, M., Elias, E.R., Chen, T.C. and Holick, M.F. (1995) Markedly inhibited 7-dehydrocholesterol-delta 7-reductase activity in liver microsomes from Smith-Lemli-Opitz homozygotes. *J. Clin. Invest.* 96: 1779–1785.

Sidransky, D. (1996) Is human patched the gatekeeper of common skin cancers? *Nat. Genet.* 14: 7–8.

Smyth, I., Narang, M.A., Evans, T., Heimann, C., Nakamura, Y., Chenevix-Trench, G., Pietsch, T., Wicking, C. and Wainwright, B.J. (1999) Isolation and characterization of human patched 2 (PTCH2), a putative tumour suppressor gene in basal cell carcinoma and medulloblastoma on chromosome 1p32. *Hum. Molec. Genet.* 8: 291–297.

Springate, J.E. (1986) The nevoid basal cell nevoid basal cell carcinoma syndrome. *J. Pediatr. Surg.* 21: 908–910.

St-Jacques, B., Hammerschmidt, M. and McMahon, A.P. (1999) Indian hedgehog signaling regulates proliferation and differentiation of chondrocytes and is essential for bone formation. *Genes Devel.* 13: 2072–2086.

Stone, D.M., Hynes, M., Armanini, M. *et al.* (1996) The tumour-suppressor gene patched encodes a candidate receptor for sonic hedgehog. *Nature* **384**: 129–134.

Strong, L. (1977) Genetic and environmental interactions. *Cancer* **40**: 1861–1866.

Su, L.K., Vogelstein, B. and Kinzler, K.W. (1993) Association of the APC tumor suppressor protein with catenins. *Science* **262**: 1734–1737.

Thiagalingam, S., Lengauer, C., Leach, F. *et al.* (1996) Evaluation of candidate tumour suppressor genes on chromosome 18 in colorectal cancers. *Nat. Genet.* **13**: 343–346.

Trofatter, J.A., MacCollin, M.M., Rutter, J.L. *et al.* (1993) A novel moesin-, ezrin-, radixin-like gene is a candidate for the neurofibromatosis 2 tumor suppressor. *Cell* **72**: 791–800.

Unden, A.B., Holmberg, E., Lundh-Rozell, B. *et al.* (1996) Mutations in the human homologue of Drosophila patched (PTCH) in basal cell carcinomas and the Gorlin syndrome: different in vivo mechanisms of PTCH inactivation. *Cancer Res.* **56**: 4562–4565.

van den Heuvel, M. and Ingham, P.W. (1996) Smoothened encodes a receptor-like serpentine protein required for hedgehog signalling. *Nature* **382**: 547–551.

van der Riet, P., Karp, D., Farmer, E. *et al.* (1994) Progression of basal cell carcinoma through loss of chromosome 9q and inactivation of a single p53 allele. *Cancer Res.* **54**: 25–27.

van der Schroeff, J.G., Evers, L.M., Boot, A.J. and Bos, J.L. (1990) Ras oncogene mutations in basal cell carcinomas and squamous cell carcinomas of human skin. *J. Invest. Dermatol.* **94**: 423–425.

Vortkamp, A., Gessler, M. and Grzeschik, K-H. (1991) GLI3 zinc-finger gene interrupted by translocations in Greig syndrome families. *Nature* **352**: 539–540.

Vortkamp, A., Lee, K., Lanske, B., Segre, G.V., Kronenberg, H.Mm. and Tabin, C.J. (1996) Regulation of rate of cartilage difference by indian hedgehog and PTH-related protein. *Science* **273**: 613–621.

Wicking, C., Shanley, S., Smyth, I. *et al.* (1997) Most germ-line mutations in the nevoid basal cell carcinoma syndrome lead to a premature terminationof the patched protein, and no genotype-phenotype correlations are evident. *Am. J. Hum. Genet.* **60**: 21–26.

Wolter, M., Reifenberger, J., Sommer, C. *et al.* (1997). Mutations in the human homologue of the *Drosophila* segment polarity gene patched (PTCH) in sporadic basal cell carcinomas of the skin and primitive neuroectodermal tumors of the central nervous system. *Cancer Res.* **57**: 2581–2585.

Wicking, C., Smyth, I. and Bale, A.E. (1999) The hedgehog signalling pathway in tumorigenesis and development. *Oncogene* **18**: 7844–7851.

Xie, J., Murone, M., Luoh, S.-M. *et al.* (1998). Activating Smoothened mutations in sporadic basal-cell carcinoma. *Nature* **391**: 90–92.

Ziegler, A., Leffell, D.J., Kunala, S. *et al.* (1993) Mutation hotspots due to sunlight in the p53 gene of nonmelanoma skin cancers. *Proc. Natl Acad. Sci. USA* **90**: 4216–4220.

15

X-linked immunodeficiencies

Hubert B. Gaspar and Christine Kinnon

1. Introduction

The primary immunodeficiencies are a heterogeneous group of inherited conditions characterized by an increased susceptibility to infection. Defects in humoral and cellular immune function result in varying degrees of severity. They range from the severe combined immunodeficiencies (SCIDs) in which there are defects in both humoral and cellular function, to pure humoral defects such as X-linked agammaglobulinemia (XLA) and neutrophil defects, for example, chronic granulomatous disease (CGD). The last decade has seen an enormous advance in the understanding of the molecular basis of primary immunodeficiencies and the genes for most of the well-defined syndromes have now been identified. The X chromosome has been a rich source of genes which are defective in immunodeficiencies with seven genes having now been identified (see *Table 1*). The study and analysis of these genetic defects has also been important in understanding the basic mechanisms in the functioning of the normal immune system. This chapter will describe the X-linked immunodeficiencies, outlining the nature of the diseases, the defective genes and our present state of understanding of the molecular pathogenesis of the clinical and immunological phenotypes.

2. X-linked severe combined immunodeficiency

2.1 Phenotype

Severe combined immunodeficiencies (SCIDs) are a heterogeneous group of inherited disorders characterized by a profound reduction or absence of T-lymphocyte function (reviewed in Fischer *et al.*, 1997). They arise from a variety of molecular defects which affect lymphocyte development and function. The most common form of SCID is the X-linked form (X-SCID) which accounts for 40–50% of all cases. Clinically it is characterized by severe and recurrent infections and a high frequency of opportunistic infections. The clinical presentation in X-SCID is similar to patients with autosomal forms of SCID and it is difficult to distinguish

Genotype to Phenotype second edition, edited by S. Malcolm and J. Goodship.
© 2001 BIOS Scientific Publishers Ltd, Oxford.

Table 1. X-linked immunodeficiencies

Disorder (year of definition of molecular basis)	Chromosomal location	Gene	Function/defect
X-linked chronic granulomatous disease (1986)	Xp21	gp91*phox*	Component of phagocyte NADPH oxidase-phagocytic respiratory burst
Properdin deficiency (1992)	Xp21	properdin	Terminal complement component
X-linked agammaglobulinemia (1993)	Xq22	Bruton's tyrosine kinase (Btk)	Intracellular signalling pathways essential for pre-B-cell maturation
X-linked severe combined immunodeficiency (1993)	Xq13	common γ chain (γ_c)	Component of IL-2,4,7,9,15 cytokine receptors; T and NK cell development, T- and B-cell function
X-linked Hyper-IgM syndrome (CD40 ligand deficiency) (1993)	Xq26	CD40 ligand (CD154)	Isotype switching, T-cell function
Wiskott–Aldrich syndrome (1994)	Xp11	WASP	Cytoskeletal architecture formation, immune cell motility and trafficking
X-linked lymphoproliferative (Duncan's) syndrome (1998)	Xq25	SAP	Regulation of T-cell responses to EBV and other viral infections

between the different forms of SCID on the basis of presentation alone. Nearly all cases of SCID have very low or absent numbers of T-lymphocytes. Patients are then grouped into those who have B-lymphocytes (T-B+ SCID) and those who do not (T-B- SCID). The phenotype can be further divided on the basis of the presence or absence of natural killer (NK) cells. In the classical forms of X-SCID there is a profound abnormality in T and NK cell development, giving rise to a characteristic T-B+NK- form of SCID.

2.2 Genetic defect in X-SCID

The X-SCID gene locus was mapped by linkage analysis to Xq12–13 and in 1993 it was shown that the γc gene was located precisely to this critical region (Russell *et al.*, 1995). Subsequent analysis of a number of unrelated patients with T-B+NK- SCID demonstrated nonsense mutations in γc, thus confirming γc as the gene responsible for X-SCID (Noguchi *et al.*, 1993a).

The γc gene is organized into eight exons and spans 4.5 kb of genomic DNA in Xq13.1. The coding sequence of 1,124 nucleotides gives rise to a 232 amino acid transmembrane glycoprotein which is expressed constitutively in all hematolymphoid cells (Takeshita *et al.*, 1992). The protein has a number of structural motifs characteristic of cytokine receptor superfamily members. Four conserved cysteine residues are found at the extracellular amino-terminal end; a juxtamembrane conserved extracellular motif, the WSXSW box, found in all cytokine receptors is encoded in exon 5; the highly hydrophobic transmembrane domain of 29 amino acids occupies most of exon 6 and the proximal intracellular domain in exon 7 has a signalling sequence called Box 1/Box 2 which has homology to the SH2 domains of Src family tyrosine kinases.

Genetic analysis of boys with X-SCID has now identified over 160 patients with mutations in γc. Although mutations have been found in all the exons of γc, they are not evenly distributed. Exon 5 is the site of over 25% of all mutations followed by exons 3 and 4 with 15% each. Only very few numbers of mutations have been found in exon 8. A number of hotspots for mutation have also been noted within CpG dinucleotides, the most prominent being in exon 5 where 16 independent missense mutations have been reported (Pepper et al., 1995). The most frequently encountered defects are missense mutations, resulting in non-conservative amino acid substitutions, followed by nonsense mutations and together these account for approximately two thirds of all X-SCID mutations.

The consequences of the different mutations on γc protein expression and function are variable. Missense mutations in the extracellular domain of the protein can result in normal, trace or absent expression of γc but missense defects in the intracellular domains appear to result in intact protein expression (Puck et al., 1997). These results were obtained from the analysis of Epstein–Barr virus (EBV) immortalized B-cell lines from X-SCID patients but recent studies from the analysis of primary cells suggest that the majority of mutations result in absent or highly abnormal gc expression (Gilmour et al., 2001).

2.3 Molecular pathogenesis of X-SCID

γc was thought initially to be a component only of the interleukin 2 cytokine receptor (IL-2R) complex which plays a role in the activation and proliferation of T- and B-cells. Thus, this did not offer a satisfactory explanation for the more complex T-B+NK-immunophenotype of X-SCID. Subsequent realization that it was also an essential component of the additional high affinity receptors for cytokines IL-4,-7,-9 and -15 has since allowed a greater understanding of the lineage specific abnormalities in X-SCID (Giri et al., 1994; Kimura et al., 1995; Noguchi et al., 1993b; Russell et al., 1993; see Figure 1). Murine 'knockout' models, in vitro cellular studies and analysis of patients with a T-B+NK+ SCID phenotype have all shown that functional IL-7/IL-7R signalling is essential for normal T-cell development in the thymus (Corcoran et al., 1996; Puel et al., 1998; Suzuki et al., 1997). Studies with mice deficient in IL-15Rα or IL-15 have identified the dominant role of IL-15 in NK cell development and survival (Kennedy et al., 2000; Lodolce et al., 1998).

The downstream consequences of γc activation have now been carefully defined. Stimulation of the receptor complex by the relevant cytokine results in the heterodimerization of the receptor subunits and tyrosine phosphorylation of the JAK3 molecule; a cytoplasmic kinase which binds specifically to the γc subunit. Tyrosine phosphorylated JAK3 in turn phosphorylates one of the

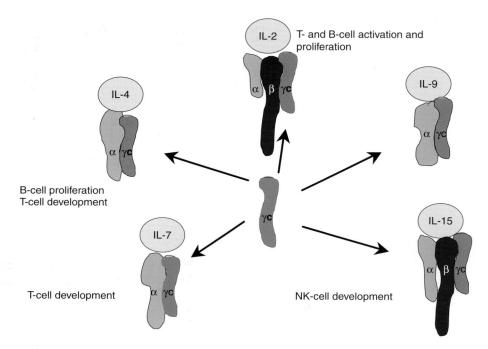

Figure 1. The common gamma (γc) chain is a component of the high affinity receptors for interleukin-2, -4, -7, -9 and -15.

STAT (Signal Tranducers and Activators of Transcription) family of transcription factors which then dimerises and translocates to the nucleus where it binds to specific sites to initiate transcriptional events (reviewed in Ihle, 1995; Taniguchi, 1995; see *Figure 2*). The specificity of γc binding to JAK3 rather than to other JAK molecules is demonstrated by the lack of JAK3 tyrosine phosphorylation in patients with X-SCID and also by the identification of mutations in JAK3 in patients with an autosomal recessive form of T-B+NK- SCID (Macchi *et al.*, 1995; Russell *et al.*, 1995). This latter finding implies that any disruption to the specific interaction between these two molecules can lead to the T-B+NK- immunophenotype. Further downstream, all γc cytokine stimulation pathways (except IL-4) activate the same members of the STAT family of molecules, STAT 3 and STAT 5, whereas IL-4 activates STAT6 (Leonard and O'Shea, 1998).

2.5 Atypical presentation of X-SCID

Atypical forms of X-SCID, in terms of both immunological and clinical phenotype have been described. Certain individuals have been reported who, despite a mutation in the γc gene, have some residual T-cell function and thus present with a less severe clinical phenotype. In one pedigree, this was associated with a splice site mutation that generated two transcripts: one truncated and one normal sized, which accounted for 80% and 20% of the total γc mRNA,

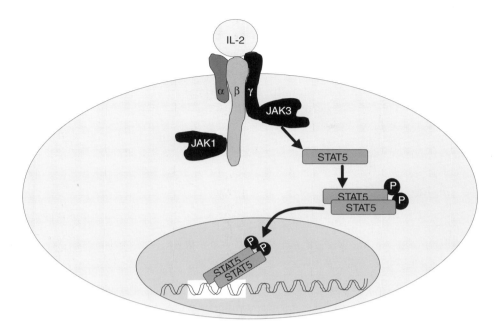

Figure 2. Signalling through the IL-2R. On activation of the IL-2 cytokine receptor complex, the JAK3 molecule, which binds selectively to γc, is activated and subsequently phosphorylates STAT5 (Signal Transducer and Activators of Transcription). Phosphorylated STAT5 dimerises and then translocates to the nucleus where it binds to specific sites to initiate transcriptional events.

respectively. IL-2 binding to high affinity receptor complexes was severely reduced and T-cells from affected individuals showed impaired *in vitro* stimulation and a restricted T-cell receptor repertoire (Di Santo *et al.*, 1994).

One important atypical X-SCID case has been described which underlines the effect of the γc defect on lymphocyte development (Stephan *et al.*, 1996). A patient with a T+B+NK- SCID phenotype with poor T-cell function and an X-linked family history was found to lack γc expression in his B-cells. Analysis of genomic DNA derived from B-cells, monocytes and neutrophils demonstrated a missense mutation in exon 3 of γc. However, analysis of his T-cells displayed normal γc surface expression and wild-type sequence. These results are explained by a reversion of the γc mutation in a committed T-cell precursor which gives rise to a pool of mature T-cells. This case confirms previous experimental data that γc expression and signalling are essential for T-lineage development and suggests that a significant growth advantage was conferred to the reverted T-cell precursors. Such data also suggests that introduction of γc into lymphoid precursors by gene transfer even at relatively low frequency, could have significant therapeutic benefit. This has been borne out in a recent clinical trial of somatic gene therapy for X-SCID, where reconstitution of γc expression in hematopoietic and lymphoid progenitor cell populations has given significant clinical benefit (Cavazzana-Calvo *et al.*, 2000).

3. X-linked agammaglobulinemia (XLA)

3.1 Phenotype

The phenotype of XLA is characterized in its classical form by the almost complete absence of immunoglobulin of all isotypes and the profound reduction of B-lymphocytes in the peripheral circulation. Examination of bone marrow from affected individuals demonstrates that there is an arrest in B-cell development. The site of the block is variable and is most commonly at the pre-B-cell stage but earlier maturation arrest has also been described (see *Figure 3*).

3.2 Genetic defect in XLA

Btk was identified as the molecular defect in XLA by two groups; one using a positional cloning approach and the other in a search for novel protein kinases expressed in B-lymphocytes (Tsukada *et al.*, 1993; Vetrie *et al.*, 1993). The gene is located on the long arm of the X chromosome at Xq21.3 and the human gene encompasses 37.5 kb. Btk is organized into 19 exons (Ohta *et al.*, 1994; Rohrer *et al.*, 1994; Sideras *et al.*, 1994), including a 5′ untranslated region (exon 1), and encodes a 659 amino acid protein that is expressed throughout the myeloid and B-cell lineages (Genevier *et al.*, 1994). The Btk protein belongs to a family of cytoplasmic tyrosine kinases that are related to, but distinct from, the Src family kinases. The protein is characterized by its modular structure which includes an amino-terminal pleckstrin homology (PH) domain followed by a tec-homology (TH) domain, a Src homology SH3 domain, an SH2 domain and a C-terminal kinase SH1 domain. Catalytic activity resides in the kinase domain while the

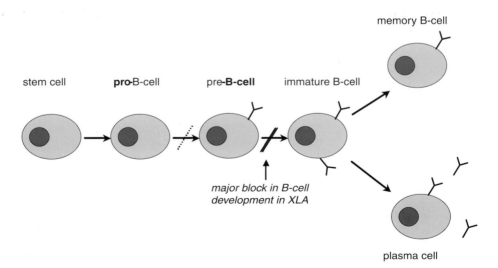

Figure 3. B-cell maturation highlighting the role of Btk. In XLA, defects in Btk result in a block in B-cell maturation at the pre-B-cell to immature B-cell transition. The site of the block is variable and earlier maturation arrest at the pro-B-cell stage has been described.

other domains are necessary for protein–protein interactions. Other members of this family include Tec, Itk and Bmx, all of which are also expressed in hematopoietic tissues.

Identification of the gene has led to mutational analysis being undertaken by a number of groups world-wide. Over 500 unique mutations in Btk have now been identified and an international database of mutations and clinical information has now been established (BTKbase and all other primary immunodeficiency data-bases can be found at: http://www.uta.fi/imt/bioinfo/mutdatbas.html#idmdb) (Vihinen et al., 1998). Various different types of genetic abnormalities have been found in the Btk gene. One third of mutations are missense mutations and these have been found predominantly in the kinase domain of the gene, although they have been found in all domains except the SH3 domain (Vihinen et al., 1996). Premature stop codons, deletions and insertions have also been described throughout the gene. Although the large majority of mutations are found in the coding region, 12% of mutations have been found to affect splice site recognition sequence. Only one mutation has been found in the promoter region and this is thought to affect transcriptional regulation of the gene. Approximately 5–10% of mutations result in gross alterations of the Btk gene. In addition to large deletions, an inversion, a duplication and a retroviral insertion have been identified (Rohrer et al., 1999).

Soon after the identification of the defect in Btk in humans, the X-linked immunodeficiency (xid) mouse was found to have a single amino acid substitution (R28C) in the PH domain of Btk (Rawlings et al., 1993). However, the immunophenotype of the xid mouse is very different from the human form. Xid mice have a much less severe reduction in the number of B-cells, they make rela-tively normal serum concentrations of most immunoglobulin types, but markedly reduced IgG3 and IgM concentrations, and they are unable to make antibody responses to T-cell independent antigens (Wicker and Scher, 1986). In both species defects in Btk are associated with an immature B-cell phenotype (Conley, 1985). The differences between the murine and human phenotypes cannot be explained by the nature of the xid genetic defect, as patients with severe pheno-types have been described with R28C mutations (Conley et al., 1998). Btk-null mice created by homologous recombination also have a B-cell phenotype that is identical to the xid mouse (Khan et al., 1995). The most likely explanation for these differences in phenotypes lies in the intrinsic differences in B-cell devel-opment and function between the two species.

3.3 Btk function and its role in B-cell development

Tyrosine kinases have been studied in many hematopoietic cell lineages and have been shown to act as signal transduction molecules mediating cell surface receptor activation events to downstream pathways. Cross-linking of a number of cell surface receptors, including IL-5, IL-6, CD38, FcRε and perhaps most impor-tantly surface IgM, on B-cells results in the recruitment of cytosolic Btk to the plasma membrane and activation of Btk by tyrosine phosphorylation (Li et al., 1997). Btk activation and downstream consequences are illustrated in Figure 4. The process of Btk activation is initiated by phosphorylation of tyrosine 551

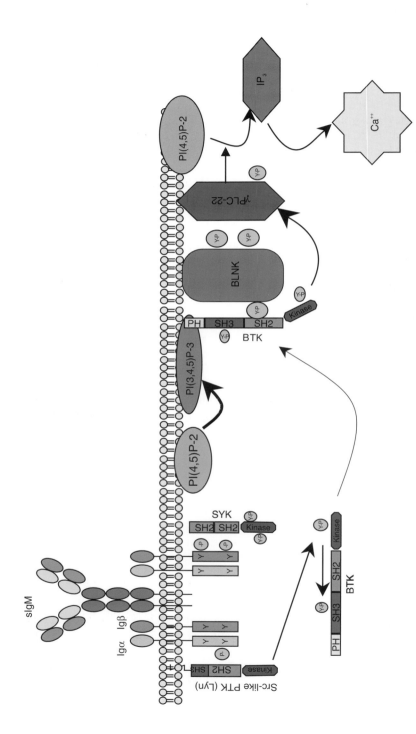

Figure 4. Signalling through the BCR. Following BCR activation, Btk is tyrosine phosphorylated on Y551 by the action of Lyn resulting in autophosphorylation of its SH3 domain. Btk is then recruited to the cell membrane through a PH domain interaction with PIP_3 and is then brought into proximity with PLC-γ2 by association with the adapter protein BLNK. Btk tyrosine phosphorylation of PLC-γ2 leads to conversion of $PIP2$ to IP_3 resulting in the release of intracellular calcium stores.

(Y551) in the kinase domain of Btk by a Src family kinase, most likely Lyn, followed by Btk autophosphorylation at Y223 in the SH3 domain (Park *et al.*, 1996; Rawlings *et al.*, 1996). Btk is then recruited to the cell membrane, a process which is dependent upon an intact PH domain and its association with phos-phatidylinositol-3,4,5-triphosphate (PIP$_3$; Fukuda *et al.*, 1996). At the cell membrane activated Btk interacts with phospholipase γ2 (PLC-γ2) leading to enhanced tyrosine phosphorylation of PLC-γ2 and resulting in accumulation of inositol 1,4,5-triphosphate (IP3) and release of calcium from internal stores. An adapter molecule BLNK (B-cell linker protein) has been identified that mediates Btk interactions with PLC-γ2 (Fu *et al.*, 1998). Following B-cell receptor (BCR) cross-linking, activated Syk phosphorylates BLNK on a number of tyrosine residues. BLNK is then able to bind via phosphorylated tyrosine motifs to the PLC-γ2 SH2 domains and also to the SH2 domain of Btk. Thus, BLNK may act to bring Btk in close proximity with PLC-γ2 and thus allow Btk phosphorylation of PLC-γ2. This is not an absolute mechanism, since Btk is able to bind to unphos-phorylated BLNK and also since PLC-γ2 phosphorylation can occur in the absence of BLNK. However, optimal function is only seen if phosphorylated BLNK is present. The crucial role of BLNK in B-cell development is further illustrated by the phenotypes of human and murine BLNK mutants which show distinct similarity to the respective Btk mutant phenotypes (Minegishi *et al.*, 1999). Despite the data available, a relationship with cellular events and specifi-cally the mechanism of B-cell developmental arrest has yet to be established. Clearly, Btk participates in a signal cascade responsible for the mobilization of calcium but this can initiate several types of transcriptional events and growth responses such as proliferation and apoptosis.

The central question of why the Btk gene defect results in B-cell development arrest remains unresolved. There is however evidence that BCR expression and signalling is necessary for positive selection of B-cells and prevention of death by apoptosis. Xid mice exhibit a progressive loss of splenic and re-circulating B-cells. Moreover, some of the xid defects can be rescued by over-expression of the anti-apoptotic proteins Bcl-2 and Bcl$_{XL}$ (Woodland *et al.*, 1996). This, together with evidence for Btk in regulation of Bcl$_{XL}$ expression, suggests that Btk may be important in mediating anti-apoptotic signals during crucial stages in B-cell development.

3.4 Atypical forms of XLA

The identification of the genetic abnormality in XLA has allowed unambiguous assignment of molecular defects to individuals with abnormalities in antibody production. Thus, although the majority of Btk deficient patients display the clas-sical immunophenotype, as many as 20% of patients with mutations in Btk have delayed onset of symptoms or higher concentrations of serum immunoglobulins than expected (Hashimoto *et al.*, 1999; Kornfeld *et al.*, 1994).

Despite the large amount of mutation and phenotypic data available the reasons for clinical and immunological heterogeneity in XLA are unclear and it has not been possible to make definitive genotype–phenotype correlations. The strongest argument for this comes from reports of considerable variation in clinical severity even within the same pedigree (Gaspar *et al.*, 2000; Kornfeld *et al.*, 1997), suggesting that factors other than the genetic mutation may influence the clinical course. It has been suggested that, as in other immunodeficiencies, the presence of cytokine gene polymorphisms or defects in innate immunity, may be important but as yet there is no evidence to refute or confirm these ideas.

4. Wiskott–Aldrich syndrome

4.1 Phenotype

The Wiskott–Aldrich syndrome (WAS) is a rare inherited X-linked recessive disease characterized by immune dysregulation and microthrombocytopenia (reviewed in Thrasher and Kinnon, 2000). Clinical manifestations of the immune disorder include susceptibility to pyogenic, viral and opportunistic infection, and eczema. The immunological manifestations are progressive, with quantitative and qualitative deficiency of T-lymphocytes (particularly CD8 cells) developing during early childhood, defects in proliferative and delayed type hypersensitivity responses, and deficient production of antibodies to both polysaccharide and protein antigens. In its less severe form, known as X-linked thrombocytopenia (XLT), mutations in the same gene produce the characteristic platelet abnormality but with minimal immunological disturbance (Villa et al., 1995). However, the natural history of XLT is less well defined than for WAS and for some patients immune dysregulation may develop over time.

4.2 Genetic defect in WAS

The WAS gene is now known to encode a 502 amino acid proline-rich intracellular protein, WASP (Derry et al., 1994). WASP is expressed exclusively in hematopoietic cells and belongs to a recently defined family of more widely expressed proteins involved in transduction of signals from receptors on the cell surface to the actin cytoskeleton (reviewed in Ramesh et al., 1999). Other members of the family include neural (N)-WASP, SCAR (suppressor of G-protein coupled cyclic-AMP receptor, isolated from Dictyostelium), four human SCAR proteins (hsSCAR1–4), other homologs of SCAR (in mouse, Caenorhabditis elegans, and Drosophila), and the WASP-related S. cerevisiae protein Las17p/Bee1p (Naqvi et al., 1998). WASP, N-WASP, SCAR, and Las17p/Bee1p are organized into modular domains defined by sequence homology and binding interactions with other signalling molecules. They all possess C-terminal polyproline (P), Wiskott-homology (WH)-2 and acidic (A) domains.

Over 100 unique mutations have been reported in WAS with mutations found throughout the gene. The distribution of mutations is uneven, with most occurring in the first four exons of the N-terminal region. Genotype–phenotype studies suggested that missense mutations located in the first three exons generally result in a mild disease phenotype, including X-linked thrombocytopenia, while more complex mutations and missense mutations in the remainder of the gene may generally result in a more severely affected phenotype with low or absent levels of WASP expression (Zhu et al., 1997). This led to the conclusion that mutations which affect the C terminus of the WAS gene have a more disruptive effect on WASP protein expression. Immunoblot analysis of WAS patient mononuclear cells showed that patients with the classical phenotype appear to lack protein expression (MacCarthy-Morrogh *et al.*, 1998). For accurate prediction of clinical course investigation of protein and/or transcription levels in primary cells may provide a more accurate assessment than mutation analysis.

WASP is implicated in the regulation of the actin cytoskeleton by studies which showed WASP clustering physically with polymerized actin and acting as a direct effector molecule for the Rho-like GTPase Cdc42 (Aspenstrom *et al.*, 1996; Symons *et al.*, 1996). Cdc42 has been implicated as both a regulator of actin-containing filopodial extensions and in the development of cell polarity (Nobes and Hall, 1995; Stowers *et al.*, 1995). In addition, the C-terminal portion of WASP has been shown to interact directly with the actin related protein (Arp)2/3 complex, which suggests a critical regulatory role for WASP in the nucleation and branching of actin filaments especially in lamellipodia and at the base of filopodia (Machesky and Insall, 1998). This complex has also been shown to enhance the nucleation of new actin filaments, a process which is considerably enhanced when bound to WASP (see *Figure 5*).

4.3 Immune cell defects in WAS

Cytoskeletal architecture abnormalities arising from defects in the WAS gene have profound effects upon multi-lineage cellular function that may explain the

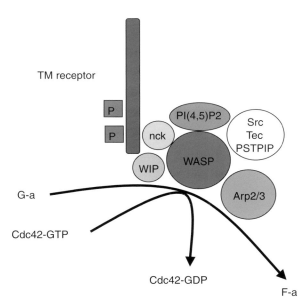

Figure 5. Formation of specialized cytoskeletal structures is dependent on the regulated assembly of a signalling complex involving WASP. Following signalling through a transmembrane (TM) receptor, there is receptor phosphorylation (P) and formation of a complex involving WASP and Cdc42, Src and Tec family tyrosine kinases, proline-serine-threonine phosphatase-interacting protein (PSTPIP), WASP interacting protein (WIP), PI(4,5)P2 and adaptor proteins such as Nck, which activates the Arp2/3 complex and the polymerization of new actin filaments (F-a) from monomeric actin (G-a).

immunopathology of the WAS phenotype. Chemotactic responses of macrophages from WAS patients *in vitro* are markedly abrogated (Zicha *et al.*, 1998) and dendritic cells (DC) fail to develop normal polarized morphologies and are qualitatively and quantitatively dysmotile when stimulated *in vitro* (Binks *et al.*, 1998). Significantly, they were also shown to exhibit major abnormalities in the distribution of the peripheral filamentous actin cytoskeleton and to lack condensations of actin on the ventral surface of the cell known as podosomes. These data suggest that WAS macrophages and DCs are compromised in their ability to develop polarized structures and are unable to respond to chemical stimuli in a directional manner. FcγR-mediated phagocytosis is impaired in WAS macrophages, in terms of the kinetics of particle uptake and in regulated formation of the phagocytic cup (Lorenzi *et al.*, 2000). This suggests impaired clearance of pathogens by macrophages and presentation of foreign antigens to the cell mediated immune system by DCs may also be contributing to immune dysfunction in WAS.

These studies suggest that a significant component of the immunological dysregulation in WAS may arise from the inability of antigen presenting cells to traffic normally between non lymphoid tissues and T-lymphocyte rich areas of lymphoid organs. The regulation of peripheral tolerance and the surveillance of tissues for viral infection or tumor cells may also be compromised by the inefficiency of cell trafficking, phagocytosis and cross presentation. Similar defects of T- and B-lymphocyte homing and activation may also exist.

5. X-linked lymphoproliferative syndrome

5.1 Clinical phenotype

X-linked lymphoproliferative disease (XLP) is a rare disorder characterized by a dysregulated immune response to EBV (Purtilo, 1981). First recognized over 25 years ago, more than 80 kindreds have now been identified world-wide. The most usual presentation is with fulminant, often fatal acute EBV infection, but its phenotypic expression is highly variable. Other presentations include acquired dysgammaglobulinemia, B-cell lymphoma, aplastic anemia, vasculitis and lymphomatoid granulomatosis (Seemayer *et al.*, 1995). In XLP patients, the primary cause of death (mortality is 100% by the age of 40 years) is hepatic necrosis and bone marrow failure by extensive tissue destruction from an uncontrolled cytotoxic T-cell response. A number of reports have now documented an XLP syndrome in association with viral triggers other than EBV suggesting that the XLP defect is not specifically to the control of EBV infection but is required more generally for regulation of immune cell activation (Brandau *et al.*, 1999).

The lack of EBV infection in all patients, the variability in the clinical phenotype and the lack of a reliable diagnostic test has made firm diagnosis of XLP extremely difficult. Historically, the diagnosis was based on a typical history with a suggestive pedigree, and sometimes the finding of elevated anti-EBV antibody levels in obligate female carriers. Unambiguous diagnosis is now possible following the identification of the genetic defect.

5.2 Genetic defect in XLP

The gene responsible for XLP was localized to Xq25 in 1990 by demonstration of an Xq25 deletion in an affected family (Sanger *et al.*, 1990), but it was not until 1998 that the gene was finally identified as that coding for SAP (signalling lymphocyte activating molecule-associated protein; Coffey *et al.*, 1998; Sayos *et al.*, 1998). Signalling lymphocyte activating molecule (SLAM) is a B- and T-cell surface marker which forms a receptor-ligand pair, triggering of which co-activates B- and T-lymphocyte responses. Using the cytoplasmic tail of SLAM as bait in a yeast two-hybrid system and a human T-cell clone library, Sayos *et al.* (1998) identified SAP, an SH2-domain containing molecule with a short tail of 26 amino acids. The SAP gene is organized into four exons and human SAP exists as two RNA species of 2.5 kilobases and 0.9 kilobases. An alternative form of human SAP mRNA has been identified in which a 55 nucleotide deletion of exon 3 occurs due to an infrequently used cryptic splice acceptor site in exon 3. SAP has homology with other SH2-domain-containing proteins including SHIP, EAT-2 and Abl. The 26 amino acid tail, however, has no homology with any other known protein domain.

SAP is primarily expressed in T-cells and is found in all major T-cell subsets. SAP expression has also been shown by northern and western blot analysis in monocytes, human spleen and thymus (Nichols *et al.*, 1998). Primary human and murine B-cells do not appear to express SAP at high levels.

5.3 Function of SAP protein

Functional *in vitro* studies and crystal structure modelling of the SAP protein provide convincing evidence that SAP binds to specific tyrosine motifs in the cytoplasmic tail of SLAM. Although SH2 domain containing proteins predominantly bind to phosphorylated tyrosines, SAP has high affinity for both unphosphorylated and phosphorylated SLAM tyrosine residues. Functionally this indicates that SAP may act to block the phosphorylation of tyrosine motifs in SLAM or by competitively binding to phosphorylated tyrosines it acts to block the binding to SLAM of other SH2 domain containing signalling proteins such as SHP-2 (Sayos *et al.*, 1998). The function of SAP in NK cells has also been explored and demonstrates similarities to its role in T-lymphocytes.

5.4 Cellular defects in XLP

Mutations of SAP in XLP patients are postulated to dysregulate effector mechanisms in cytotoxic T-cells and NK cells thereby resulting in a dysregulated response to EBV infection and the inability of effector cells to control EBV transformed B-cell proliferation. A number of cellular studies appear to support this hypothesis. EBV specific cytotoxic T-cells from XLP patients have shown defects in killing and defective production of cytokines (Yasuda *et al.*, 1991) and more recent evidence suggests that NK cell killing is severely inhibited (Parolini *et al.*, 2000; Tangye *et al.*, 2000).

5.5 Variability in the XLP phenotype

The availability of mutation analysis for individuals with suspected XLP has identified a number of atypical cases. Males previously labelled as common

variable immunodeficiency (CVID) or XLA have now been show to have mutations in the SAP gene (Gilmour *et al.*, 2000). In addition two boys with non EBV
driven non Hodgkins lymphoma were also found to have XLP (Strahm *et al.*,
2000). These cases suggest that the true incidence of XLP may be signficantly
underestimated and underlines the need for more extensive screening of patients
with CVID or B-cell lymphoma for XLP. At present there is no evidence for a
correlation between the genetic defect and the clinical course of the patient
(Sumegi *et al.*, 2000).

6. X-linked chronic granulomatous disease

6.1 Clinical phenotype

Chronic granulomatous disease (CGD) is a condition characterized by defective
phagocyte function that can arise from a number of different molecular defects.
Phagocytic cells in children with CGD ingest organisms normally but fail to kill
them as a result of defects in the superoxide-producing pathway (Goldblatt and
Thrasher, 2000). The importance of the system to host immunity is exemplified by
the clinical phenotype which although variable results in significant morbidity
and mortality due to susceptibility to bacterial and fungal infection. The clinical
syndrome is most often characterized by recurrent life-threatening *Staphylococcus
aureus*, *Proteus* or *Pseudomonas* sepsis, hypergammaglobulinemia and widespread
chronic granulomatous infiltration.

6.2 Genetic defect in X-CGD

The gene which causes X-CGD was the first human disease causing gene to be
identified by a positional cloning strategy (Royer-Pokora *et al.*, 1986). The gene,
CYBB, maps to Xp21.1 and encodes a 570 amino acid transmembrane glyco-
porotein, gp91-*phox*, a component of the NADPH oxidase complex of phagocytic
cells. Over two thirds of all cases of CGD are X-linked. Mutations, including a
relatively large proportion of deletions, have now been identified in many X-CGD
families (Roos *et al.*, 1996) and can be used for unambiguous carrier determination
and prenatal diagnosis. In all CGD families where genetic analysis is not possible,
carrier detection can be performed using the nitroblue tetrazolium (NBT) test and
prenatal diagnosis can also be performed using this test on fetal blood samples.
Similar assays are available with flow cytometric analysis. Without genetic
analysis more accurate diagnosis depends on investigation of the relevant proteins
presence or absence in the neutrophils by immunoblot analysis.

6.3 Molecular pathogenesis of X-CGD

Neutrophils from CGD patients fail to exhibit a characteristic increase of
oxidative metabolism, the 'respiratory burst', during phagocytosis. This results
from molecular lesions in the genes encoding a phagocyte-specific multi-
component enzyme system, the NADPH-oxidase. NADPH oxidase is responsible
for the production of superoxide and is known to consist of at least six specific

components, two of which are the membrane bound flavocytochrome b_{558} components, gp91-*phox* and p22-*phox* and four are cytosolic, p40-*phox*, p47-*phox*, p67-*phox* and p21 *rac* (reviewed in (Goldblatt and Thrasher, 2000). When the cell is activated two cytosolic components, p47-*phox* and p67-*phox*, translocate to the membrane forming an active complex with the flavocytochrome which renders it permissive for electron transport and leads to reduction of molecular oxygen to derivative free radical anions and other microbicidal compounds. Molecular defects in all components, other than p40-*phox*, have been found to give rise to CGD (Roos *et al.*, 1996).

7. X-linked hyper IgM syndrome

7.1 Clinical phenotype

X-linked hyper IgM syndrome (XHM) has been recognized since 1966, its characteristic immunological phenotype including very low IgG and IgA levels, normal or elevated IgM and normal lymphocyte subpopulations. T-cell function by normal laboratory parameters is also usually normal. The clinical phenotype is one of susceptibility to bacterial infection, but, for many years before the responsible gene was identified, it was known that boys affected by XHM were susceptible to opportunistic infection with organisms such as *Pneumocystis carinii* (Banatvala *et al.*, 1994) and *Cryptosporidium parvum*. Boys with XHM are particularly susceptible to liver disease and liver/gastrointestinal malignancy, with a link between chronic *cryptosporidial* infection and sclerosing cholangitis (Hayward *et al.*, 1997). The European Society for Immunodeficiency database for XHM deficiency currently contains clinical and molecular data from over 100 affected males. The actuarial survival at 25 years is only 25%, and the incidence of liver disease by the age of 20 years is 80% (Levy *et al.*, 1997).

7.2 Genetic defect in XHM

The clinical features of XHM indicate that the genetic defect affects both T- and B-cell function. It was therefore not surprising when the gene responsible for XHM was identified in 1993 as that coding for CD40 ligand (CD40L, now renamed CD154), a surface molecule present on activated T-cells (Korthauer *et al.*, 1993). Mutations in CD154 are scattered throughout the length of the gene although they are more common in the extracellular domain (Katz *et al.*, 1996). There is so far little apparent genotype–phenotype correlation, and there is considerable intra-familial variation in clinical phenotype, so predictions of severity or likelihood of liver disease are not possible. Some of the CD154 missense mutations have been shown to prevent binding to CD40 while others affect folding of the monomer or trimer formation and may therefore result in the inability to express the mutant CD154 molecule on the surface.

Knowledge of the molecular basis of XHM allows accurate diagnosis not only in boys who have a typical phenotype or a positive family history, but also in some cases of previously undefined hypogammaglobulinemia. Several adult males diagnosed as affected by CVID have recently been found to have CD154 deficiency

(Mouthon *et al.*, 1999). Although these individuals may not develop the life-threatening complications of CD154 deficiency, other affected family members could have a more severe phenotype. Carrier detection and prenatal or early diagnosis is therefore important for at risk relatives.

7.3 Molecular pathogenesis of XHM

CD154 is a 261 amino acid long, type II transmembrane protein, with a short intra-cytoplasmic tail, a transmembrane region and an extracellular domain that shares homology to tumor necrosis factor-alpha (TNF-α) (reviewed in Notarangelo and Hayward, 2000). Similar to other TNF family members, CD154 is expressed as a homotrimer and molecular modelling indicates that residues 189–209 are involved in binding to CD40. The cytoplasmic tail has a phosphorylation site which may transduce signals within the T-cell. The CD40 molecule to which CD154 binds is a member of the TNF receptor superfamily and is normally present on the surface of B-cells, mononuclear phagocytes and DCs. The cytoplasmic domain of CD40 binds members of the TNF receptor associated factor (TRAFs) proteins and further signalling is mediated by JNK and NF-kB resulting in activation of pathways leading to a variety of cellular events. CD40–CD154 interaction is obviously of critical importance to immunoglobulin gene switching but it has also been shown that this interaction can also lead to apoptosis depending on the target cell type and state of activation (Hess and Engelmann, 1996).

The function of CD154 in T-cells is less clearly understood. CD154 can act as a T-cell growth factor and acts as co-stimulatory molecule for α/β and γ/δ T-cells. T-cell proliferative responses to a range of antigens have been shown to be defective in CD154 deficiency (Ameratunga *et al.*, 1997). CD154 is expressed on the surface of biliary epithelial cells, and CD40-CD154 interaction here is thought to be involved in control of intracellular pathogens such as *Cryptosporidium*. CD154 knockout mice show increased susceptibility to *Pneumocystis carinii* and *Cryptosporidium* and their immunity to leishmania organisms is also impaired. Although they are capable of cytotoxic T-lymphocyte responses, they do not make long lived cytotoxic memory cells (Cosyns *et al.*, 1998). The exact mechanism of bilary damage in XHM humans and mice is as yet unknown.

8. Properdin deficiency

Properdin deficiency is characterized by absence of extracellular properdin, a positive regulator of the alternative pathway of complement activation. Properdin promotes alternative pathway activation by stabilizing the alternative pathway C3 convertase, C3bB (Pangburn and Muller-Eberhard, 1984). Properdin deficiency is associated with increased susceptibility to severe meningococcal disease (reviewed in Figueroa and Densen, 1991). The infections are mostly caused by rare variants of *Neisseria meningitidis*. The reported fatality rate in properdin-deficient patients is high and survivors rarely have recurrent infections, which is in contrast to findings in other complement deficiencies. Three distinct properdin deficiency phenotypes have been identified, all are X-linked. Properdin deficiency type I, the most common, is characterized by no circulating properdin, while type II is characterized by low concentrations of properdin and type III by properdin dysfunction. Genetic defects have been identified in a relatively small number of families. Point mutations have been found to be associated with all three phenotypes (Fredrikson *et al.*, 1996; Westberg *et al.*, 1995).

References

Ameratunga, R., Lederman, H.M., Sullivan, K.E. *et al.* (1997) Defective antigen-induced lymphocyte proliferation in the X-linked hyper-IgM syndrome. *J. Pediatr.* **131**: 147–150.

Aspenstrom, P., Lindberg, U. and Hall, A. (1996) Two GTPases, Cdc42 and Rac, bind directly to a protein implicated in the immunodeficiency disorder Wiskott–Aldrich syndrome. *Curr. Biol.* **6**: 70–75.

Banatvala, N., Davies, J., Kanariou, M., Strobel, S., Levinsky, R. and Morgan, G. (1994) Hypogammaglobulinaemia associated with normal or increased IgM (the hyper IgM syndrome): a case series review. *Arch. Dis. Child* **71**: 150–152.

Binks, M., Jones, G.E., Brickell, P.M., Kinnon, C., Katz, D.R. and Thrasher, A.J. (1998) Intrinsic dendritic cell abnormalities in Wiskott–Aldrich syndrome. *Eur. J. Immunol.* **28**: 3259–3267.

Brandau, O., Schuster, V., Weiss, M. *et al.* (1999) Epstein-barr virus-negative boys with non-hodgkin lymphoma are mutated in the SH2D1A gene, as are patients with X-linked lymphoproliferative disease (XLP). *Hum. Mol. Genet.* **8**: 2407–2413.

Cavazzana-Calvo, M., Hacein-Bey, S., de Saint, B.G. *et al.* (2000) Gene therapy of human severe combined immunodeficiency (SCID)-X1 disease. *Science* **288**: 669–672.

Coffey, A.J., Brooksbank, R.A., Brandau, O. *et al.* (1998) Host response to EBV infection in X-linked lymphoproliferative disease results from mutations in an SH2-domain encoding gene *Nat. Genet.* **20**: 129–135.

Conley, M.E. (1985) B-cells in patients with X-linked agammaglobulinemia. *J. Immunol.* **134**: 3070–3074.

Conley, M.E., Mathias, D., Treadaway, J., Minegishi, Y. and Rohrer, J. (1998) Mutations in btk in patients with presumed X-linked agammaglobulinemia. *Am. J. Hum. Genet.* **62**: 1034–1043.

Corcoran, A.E., Smart, F.M., Cowling, R.J., Crompton, T., Owen, M.J. and Venkitaraman, A.R. (1996) The interleukin-7 receptor alpha chain transmits distinct signals for proliferation and differentiation during B lymphopoiesis. *EMBO J.* **15**: 1924–1932.

Cosyns, M., Tsirkin, S., Jones, M., Flavell, R., Kikutani, H. and Hayward, A.R. (1998) Requirement of CD40-CD40 ligand interaction for elimination of Cryptosporidium parvum from mice. *Infect. Immun.* **66**: 603–607.

Derry, J.M., Ochs, H.D. and Francke, U. (1994) Isolation of a novel gene mutated in Wiskott–Aldrich syndrome. *Cell* **79**: 639–644.

Di Santo, J.P., Rieux-Laucat, F., Dautry-Varsat, A., Fischer, A. and de Saint Basile, G. (1994) Defective human interleukin 2 receptor gamma chain in an atypical X chromosome-linked severe combined immunodeficiency with peripheral T-cells. *Proc.Natl Acad. Sci. USA* **91**: 9466–9470.

Figueroa, J.E. and Densen, P. (1991) Infectious diseases associated with complement deficiencies. *Clin. Microbiol. Rev.* **4**: 359–395.

Fischer, A., Cavazzana-Calvo, M., De Saint Basile, G., DeVillartay, J.P., DiSanto. J.P., Hivroz, C., Rieux-Laucat, F. and Le Deist, F. (1997) Naturally occurring primary deficiencies of the immune system. *Ann. Rev. Immunol.* **15**: 93–124.

Fredrikson, G.N., Westberg, J., Kuijper, E.J., Tijssen, C.C., Sjoholm, A.G., Uhlen, M. and Truedsson, L. (1996) Molecular characterization of properdin deficiency type III: dysfunction produced by a single point mutation in exon 9 of the structural gene causing a tyrosine to aspartic acid interchange. *J. Immunol.* **157**: 3666–3671.

Fu, C., Turck, C.W., Kurosaki, T. and Chan, A.C. (1998) BLNK: a central linker protein in B-cell activation. *Immunity* **9**: 93–103.

Fukuda, M., Kojima, T., Kabayama, H. and Mikoshiba, K. (1996) Mutation of the pleckstrin homology domain of Bruton's tyrosine kinase in immunodeficiency impaired inositol 1,3,4,5-tetrakisphosphate binding capacity. *J. Biol. Chem.* **271**: 30303–30306.

Gaspar, H.B., Ferrando, M., Caragol, I. *et al.* (2000) Kinase mutant Btk results in atypical X-linked agammaglobulinaemia phenotype. *Clin. Exp. Immunol.* **120**: 346–350.

Genevier, H.C., Hinshelwood, S., Gaspar, H.B. *et al.* (1994) Expression of Bruton's tyrosine kinase protein within the B-cell lineage. *Eur. J. Immunol.* **24**: 3100–3105.

Gilmour, K.C., Cranston, T., Jones, A., Davies, E.G., Goldblatt, D., Thrasher, A., Kinnon, C., Nichols, K.E. and Gaspar, H.B. (2000) Diagnosis of X-linked lymphoproliferative disease by analysis of SLAM- associated protein expression. *Eur. J. Immunol.* **30**: 1691–1697.

Gilmour, K.C., Cranston, T., Loughlin, S. *et al.* (2000) Rapid protein based assays for the diagnosis of T-B+ severe combined immunodeficiency. *Br. J. Haemotol.* **112**: 671–676.

Giri, J.G., Ahdieh, M., Eisenman, J. *et al.* (1994) Utilization of the beta and gamma chains of the IL-2 receptor by the novel cytokine IL-15. *EMBO J.* **13**: 2822–2830.

Goldblatt, D. and Thrasher, A.J. (2000) Chronic granulomatous disease immunodeficiency review. *Clin. Exp. Immunol.* **122**: 1–9.

Hashimoto, S., Miyawaki, T., Futatani, T. *et al.* (1999) Atypical X-linked agammaglobulinemia diagnosed in three adults. *Intern. Med.* **38**: 722–725.

Hayward, A.R., Levy, J., Facchetti, F., Notarangelo, L., Ochs, H.D., Etzioni, A., Bonnefoy, J.Y., Cosyns, M. and Weinberg, A. (1997) Cholangiopathy and tumors of the pancreas, liver, and biliary tree in boys with X-linked immunodeficiency with hyper-IgM. *J. Immunol.* **158**: 977–983.

Hess, S. and Engelmann, H. (1996) A novel function of CD40: induction of cell death in transformed cells. *J. Exp. Med.* **183**: 159–167.

Ihle, J.N. (1995) Cytokine receptor signalling. *Nature* **377**: 591–594.

Katz, F., Hinshelwood, S., Rutland, P., Jones, A., Kinnon, C. and Morgan, G. (1996) Mutation analysis in CD40 ligand deficiency leading to X-linked hypogammaglobulinemia with hyper IgM syndrome. *Hum. Mutat.* **8**: 223–228.

Kennedy, M.K., Glaccum, M., Brown, S.N. *et al.* (2000) Reversible defects in natural killer and memory CD8 T cell lineages in interleukin 15-deficient mice. *J. Exp. Med.* **191**: 771–780.

Khan, W.N., Alt, F.W., Gerstein, R.M. *et al.* (1995) Defective B-cell development and function in Btk-deficient mice. *Immunity* **3**: 283–299.

Kimura, Y., Takeshita, T., Kondo, M., Ishii, N., Nakamura, M., Van Snick, J. and Sugamura, K. (1995) Sharing of the IL-2 receptor gamma chain with the functional IL-9 receptor complex. *Int. Immunol.* **7**: 115–120.

Kornfeld, S.J., Good, R.A. and Litman, G.W. (1994) Atypical X-linked agammaglobulinemia. *N. Engl. J. Med.* **331**: 949–951.

Kornfeld, S.J., Haire, R.N., Strong, S.J., Brigino, E.N., Tang, H., Sung, S.S., Fu, S.M. and Litman, G.W. (1997) Extreme variation in X-linked agammaglobulinemia phenotype in a three-generation family. *J. Allergy Clin. Immunol.* **100**: 702–706.

Korthauer, U., Graf, D., Mages, H.W. *et al.* (1993) Defective expression of T-cell CD40 ligand causes X-linked immunodeficiency with hyper-IgM. *Nature* **361**: 539–541.

Leonard, W.J. and O'Shea, J.J. (1998) Jaks and STATs: biological implications. *Annu. Rev. Immunol.* **16**: 293–322.

Levy, J., Espanol-Boren, T., Thomas, C. *et al.* (1997) Clinical spectrum of X-linked hyper-IgM syndrome. *J. Pediatr.* **131**: 47–54.

Li, T., Rawlings, D.J., Park, H., Kato, R.M., Witte, O.N. and Satterthwaite, A.B. (1997) Constitutive membrane association potentiates activation of Bruton tyrosine kinase. *Oncogene* **15**: 1375–1383.

Lodolce, J.P., Boone, D.L., Chai, S., Swain, R.E., Dassopoulos, T., Trettin, S. and Ma, A. (1998) IL-15 receptor maintains lymphoid homeostasis by supporting lymphocyte homing and proliferation. *Immunity* **9**: 669–676.

Lorenzi, R., Brickell, P.M., Katz, D.R., Kinnon, C. and Thrasher, A.J. (2000) Wiskott–Aldrich syndrome protein is necessary for efficient IgG-mediated phagocytosis. *Blood* **95**: 2943–2946.

MacCarthy-Morrogh, L., Gaspar, H.B., Wang, Y.C., Katz, F., Thompson, L., Layton, M., Jones, A.M. and Kinnon, C. (1998) Absence of expression of the Wiskott–Aldrich syndrome protein in peripheral blood cells of Wiskott–Aldrich syndrome patients. *Clin. Immunol. Immunopathol.* **88**: 22–27.

Macchi, P., Villa, A., Giliani, S., Sacco, M.G., Frattini, A., Porta, F., Ugazio, A.G., Johnston, J.A., Candotti, F. and O'Shea, J.J. (1995) Mutations of Jak-3 gene in patients with autosomal severe combined immune deficiency (SCID). *Nature* **377**: 65–68.

Machesky, L.M. and Insall, R.H. (1998). Scar1 and the related Wiskott-Aldrich syndrome protein, WASP, regulate the actin cytoskeleton through the Arp2/3 complex. *Curr. Biol.* **8**: 1347–1356.

Minegishi, Y., Rohrer, J., Coustan-Smith, E., Lederman, H.M., Pappu, R., Campana, D., Chan, A.C., and Conley, M.E. (1999). An essential role for BLNK in human B-cell development. *Science* **286**: 1954–1957.

Mouthon, L., Cohen, P., Larroche, C., Andre, M.H., Royer, I., Casassus, P. and Guillevin, L. (1999) Common variable immunodeficiency: one or multiple illnesses? 3 clinical cases. *Ann. Med. Interne (Paris)* **150**: 275–282.

Naqvi, S.N., Zahn, R., Mitchell, D.A., Stevenson, B.J. and Munn, A.L. (1998) The WASp homologue Las17p functions with the WIP homologue End5p/verprolin and is essential for endocytosis in yeast. *Curr. Biol.* **8**: 959–962.

Nichols, K.E., Harkin, D.P., Levitz, S. *et al.* (1998) Inactivating mutations in an SH2 domain-encoding gene in X-linked lymphoproliferative syndrome. *Proc. Natl Acad. Sci. USA* **95**: 13765–13770.

Nobes, C.D. and Hall, A. (1995) Rho, rac, and cdc42 GTPases regulate the assembly of multimolecular focal complexes associated with actin stress fibers, lamellipodia, and filopodia. *Cell* **81**: 53–62.

Noguchi, M., Yi, H., Rosenblatt, H.M., Filipovich, A.H., Adelstein, S., Modi, W.S., McBride, O.W. and Leonard, W.J. (1993a) Interleukin-2 receptor gamma chain mutation results in X-linked severe combined immunodeficiency in humans. *Cell* **73**: 147–157.

Noguchi, M., Nakamura, Y., Russell, S.M., Ziegler, S.F., Tsang, M., Cao, X. and Leonard, W.J. (1993b) Interleukin-2 receptor gamma chain: a functional component of the interleukin-7 receptor. *Science* **262**: 1877–1880.

Notarangelo, L.D. and Hayward, A.R. (2000) X-linked immunodeficiency with hyper-IgM (XHIM). *Clin. Exp. Immunol.* **120**: 399–405.

Ohta, Y., Haire, R.N., Litman, R.T. *et al.* (1994) Genomic organization and structure of Bruton agammaglobulinemia tyrosine kinase: localization of mutations associated with varied clinical presentations and course in X chromosome-linked agammaglobulinemia. *Proc. Natl Acad. Sci. USA* **91**: 9062–9066.

Pangburn, M.K. and Muller-Eberhard, H.J. (1984) The alternative pathway of complement. *Springer Semin. Immunopathol.* **7**: 163–192.

Park, H., Wahl, M.I., Afar, D.E., Turck, C.W., Rawlings, D.J., Tam, C., Scharenberg, A.M., Kinet, J.P. and Witte, O.N. (1996) Regulation of Btk function by a major autophosphorylation site within the SH3 domain. *Immunity* **4**: 515–525.

Parolini, S., Bottino, C., Falco, M. *et al.* (2000) X-linked lymphoproliferative disease. 2B4 molecules displaying inhibitory rather than activating function are responsible for the inability of natural killer cells to kill Epstein-Barr virus-infected cells. *J. Exp. Med.* **192**: 337–346.

Pepper, A.E., Buckley, R.H., Small, T.N. and Puck, J.M. (1995) Two mutational hotspots in the interleukin-2 receptor gamma chain gene causing human X-linked severe combined immunodeficiency. *Am. J. Hum. Genet.* **57**: 564–571.

Puck, J.M., Pepper, A.E., Henthorn, P.S. *et al.* (1997) Mutation analysis of IL2RG in human X-linked severe combined immunodeficiency. *Blood* **89**: 1968–1977.

Puel, A., Ziegler, S.F., Buckley, R.H. and Leonard, W.J. (1998) Defective IL7R expression in T(-)B(+)NK(+) severe combined immunodeficiency. *Nat. Genet.* **20**: 394–397.

Purtilo, D.T. (1981) X-linked lymphoproliferative syndrome. An immunodeficiency disorder with acquired agammaglobulinemia, fatal infectious mononucleosis, or malignant lymphoma. *Arch. Pathol. Lab. Med.* **105**: 119–121.

Ramesh, N., Anton, I.M., Martinez, Q.N. and Geha, R.S. (1999) Waltzing with WASP. *Trends. Cell Biol.* **9**: 15–19.

Rawlings, D.J., Saffran, D.C., Tsukada, S. *et al.* (1993) Mutation of unique region of Bruton's tyrosine kinase in immunodeficient XID mice. *Science* **261**: 358–361.

Rawlings, D.J., Scharenberg, A.M., Park, H., Wahl, M.I., Lin, S., Kato, R.M., Fluckiger, A.C., Witte, O.N. and Kinet, J.P. (1996) Activation of BTK by a phosphorylation mechanism initiated by SRC family kinases. *Science* **271**: 822–825.

Rohrer, J., Minegishi, Y., Richter, D., Eguiguren, J. and Conley, M.E. (1999) Unusual mutations in Btk: an insertion, a duplication, an inversion, and four large deletions. *Clin. Immunol.* **90**: 28–37.

Rohrer, J., Parolini, O., Belmont, J.W., Conley, M.E. and Parolino, O. (1994) The genomic structure of human BTK, the defective gene in X-linked agammaglobulinemia. *Immunogenetics* **40**: 319–324.

Roos, D., de Boer, M., Kuribayashi, F. *et al.* (1996) Mutations in the X-linked and autosomal recessive forms of chronic granulomatous disease. *Blood* 87: 1663–1681.

Royer-Pokora, B., Kunkel, L.M., Monaco, A.P., Goff, S.C., Newburger, P.E., Baehner, R.L., Cole, F.S., Curnutte, J.T. and Orkin, S.H. (1986) Cloning the gene for the inherited disorder chronic granulomatous disease on the basis of its chromosomal location. *Cold Spring Harb. Symp. Quant. Biol.* 51 Pt 1: 177–183.

Russell, S.M., Keegan, A.D., Harada, N. *et al.* (1993) Interleukin-2 receptor gamma chain: a functional component of the interleukin-4 receptor. *Science* 262: 1880–1883.

Russell, S.M., Tayebi, N., Nakajima, H. *et al.* (1995) Mutation of Jak3 in a patient with SCID: essential role of Jak3 in lymphoid development. *Science* 270: 797–800.

Sanger, W.G., Grierson, H.L., Skare, J., Wyandt, H., Pirruccello, S., Fordyce, R. and Purtilo, D.T. (1990) Partial Xq25 deletion in a family with the X-linked lymphoproliferative disease (XLP). *Cancer Genet. Cytogenet.* 47: 163–169.

Sayos, J., Wu, C., Morra, M., Wang, N. *et al.* (1998) The X-linked lymphoproliferative-disease gene product SAP regulates signals induced through the co-receptor SLAM. *Nature* 395: 462–469.

Seemayer, T.A., Gross, T.G., Egeler, R.M., Pirruccello, S.J., Davis, J.R., Kelly, C.M., Okano, M., Lanyi, A. and Sumegi, J. (1995) X-linked lymphoproliferative disease: twenty-five years after the discovery. *Pediatr. Res.* 38: 471–478.

Sideras, P., Muller, S., Shiels, H. *et al.* (1994) Genomic organization of mouse and human Bruton's agammaglobulinemia tyrosine kinase (Btk) loci. *J. Immunol.* 153: 5607–5617.

Stephan, V., Wahn, V., Le-Deist, F., Dirksen, U., Broker, B., Muller, F., I, Horneff, G., Schroten, H., Fischer, A. and De-Saint-Basile, G. (1996) Atypical X-linked severe combined immunodeficiency due to possible spontaneous reversion of the genetic defect in T cells. *N. Engl. J. Med* 335: 1563–1567.

Stowers, L., Yelon, D., Berg, L.J. and Chant, J. (1995) Regulation of the polarization of T cells toward antigen-presenting cells by Ras-related GTPase CDC42. *Proc. Natl Acad. Sci. USA* 92: 5027–5031.

Strahm, B., Rittweiler, K., Duffner, U. *et al.* (2000) Recurrent B-cell non-Hodgkin's lymphoma in two brothers with X-linked lymphoproliferative disease without evidence for Epstein-Barr virus infection. *Br. J. Haematol.* 108: 377–382.

Sumegi, J., Huang, D., Lanyi, A. *et al.* (2000) Correlation of mutations of the SH2D1A gene and epstein-barr virus infection with clinical phenotype and outcome in X-linked lymphoproliferative disease. *Blood* 96: 3118–3125.

Suzuki, H., Duncan, G.S., Takimoto, H. and Mak, T.W. (1997) Abnormal development of intestinal intraepithelial lymphocytes and peripheral natural killer cells in mice lacking the IL-2 receptor beta chain. *J. Exp. Med.* 185: 499–505.

Symons, M., Derry, J.M., Karlak, B., Jiang, S., Lemahieu, V., Mccormick, F., Francke, U. and Abo, A. (1996) Wiskott-Aldrich syndrome protein, a novel effector for the GTPase CDC42Hs, is implicated in actin polymerization. *Cell* 84: 723–734.

Takeshita, T., Asao, H., Ohtani, K., Ishii, N., Kumaki, S., Tanaka, N., Munakata, H., Nakamura, M. and Sugamura, K. (1992) Cloning of the gamma chain of the human IL-2 receptor. *Science* 257: 379–382.

Tangye, S.G., Lazetic, S., Woollatt, E., Sutherland, G.R., Lanier, L.L. and Phillips, J.H. (1999) Cutting edge: human 2B4, an activating NK cell receptor, recruits the protein tyrosine phosphatase SHP-2 and the adaptor signaling protein SAP. *J. Immunol.* 162: 6981–6985.

Tangye, S.G., Phillips, J.H., Lanier, L.L. and Nichols, K.E. (2000) Functional requirement for SAP in 2B4-mediated activation of human natural killer cells as revealed by the X-linked lymphoproliferative syndrome. *J. Immunol.* 165: 2932–2936.

Taniguchi, T. (1995) Cytokine signaling through nonreceptor protein tyrosine kinases. *Science* 268: 251–255.

Thrasher, A.J. and Kinnon, C. (2000) The Wiskott-Aldrich syndrome. *Clin. Exp. Immunol.* 120: 2–9.

Tsukada, S., Saffran, D.C., Rawlings, D.J. *et al.* (1993). Deficient expression of a B-cell cytoplasmic tyrosine kinase in human X-linked agammaglobulinemia. *Cell* 72: 279–290.

Vetrie, D., Vorechovsky, I., Sideras, P. *et al.* (1993) The gene involved in X-linked agammaglobulinaemia is a member of the src family of protein-tyrosine kinases. *Nature* 361: 226–233.

Vihinen, M., Brandau, O., Branden, L.J. *et al.* (1998) BTKbase, mutation database for X-linked agammaglobulinemia (XLA). *Nucleic Acids. Res.* 26: 242–247.

Vihinen, M., Iwata, T., Kinnon, C., Kwan, S.P., Ochs, H.D., Vorechovsky, I. and Smith, C.I. (1996) BTKbase, mutation database for X-linked agammaglobulinemia (XLA). *Nucleic Acids. Res.* **24**: 160–165.

Villa, A., Notarangelo, L., Macchi, P. *et al.* (1995) X-linked thrombocytopenia and Wiskott–Aldrich syndrome are allelic diseases with mutations in the WASP gene. *Nat. Genet.* **9**: 414–417.

Westberg, J., Fredrikson, G.N., Truedsson, L., Sjoholm, A.G. and Uhlen, M. (1995) Sequence-based analysis of properdin deficiency: identification of point mutations in two phenotypic forms of an X-linked immunodeficiency. *Genomics* **29**: 1–8.

Wicker, L.S. and Scher, I. (1986) X-linked immune deficiency (xid) of CBA/N mice. *Curr. Top. Microbiol. Immunol.* **124**: 87–101.

Woodland, R.T., Schmidt, M.R., Korsmeyer, S.J. and Gravel, K.A. (1996) Regulation of B-cell survival in xid mice by the proto-oncogene bcl-2. *J. Immunol.* **156**: 2143–2154.

Yasuda, N., Lai, P.K., Rogers, J. and Purtlo, D.T. (1991) Defective control of Epstein–Barr virus-infected B-cell growth in patients with X-linked lymphoproliferative disease. *Clin. Exp. Immunol.* **83**: 10–16.

Zhu, Q., Watanabe, C., Liu, T., Hollenbaugh, D., Blaese, R.M., Kanner, S.B., Aruffo, A. and Ochs, H.D. (1997) Wiskott-Aldrich syndrome/X-linked thrombocytopenia: WASP gene mutations, protein expression, and phenotype. *Blood* **90**: 2680–2689.

Zicha, D., Allen, W.E., Brickell, P.M., Kinnon, C., Dunn, G.A., Jones, G.E. and Thrasher, A.J. (1998) Chemotaxis of macrophages is abolished in the Wiskott-Aldrich syndrome. *Br. J. Haematol.* **101**: 659–665.

16

The ubiquitin–proteasome system and genetic diseases: Protein degradation gone awry

S. Russ Price and William E. Mitch

1. Introduction

Protein degradation is a critical process for the growth and function of all cells. It eliminates abnormal proteins, halts regulatory processes, and supplies amino acids for cellular remodeling. When protein substrates of proteolytic pathways are not recognized, or there is mistiming of proteolysis, profound changes in cell function will occur. Based on these potential problems, it is not surprising that genetic alterations in proteolytic enzymes/cofactors or the protein substrates that render them more or less susceptible to degradation are responsible for inherited disorders.

Multiple pathways exist for degrading proteins. Lysosomes contain proteases that have optimal activity at an acidic pH (e.g. cathepsins) and degrade proteins engulfed by endocytosis. Traditionally, degradation by this pathway has been thought to be nonspecific but there is growing evidence that some proteins can be specifically targeted to the lysosome for degradation (Agarraberes et al., 1997; Cuervo et al., 1997). A second proteolytic pathway involves calcium-dependent proteases (e.g. calpains). The cellular functions of these proteases are believed to play a role in cytoskeletal reorganization. Finally, there are the energy requiring proteolytic systems. The best described system is the ubiquitin–proteasome pathway which requires ATP and degrades the bulk of cellular and some membrane proteins (Bailey et al., 1996; Rock et al., 1994).

2. The ubiquitin–proteasome pathway

2.1 Ubiquitin

Proteins degraded by the very large (>1500 kDa) 26S proteasome complex must first undergo a modification process which targets the substrate proteins by

Genotype to Phenotype second edition, edited by S. Malcolm and J. Goodship.

linking it to a small protein, ubiquitin (*Figure 1*). Ubiquitin is a member of the heat-shock protein family. It is found in all cells and is one of the most evoluntionarily conserved proteins known. Multiple genes encode precursors of ubiquitin, a protein of 76 amino acids. Two genes, *UbC* and *UbB*, have been localized to chromosome band 12q24.3 and 17p11.1–17p12, respectively and encode chains of consecutive head-to-tail copies of the ubiquitin. These chains are proteolytically processed to yield monomeric ubiquitin (Schlesinger and Bond, 1987). Other genes encode ubiquitin-fusion proteins consisting of the ubiquitin peptide fused to another protein (e.g. ribosomal proteins (Baker and Board, 1991)). These fusion proteins are proteolytically cleaved to generate 'free' ubiquitin. Thus, multiple genes contribute to the pool of ubiquitin monomers.

2.2 Ubiquitin conjugation enzymes

The process of linking ubiquitin to substrate proteins occurs before their degradation by the proteasome (reviewed in Hershko and Ciechanover, 1998). Ubiquitination involves a series of complex steps: initially, an E1 ubiquitin-activating enzyme uses ATP to form an E1-ubiquitin thiolester through the carboxyl terminal glycine of ubiquitin (*Figure 1*). A single E1 enzyme is responsible for ubiquitin activation in all mammalian cells. After ubiquitin activation, one of several E2 ubiquitin-carrier proteins (also called ubiquitin-conjugating enzymes or UBCs) transfers the activated ubiquitin to the protein substrate which has been bound to a member of the E3 ubiquitin-protein ligase family (*Figure 1*). These E3 enzymes bind specific proteins and interact with specific E2s. Thus, selectivity of the proteolytic process occurs at the level of ubiquitin activation and conjugation and possibly at other steps.

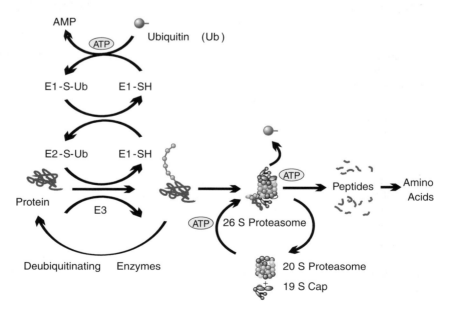

Figure 1. Ubiquitin–proteasome proteolytic pathway.

In most cases, ubiquitin is linked to the substrate protein through an isopeptide bond between the ε-amino group of a lysine residue in the target protein and the carboxyl terminal glycine of ubiquitin. Cycles of these reactions link additional ubiquitins to lysines within ubiquitins added previously. Typically, ubiquitin chains are linked through K48 in ubiquitin but alternate linkages through K63 or K29 have been described (Wilkinson, 2000). It is possible that these other linkages have distinct functions but to date, no specific roles have been identified.

Various E3 ubiquitin ligases provide selectivity to the ubiquitination process by serving as docking proteins that bring the substrate protein and the E2 carrier protein with activated ubiquitin together. In some instances, accessory proteins required for ubiquitin conjugation interact with specific E3 ubiquitin ligases. Much of our knowledge about the functions of various E2 and E3 proteins has been gained by studying the ubiquitination process in yeast; homologs of many of these yeast proteins have been identified in higher eukaryotes including humans. The E3 ubiquitin ligases are grouped into families based on structural similarities and the functional classes of substrates they recognize.

E3α. The 'N-end Rule' determines the rate of degradation of proteins based on their NH_2-terminal residues. Proteins with basic or bulky hydrophobic N-terminal amino acids are degraded very rapidly (e.g. 2–3 minutes) whereas proteins with stabilizing residues have half-lives measured in many hours. Selectivity is achieved because the E3α ligase recognizes proteins with 'destabilizing' residues and presents them to the $E2_{14k}$ carrier protein which completes the ubiquitination reaction (Kwon *et al.*, 1998). Subsequently, these proteins are degraded by the 26S proteasome. The specificity of these reactions was worked out in yeast and little is known about the *in vivo* substrates in eukaryotic organs. However, recent reports indicate the N-end Rule pathway is responsible for ubiquitination of muscle proteins and the system is activated in conditions that cause loss of muscle mass (Lecker *et al.*, 1999; Solomon *et al.*, 1998).

HECT domain ubiquitin ligases. These E3 ligases are referred to as the *hect* domain proteins because they have structural motifs that are **h**omologous to the **E**6-AP **C**-**t**erminus. The prototypical family member is E6-AP, a 100-kDa protein that is an accessory protein required for ubiquitination and degradation of the p53 tumor suppressor protein in cells infected by the human papilloma virus. The E6-AP ligase does not directly interact with p53. Rather, E6-AP serves as an intermediate by accepting the activated ubiquitin thiolester on a cysteine near its C-terminus and this E6-AP-ubiquitin complex interacts with the accessory E6 protein encoded by the virus. Some reports indicate that E6-AP can function independently of the E6 viral protein to transfer ubiquitin to other protein substrates.

The biochemical action of other HECT E3 ligase family members involves conjugation of an activated ubiquitin on a cysteine near their C-termini; the activated ubiquitin is then transferred to a target protein. Substrate recognition features of the HECT ligases include a highly variable N-terminal domain and an internal WW domain that interacts with a hydrophobic PPxY sequence (PY motif) in the target proteins. Substrates of the HECT E3 ligases (other than the p53 tumor suppressor) include the epithelial sodium channel and components of the TGF-β/SMAD signal transduction pathway (Attisano and Wrana, 2000; Zhu *et al.*, 1999).

Anaphase promoting complex (APC). The APC of proteins (also referred to as the cyclosome) acts to conjugate ubiquitin to proteins that regulate important mitotic events (e.g. type A and B cyclins). These APC E3 ligase complexes are activated at specific times during mitosis to degrade cyclins and hence, regulate the levels of proteins that initiate anaphase and exit from telophase. Proteins that are recognized by the APC complex generally contain a degenerate nine amino acid pattern, **R**-A/T-A-**L**-G-X-I/V-G/T-N (the invariant residues are indicated in bold and are underlined). Changes in the phosphorylation/dephosphorylation of some APC complex subunits are important for activation/inactivation. This scheme provides cells with a mechanism to regulate the degradation of substrate proteins rapidly.

Skp-1-cullin-F-box (SCF) complex. The SCF family of E3 ligases recognizes cell-cycle regulatory and other signaling proteins. Like the APC complexes, the SCFs are multi-subunit complexes of proteins that include one of a family of variable F-box proteins. The F-box proteins function to recognize substrate proteins and also bind the Skp-1 protein (Bai *et al.*, 1996). Skp-1 links F-box proteins to other components of the SCF E3 ligase complex and these other components recruit specific E2 conjugating enzymes. Typically, substrate proteins of the SCF complex are phosphorylated before they undergo conjugation to ubiquitin.

One SCF complex contains the F-box protein, βTrCP, and is responsible for conjugating ubiquitin to two important signaling molecules, β-catenin and IκBα (Hart *et al.*, 1999; Hatakeyama *et al.*, 1999; Kroll *et al.*, 1999; Maniatis, 1999;). β-catenin is a cell–cell adhesion protein that also participates in cytosolic signal transduction linked to cellular developmental systems (Polakis, 1999, 2000). Normal β-catenin function is important because it prevents many forms of cancer (see below).

IκB is the inhibitory protein of NFκB, a transcription factor activated by inflammatory cytokines and other stress signals. IκB interacts with NFκB in the cytosol, rendering it inactive as a transcription factor because it can not translocate to the nucleus (reviewed in Karin and Ben Neriah, 2000). In fact, signals that activate NFκB initially stimulate a kinase complex (IκB kinase) that phoshorylates IκB on Ser 32 or Ser 36. The phosphorylated IκB can then be recognized by the SCF^βTrCP E3 complex that conjugates ubiquitin to K21 and K22 of IκB. The sequential phosphorylation and ubiquitin conjugation of IκB triggers its degradation by the proteasome.

Von Hipple-Lindau (VHL). The VHL tumor suppressor protein associates with the B and C elongins and some cullin proteins to form E3 ligase complexes similar to the SCF ligases (*Lisztwan et al.*, 1999; Schoenfeld *et al.*, 2000). Mutations in the VHL tumor suppressor cause the rare VHL disease. Patients with this disorder (1/36 000 births) are predisposed to develop tumors in several organs by an unknown mechanism (Hes *et al.*, 2000). The only confirmed substrate for the VHL E3 ligase complex is the hypoxia-inducible factor, HIF-1, a heterodimeric transcription factor that is activated by hypoxia and rapidly degraded upon reoxygenation (Cockman *et al.*, 2000; Kamura *et al.*, 2000). It is found in many cell types and regulates the expression of a range of genes involved in metabolic adaptation, iron metabolism, angiogenesis, inflammation, and cell survival.

2.3 26S Proteasome

The 26S proteasome has over 30 different subunits and exhibits several protease activities that degrade proteins to small peptides (reviewed in Bochtler *et al.*, 1999; Voges *et al.*, 1999). It has a 20S catalytic core particle and two 19S regulatory particles (*Figure 1*). The 20S core is a barrel-shaped stack of four rings (Groll *et al.*, 1997): the two outer rings are identical and are composed of seven different α type subunits. The two inner rings contain seven different β type subunits and are also identical. Structural features of the 20S core block the random degradation of cytosolic proteins providing another level of specificity. A small opening in the outer α-subunit rings prevents proteins from inadvertently entering the catalytic chamber where the proteolytic sites are located on β subunits. Threonine residues are nucleophilic active sites for at least three distinct proteolytic activities: chymotrypsin-like, trypsin-like, and post-glutamylpeptidyl hydrolase-like. Not all proteasome complexes are identical: the form of the proteasome that generates class I MHC molecules has several β subunit substitutions that are induced by the presence of interferon-γ.

A 19S particle is attached to each outer α-subunit ring of the 20S core. It functions as the recognition component of the 26S complex and, therefore, prevents proteins from being degraded in a random fashion. One or more of the 19S subunits recognize ubiquitin chains of four or more residues ensuring that mono-ubiquitinated or non-conjugated substrates do not get degraded. Other 19S subunits confer ubiquitin hydrolase, ATPase or chaperone functions to the cap particle. Models of proteolysis suggest that the polyubiquitin chain on the target protein is cleaved by the 19S particle. Subsequently, proteins are unfolded and guided into the 20S proteolytic core for cleavage to small peptides. The peptides are released and degraded to amino acids by exopeptidases in the cytosol. The major exception to complete degradation is the Class I MHC molecules; these molecules are transported to the cell surface (reviewed in Rock and Goldberg, 1999).

2.4 DeUbiquitinating enzymes (DUBs)

A large family of ubiquitin C-terminal hydrolases (UCHs) and isopeptidases (also known as ubiquitin-specific processing proteases or UBPs) are capable of processing ubiquitin fusion proteins to free ubiquitin (reviewed in Wilkinson, 2000). They also cleave mono-ubiquitin (or short ubiquitin chains) from proteins and can disassemble polyubiquitin chains that are released from substrates before degradation in the proteasome. These enzymes are important because they recycle ubiquitin for use in the conjugation reaction. For example, isopeptidase T preferentially breaks down polyubiquitin chains that can inhibit the 26S proteasome. Other isopeptidases may control the turnover of specific proteins by removing ubiquitin or polyubiquitin chains from the substrate, thereby, preventing its degradation by the proteasome. Many DUB enzymes are expressed in a tissue-specific manner and could exert different roles in various cells.

3. Ubiquitin-like modifier proteins (UBLs)

A number of proteins that bear homology to ubiquitin or contain a ubiquitin-like domain within a larger protein have been identified (Hochstrasser, 2000). These UBLs are covalently linked to proteins by reactions that are analogous to the ubiquitin conjugation reactions (Tanaka *et al.*, 1998). Similarly, there are isopeptidases that cleave the isopeptidyl bond that links the ubiquitin-like modifiers to their target proteins (Desterro *et al.*, 1999; Suzuki *et al.*, 1999). In theory, a competitive interplay between the conjugation of ubiquitin and UBLs to the same targets could provide a regulatory mechanism that determines the fate of specific proteins.

A potentially prominent member of the UBL protein family is SUMO (<u>s</u>mall <u>u</u>biquitin-related <u>mo</u>difier). SUMO has been found linked to RanGAP1, a trafficking protein that helps shuttle proteins from the cytosol to the nucleus (Lee *et al.*, 1998). For example, addition of SUMO to RanGAP1 is required before RanGAP1 will associate with nuclear pores (Matunis *et al.*, 1998). Other proteins (e.g. IκB) can undergo conjugation to either SUMO or ubiquitin but conjugation of SUMO to IκB blocks ubiquitin conjugation of IκB and hence, its degradation. Thus, SUMOylation of IκB prevents the release and translocation of NFκB to the nucleus (Desterro *et al.*, 1998).

4. Genetic abnormalities involving the ubiquitin–proteasome system

This review of the ubiquitin–proteasome pathway points out the complexity of protein degradation in the ubiquitin–proteasome pathway and suggests several points where genetic abnormalities could change its activity. However, diseases are generally recognized by an abnormality in a specific organ raising the central question of how this pathway, which is present in all cells, could be dysfunctional in only one organ. Potential responses to this question are:

1. a mutation or deletion in a substrate protein expressed only in one organ (e.g. mutation of a phosphorylation site) leading to more or less degradation of the substrate;
2. dysfunction of a DNA/RNA editing system in the organ could alter components of the ubiquitin–proteasome pathway;
3. presence of a factor (transcription or other) in the organ that changes expression of components of the ubiquitin–proteasome pathway;
4. increased or decreased response to an external signal (e.g. a hormone or neural signal) that regulates activity of the ubiquitin–proteasome pathway directly or through initial steps leading to degradation of a substrate protein;
5. expression of an accessory factor that accelerates or depresses functions of the ubiquitin–proteasome pathway in an organ.

A variety of inherited diseases have been linked to dysfunctional proteolysis by the ubiquitin–proteasome system. A few selected disorders will be discussed.

4.1 Angelman syndrome

The Angelman syndrome (AS) is a rare genetic disorder that is characterized by severe mental retardation, seizures and an ataxic gait. Deletions or mutations in the 15q11-q13 region of chromosome 15 have been linked to this syndrome and the Prader–Willi syndrome. The observation that different clinical entities result from deletion of the same region has been attributed to gene imprinting in the affected region of chromosome 15 (Cassidy and Schwartz, 1998; Jiang *et al.*, 1999; Kishino *et al.*, 1997; Matsuura *et al.*, 1997). In normal individuals, the 15q11-q13 region includes the *UBE3A* gene that encodes E6-AP E3 ligase. Originally, *UBE3A* was discounted as a candidate gene for AS because both the maternal and paternal genes are expressed in most tissues. However, recent studies have revealed that the maternal copy of *UBE3A* is expressed in specific regions of the brain and central nervous system while the paternal copy of the gene is silent (Rougeulle *et al.*, 1997; Vu and Hoffman, 1997). This is the imprinting phenomenon in which there is differential expression of a gene from one parent; this phenomenon is related to the presence of an imprinting center which directs gene expression.

Approximately 70% of all patients with AS (Type I) have *de novo* deletions in the 15q11-q13 region of the maternal chromosome (Matsuura *et al.*, 1997). Approximately 3–5% of affected individuals (Type II) have unbalanced translocations or paternal uniparental disomy which may arise by chromosome nondisjunction. Note that two normal paternal copies of *UBE3A* are present in these individuals but both of these genes are silent due to imprinting. In 3–5% of patients (Type III), a mutation in the imprinting center has been identified. In 4–6% of patients (Type IV), point mutations in *UBE3A* result in a loss of E6-AP function. In the remaining ~10% of patients, the genetic basis of AS does not fall into one of the above categories raising questions about the role of E6-AP in the disease. This disorder could be an example of diseases in Category 1 in our list above.

4.2 Neurodegenerative diseases

Neurofibrillary tangles and plaques characterize Alzheimer's disease and other neurodegenerative diseases. These abnormal structures contain excessive quantities of tau protein and the amyloid β peptide. High amounts of ubiquitin protein are colocalized in these structures. The uniform presence of these proteins suggests a cause-and-effect relationship. Amyloid b peptides arise from the processing of the amyloid β precursor protein by the presenilin proteases (i.e. presenilin 1 and 2). Presenilin proteases are degraded by the proteasome. Genetic abnormalities in the genes for amyloid precursor protein or the presenilins are generally linked to the familial form of Alzheimer's disease which is characterized by an early age of onset. Circumstantial evidence suggests that mutations in the presenilins alter their degradation by the proteasome and hence may be the key to increased production of amyloid β peptides (Marambaud *et al.*, 1998a,b).

Genetically-related abnormalities of the ubiquitin conjugation reaction have not been described in familial Alzheimer's disease although a mutant form of ubiquitin, ubiquitin-B^{+1} (Ub^{+1}), is uniquely expressed in the brains of many elderly patients with the nonfamilial form of Alzheimer's (van Leeuwen *et al.*, 1998). Ub^{+1} arises during post-transcriptional editing of ubiquitin mRNA when dinucleotide deletions occur adjacent to a GAGAG motif in the polyubiquitin-B mRNA . The polyubiquitin-B gene (*Ub-B*) encodes a precursor consisting of three consecutive head to tail copies of ubiquitin which are processed to monomeric ubiquitin by a ubiquitin C-terminal hydrolase (Baker and Board, 1987). A dinu-cleotide deletion in the Ub^{+1} mRNA introduces a frameshift in one of the ubiq-uitin repeats causing a replacement of the C-terminal glycine with a 20 amino acid extension (van Leeuwen *et al.*, 1998). After translation and processing, Ub^{+1} monomers are conjugated to proteins but degradation by the proteasome is inhibited by the resulting conjugate (Lam *et al.*, 2000). Processing of Ub^{+1}-containing chains by isopeptidase T is also blocked because this ubiquitin C-terminal hydrolase requires the presence of a terminal G76 in ubiquitin. The accumulation of Ub^{+1}-containing chains competitively inhibits the 26S proteasome. Thus, Ub^{+1} acts in a dominant negative fashion to inhibit the degra-dation of important (or deleterious) proteins by the ubiquitin–proteasome system in patients with neurodegenerative disorders. Notably, Ub^{+1} has been found to associate with the tau protein in neurofibrillary tangles in brains of patients with progressive supranuclear palsy (Wang *et al.*, 2000). This finding suggests that Ub^{+1} may play a common role in other neurodegenerative disorders. These forms of neurodegenerative diseases are examples of category 2 in our list.

4.3 Malignancy and oncogenes

Signaling pathways regulate cell growth and differentiation by controlling the levels of checkpoint proteins. One regulatory step is the degradation of critical proteins by the ubiquitin–proteasome pathway at precise times during the cell cycle. Prolonged activation of signaling proteins because their degradation is suppressed can have deleterious effects on normal cell function. In fact, impaired degradation of genetically altered proteins results in their prolonged activation and this phenomena has been implicated in tumor production and oncogenesis.

The proto-oncogene, c-myc, is an example of the relationship between the ubiquitin–proteasome system and tumor progression. It is a short-lived tran-scription factor that is critical for the regulation of cell growth and proliferation and is degraded by the ubiquitin–proteasome pathway. Inactivation of c-myc prolongs cells in the G1 and G2 phases of the cell cycle (Mateyak *et al.*, 1997) and mutated forms of the protein are frequently found in cancers, especially lymphomas (Bahram *et al.*, 2000; Gregory and Hann, 2000; Salghetti *et al.*, 1999). Mutations have been identified throughout the c-myc protein but a mutation 'hot spot', T58, is located in the N-terminal transactivation domain, a region that is important in determining c-myc stability. The T58 'hotspot' can be phospho-rylated by several signaling proteins including glycogen synthase kinase-3 (GSK-3). Phosphorylation is a prerequisite for ubiquitin conjugation which, in turn, is necessary for c-myc degradation by the 26S proteasome (Bahram *et al.*, 2000; Gregory and Hann, 2000; Salghetti *et al.*, 1999). Mutations in T58 or in the N-terminal region of c-myc have been found in Burkitt's lymphoma cells and other cancers. These mutations decrease the turnover rate of c-myc by reducing the efficiency of ubiquitin conjugation. Other regions of c-myc also influence its turnover. For example, residues in the C-terminal, DNA-binding domain are important for interactions between c-myc and Miz-1, a protein that stabilizes c-myc (Salghetti *et al.*, 1999). Thus, various mutations can account for the different half-lives of c-myc in different types of cancer.

In addition to being involved in cell–cell adhesion, β-catenin functions as a transducer molecule in the Wnt signaling cascade which regulates cell growth and survival. Cytosolic β-catenin forms a complex with glycogen synthase kinase-3β (GSK3β), the adenomatous polyposis coli (APC) tumor suppressor and axin; in this form, β-catenin is inactive. As with other proteins, after phosphorylation (e.g. GSK-3β), β-catenin becomes a target for conjugation to ubiquitin by the SCF^{TFβCP} E3 ligase enzyme complex. This explains why certain mutations prevent β-catenin phosphorylation and degradation. However, there are other mutations in the NH$_2$-terminal region of β-catenin that prevent formation of the β-catenin-APC-axin-GSK-3β complex, and hence its phosphorylation and ultimately its degradation by the 26S proteasome (Morin *et al.*, 1997; Rubinfeld *et al.*, 1997). Finally, mutations in the *APC* gene can produce a truncated protein that does not interact with β-catenin. The result is reduced β-catenin degradation, increased Wnt pathway signaling and poorly regulated cell growth (Kitagawa *et al.*, 1999). Mutations in c-myc, β-catenin and APC tumor suppressor are examples of a category 1 defect in our list.

4.4 Cystic fibrosis

Cystic fibrosis is an autosomal recessive disorder resulting from the loss or malfunction of the cystic fibrosis transmembrane conductance regulator protein (CFTR), a transmembrance chloride ion channel. As many as 400 different mutations in the CFTR locus have been documented but greater than 70% of all cystic fibrosis cases involve deletion of F508 (ΔF508) in at least one allele (Kopito, 1999; Skach, 2000). CFTR is synthesized and glycosylated in the rough endoplasmic reticulum before being inserted into the membrane. Only 20–40% of wild-type channel and nearly all of the ΔF508 channel fail to become mature CFTR because they do not fold properly (Kopito, 1999). Mutant or immature forms of CFTR remain in the endoplasmic reticulum where they become conjugated to ubiquitin and degraded by the proteasome (Jensen *et al.*, 1995; Ward *et al.*, 1995). This explains why inhibitors of the proteasome (e.g. lactacystin, MG-132) lead to accumulation of ubiquitin-conjugated forms of CFTR in cells and why coexpression of a dominant negative ubiquitin with the ΔF508 CFTR prevents degradation of the chloride channel. Interestingly, CFTR ΔF508 maintains chloride channel activity even though the mutation causes its accelerated degradation. CFTR DF508 is an example of a category 1 defect in our list.

4.5 Liddle syndrome

Liddle syndrome is an autosomal dominant form of hypertension characterized by the early onset of severe hypertension, hypokalemia, metabolic alkalosis, salt sensitivity, and low values of serum aldosterone and renin. The disorder has been attributed to mutations in subunits of the epithelial sodium channel (ENaC) producing increased numbers of membrane channels that sustain channel activity (Abriel *et al.*, 1999; Bubien *et al.*, 1996; Firsov *et al.*, 1996; Schild *et al.*, 1996; Staub *et al.*, 2000). ENaC is composed of α, β, and γ subunits that contain PY motifs recognized by NEDD4, a member of the HECT E3 ubiquitin-ligase family. Replacing lysines clustered near the N-termini of α ENaC and γ ENaC proteins with arginine residues results in a decrease in ubiquitin conjugation of the protein and increased ENaC channel activity. Likewise, we found that inhibition of the proteasome can increase ENaC function in kidney cells (Malik *et al.*, 2000). Based on these findings, it appears that mutations in the PY motifs of the β and/or γ subunits in Liddle's syndrome results in channels that do not interact with NEDD4. Ubiquitin conjugation can target unassembled subunits for proteasomal degradation. However, fully-functional ENaC channels can be degraded by lysosomes even though they are monoubiquitinated (Staub *et al.*, 1997). Again, the genetic abnormalities leading to Liddle's syndrome is an example of a category 1 defect in our list.

4.6 Other defects

We have included category 3 defects in our list because interferon-γ is known to induce the production of new proteasome β-subunits that change the structural composition and peptidase activities of the 26S proteasome (Rock and Goldberg, 1999). This response causes the proteasome to make antigens more efficiently. Conceivably, there could be a defect in factors that are induced by interferon-γ that would impair this critical response to the cytokine. Other cytokines/signaling molecules could also change the components of the ubiquitin–proteasome pathway. This type of defect might reside in abnormalities of specific transcription factors but at present, no specific example of this category has been reported.

We do not know of an established example of category 4 in our list but indirect evidence we and others have gained during investigation of protein catabolic mechanisms suggests this possibility: a disease (i.e. a manifestation of an abnormal, organ-specific response) could be due to a genetic disorder that changes the response to a signal (e.g. hormones or neural responses). For example, animal models of several catabolic conditions reveal a pattern of responses that include increased degradation of muscle proteins by the ubiquitin–proteasome pathway plus high levels of the mRNAs encoding components of the pathway in muscle (reviewed in Mitch and Goldberg, 1996). For example, in metabolic acidosis or acute diabetes, the increase in mRNAs involves stimulation of transcription of genes encoding ubiquitin and subunits of the proteasome (Bailey *et al.*, 1996; Price *et al.*, 1996) and in the case of metabolic acidosis, starvation or acute diabetes, the increase in mRNAs requires glucocorticoids (Mitch *et al.*, 1999; Price *et al.*, 1994; Wing and Goldberg, 1993). Moreover, there is evidence that increased transcription of the ubiquitin gene in response to glucocorticoids occurs preferentially in muscle rather than intestine, liver, or kidney by a mechanism that involves the Sp1 transcription factor (Marinovic *et al.*, 2000). Thus, a disorder that raises glucocorticoid production could cause an increase or decrease in muscle mass if there were heightened or impaired sensitivity of skeletal muscle to glucocorticoids. Similarly, activation of muscle proteolysis and an increase in mRNAs for constituents of the ubiquitin–proteasome pathway are found in response to muscle denervation, suggesting that loss of neural activity regulates activity of the pathway as well as transcription of genes encoding pathway components (Medina *et al.*, 1991, 1995). Thus, impaired or heightened responses to neural signals could cause slower or faster rates of muscle protein breakdown and hence, changes in muscle mass. Although there is a hormonal or neural component to the proteolytic responses linked to muscle wasting in catabolic states, there are no conditions in which specific genetic abnormalities of category 4 in our list have been documented.

5. Summary

There has been an explosion of information about the biochemical mechanisms and functions of the ubiquitin–proteasome pathway. Because the pathway is complex, it could be regulated at several steps or could react to the presence of an abnormal substrate by increasing or decreasing its degradation (i.e. category 1 in our list). Most known genetic abnormalities involving the pathway are examples of our category 1 defects. We expect, however, that examples of other categories in our list will be identified as we learn more about genetic defects and about the function of the ubiquitin–proteasome pathway.

References

Abriel, H., Loffing, J., Rebhun, J.F., Pratt, J.H., Schild, L., Horisberger, J.D., Rotin, D. and Staub, O. (1999) Defective regulation of the epithelial Na+ channel by Nedd4 in Liddle's syndrome. *J. Clin. Invest.* 103: 667–673.

Agarraberes, F.A., Terlecky, S.R. and Dice, J.F. (1997) An intralysosomal hsp70 is required for a selective pathway of lysosomal protein degradation. *J. Cell Biol.* 137: 825–834.

Attisano, L. and Wrana, J.L. (2000) Smads as transcriptional co-modulators. *Curr. Opin. Cell Biol.* 12: 235–243.

Bahram, F., von der Lehr, N., Cetinkaya, C. and Larsson, L.G. (2000) c-Myc hot spot mutations in lymphomas result in inefficient ubiquitination and decreased proteasome-mediated turnover. *Blood* 95: 2104–2110.

Bai, C., Sen, P., Hofmann, K., Ma, L., Goebl, M., Harper, J.W., and Elledge, S.J. (1996) SKP1 connects cell cycle regulators to the ubiquitin proteolysis machinery through a novel motif, the F-box. *Cell* 86: 263–274.

Bailey, J.L., Wang, X., England, B.K., Price, S.R., Ding, X. and Mitch, W.E. (1996) The acidosis of chronic renal failure activates muscle proteolysis in rats by augmenting transcription of genes encoding proteins of the ATP-dependent, ubiquitin–proteasome pathway. *J. Clin. Invest.* 97: 1447–1453.

Baker, R.T. and Board, P.G. (1987) The human ubiquitin gene family: structure of a gene and pseudogene from the Ub B subfamily. *Nucleic Acids Res.* 15: 443–463.

Baker, R.T. and Board, P.G. (1991) The human ubiquitin-52 amino acid fusion protein gene shares several structural features with mammalian ribosomal protein genes. *Nucleic Acids Res.* 19: 1035–1040.

Bochtler, M., Ditzel, L., Groll, M., Hartmann, C. and Huber, R. (1999) The proteasome. *Annu. Rev. Biophys. Biomol. Struct.* 28: 295–317.

Bubien, J.K., Ismailov, I.I., Berdiev, B.K. *et al.* (1996) Liddle's disease: abnormal regulation of amiloride-sensitive Na+ channels by β-subunit mutation. *Am. J. Physiol.* **270**: C208–C213.

Cassidy, S.B. and Schwartz, S. (1998) Prader–Willi and Angelman syndromes. Disorders of genomic imprinting. *Medicine* **77**: 140–151.

Cockman, M.E., Masson, N., Mole, D.R. *et al.* (2000) Hypoxia inducible factor-alpha binding and ubiquitylation by the von hippel-lindau tumor suppressor protein. *J. Biol. Chem.* **275**: 25733–25741.

Cuervo, A.M., Dice, J.F. and Knecht, E. (1997) A population of rat liver lysosomes responsible for the selective uptake and degradation of cytosolic proteins. *J. Biol. Chem.* **272**: 5606–5615.

Desterro, J.M., Rodriguez, M.S., Kemp, G.D. and Hay, R.T. (1999) Identification of the enzyme required for activation of the small ubiquitin-like protein SUMO-1. *J. Biol. Chem.* **274**: 10618–10624.

Desterro, J.M.P., Rodriguez, M.S. and Hay, R.T. (1998) SUMO-1 modification of IκBα inhibits NF-κB activation. *Mol. Cell.* **2**: 233–239.

Firsov, D., Schild, L., Gautschi, I., Merillat, A.M., Schneeberger, E. and Rossier, B.C. (1996) Cell surface expression of the epithelial Na channel and a mutant causing Liddle syndrome: a quantitative approach. *Proc. Natl Acad. Sci. USA* **93**: 15370–15375.

Gregory, M.A. and Hann, S.R. (2000) c-Myc proteolysis by the ubiquitin–proteasome pathway: stabilization of c-Myc in Burkitt's lymphoma cells. *Mol. Cell Biol.* **20**: 2423–2435.

Groll, M., Ditzel, L., Lowe, J., Stock, D., Bochtler, M., Bartunik, H.D. and Huber, R. (1997) Structure of the 20S proteasome from yeast at 2.4 Å resolution. *Nature* **386**: 463–471.

Hart, M., Concordet, J.P., Lassot, I. *et al.* (1999) The F-box protein beta-TrCP associates with phosphorylated beta-catenin and regulates its activity in the cell. *Curr. Biol.* **9**: 207–210.

Hatakeyama, S., Kitagawa, M., Nakayama, K. *et al.* 1999) Ubiquitin-dependent degradation of IkappaBalpha is mediated by a ubiquitin ligase Skp1/Cul 1/F-box protein FWD1. *Proc. Natl Acad. Sci. USA* **96**: 3859–3863.

Hershko, A. and Ciechanover, A. (1998) The ubiquitin system. *Annu. Rev. Biochem.* **67**: 425–479.

Hes, F., Zewald, R., Peeters, T. *et al.* (2000) Genotype-phenotype correlations in families with deletions in the von Hippel-Lindau (VHL) gene. *Hum. Genet.* **106**: 425–431.

Hochstrasser, M. (2000) All in the ubiquitin family. *Science* **289**: 563–564.

Jensen, T.J., Loo, M.A., Pind, S., Williams, D.B., Goldberg, A.L. and Riordan, J.R. (1995) Multiple proteolytic systems, including the proteasome, contribute to CFTR processing. *Cell* **83**: 129–135.

Jiang, Y., Lev-Lehman, E., Bressler, J., Tsai, T.F. and Beaudet, A.L. (1999) Genetics of Angelman syndrome. *Am. J. Hum. Genet.* **65**: 1–6.

Kamura, T., Sato, S., Iwai, K., Czyzyk-Krzeska, M., Conaway, R.C. and Conaway, J.W. (2000) Activation of HIF1α ubiquitination by a reconstituted von Hippel-Lindau (VHL) tumor suppressor complex. *Proc. Natl Acad. Sci. USA* **97**: 10430–10435.

Karin, M. and Ben Neriah, Y. (2000) Phosphorylation meets ubiquitination: the control of NF-κB activity. *Ann. Rev. Immunol.* **18**: 621–663.

Kishino, T., Lalande, M. and Wagstaff, J. (1997) UBE3A/E6-AP mutations cause Angelman syndrome *Nature Genet.* **15**: 70–73.

Kitagawa, M., Hatakeyama, S., Shirane, M., Matsumoto, M., Ishida, N., Hattori, K., Nakamichi, I., Kikuchi, A. and Kakayama, K. (1999) An F-box protein, FWD1, mediates ubiquitin-dependent proteolysis of beta-catenin. *EMBO J.* **18**: 2401–2410.

Kopito, R.R. (1999) Biosynthesis and degradation of CFTR. *Physiol. Rev.* **79**: S167–S173.

Kroll, M., Margottin, F., Kohl, A., Renard, P., Durand, H., Concordet, J.P., Bachelerie, F., Arenzana-Seisdedos, F. and Benarous, R. (1999) Inducible degradation of IkappaBalpha by the proteasome requires interaction with the F-box protein h-betaTrCP. *J. Biol. Chem.* **274**: 7941–7945.

Kwon, Y.T., Reiss, Y., Fried, V.A. *et al.* (1998) The mouse and human genes encoding the recognition component of the N-end rule pathway. *Proc. Natl Acad. Sci. USA* **95**: 7898–7903.

Lam, Y.A., Pickart, C.M., Alban, A., Jamieson, C., Ramage, R., Mayer, R.J. and Layfield, R. (2000) Inhibition of the ubiquitin–proteasome system in Alzheimer's disease. *Proc. Natl Acad. Sci. USA* **97**: 9902–9906.

Lecker, S.H., Solomon, V., Price, S.R., Kwon, Y.T., Mitch, W.E. and Goldberg, A.L. (1999) Ubiquitin conjugation by the N-end rule pathway and mRNAs for its components increase in muscles of diabetic rats. *J. Clin. Invest.* **104**: 1411–1420.

Lee, G.W., Melchior, F., Matunis, M.J., Mahajan, R., Tian, Q. and Anderson, P. (1998) Modification of Ran GTPase-activating protein by the small ubiquitin-related modifier SUMO-1 requires Ubc9, an E2-type ubiquitin-conjugating enzyme homologue. *J. Biol. Chem.* **273**: 6503–6507.

Lisztwan, J., Imbert, G., Wirbelauer, C., Gstaiger, M. and Krek, W. (1999) The von Hippel-Lindau tumor suppressor protein is a component of an E3 ubiquitin-protein ligase activity. *Genes Dev.* **13**: 1822–1833.

Malik, B., Schlanger, H.-F., Bao, A.-K. and Eaton, D.C. (2000) Regulation of epithelial Na+ channels through proteasomal pathway. *FASEB J.* **14**: A103–A103.

Maniatis, T. (1999) A ubiquitin ligase complex essential for the NF-kappaB, Wnt/Wingless, and Hedgehog signaling pathways. *Genes Dev.* **13**: 505–510.

Marambaud, P., Alves, da Costa, Ancolio, K. and Checler, F. (1998a) Alzheimer's disease-linked mutation of presenilin 2 (N141I-PS2) drastically lowers APPalpha secretion: control by the proteasome. *Biochem Biophys. Res. Commun.* **252**: 134–138.

Marambaud, P., Ancolio, K., Lopez-Perez, E. and Checler, F. (1998b) Proteasome inhibitors prevent the degradation of familial Alzheimer's disease-linked presenilin 1 and potentiate A beta 42 recovery from human cells. *Mol. Med.* **4**: 147–157.

Marinovic, A.C., Mitch, W.E. and Price, S.R. (2000) Muscle-specific regulation of ubiquitin (*UbC*) transcription by glucocorticoids involves Sp1. *J. Am. Soc. Nephrol.* **11**: 624A–624A.

Mateyak, M.K., Obaya, A.J., Adachi, S. and Sedivy, J.M. (1997) Phenotypes of c-Myc-deficient rat fibroblasts isolated by targeted homologous recombination. *Cell Growth Differ.* **8**: 1039–1048.

Matsuura, T., Sutcliffe, J.S., Fang, P., Galjaard, R.J., Jiang, Y.H., Benton, C.S., Rommens, J.M. and Beaudet, A.L. (1997) De novo truncating mutations in E6-AP ubiquitin-protein ligase gene (UBE3A) in Angelman syndrome. *Nat. Genet.* **15**: 74–77.

Matunis, M.J., Wu, J. and Blobel, G. (1998) SUMO-1 modification and its role in targeting the Ran GTPase-activating protein, RanGAP1, to the nuclear pore complex. *J. Cell Biol.* **140**: 499–509.

Medina, R., Wing, S.S. and Goldberg, A.L. (1995) Increase in levels of polyubiquitin and proteasome mRNA in skeletal muscle during starvation and denervation atrophy. *Biochem. J.* **307**: 631–637.

Medina, R., Wing, S.S., Haas, A. and Goldberg, A.L. (1991) Activation of the ubiquitin-ATP-dependent proteolytic system in skeletal muscle during fasting and denervation atrophy. *Biomed. Biochim. Acta.* **50**: 347–356.

Mitch, W.E., Bailey, J.L., Wang, X., Jurkovitz, C., Newby, D. and Price, S.R. (1999) Evaluation of signals activating ubiquitin–proteasome proteolysis in a model of muscle wasting. *Am. J. Physiol.* **276**: C1132–C1138.

Mitch, W.E. and Goldberg, A.L. (1996) Mechanisms of muscle wasting: The role of the ubiquitin–proteasome pathway. *N. Engl. J. Med.* **335**: 1897–1905.

Morin, P.J., Sparks, A.B., Korinek, V., Barker, N., Clevers, H., Vogelstein, B. and Kinzler, K.W. (1997) Activation of beta-catenin-Tcf signaling in colon cancer by mutations in beta-catenin or APC. *Science* **275**: 1787–1790.

Polakis, P. (1999) The oncogenic activation of beta-catenin. *Curr. Opin. Genet. Dev.* **9**: 15–21.

Polakis, P. (2000) Wnt signaling and cancer. *Genes Dev.* **14**: 1837–1851.

Price, S.R., Bailey, J.L., Wang, X., Jurkovitz, C., England, B.K., Ding, X., Phillips, L.S. and Mitch, W.E. (1996) Muscle wasting in insulinopenic rats results from activation of the ATP-dependent, ubiquitin proteasome proteolytic pathway by a mechanism including gene transcription. *J. Clin. Invest.* **98**: 1703–1708.

Price, S.R., England, B.K., Bailey, J.L., Van Vreede, K. and Mitch, W.E. (1994) Acidosis and glucocorticoids concomitantly increase ubiquitin and proteasome subunit mRNAs levels in rat muscle. *Am. J. Physiol.* **267**: C955–C960.

Rock, K.L. and Goldberg, A.L. (1999) Degradation of cell proteins and the generation of MHC class I-presented peptides. *Annu. Rev. Immunol.* **17**: 739–779.

Rock, K.L., Gramm, C., Rothstein, L., Clark, K., Stein, R., Dick, L., Hwang, D. and Goldberg, A.L. (1994) Inhibitors of the proteasome block the degradation of most cell proteins and the generation of peptides presented on the MHC Class I molecules. *Cell.* **78**: 761–771.

Rougeulle, C., Glatt, H. and Lalande, M. (1997) The Angelman syndrome candidate gene, UBE3A/E6-AP, is imprinted in brain. *Nat. Genet.* **17**: 14–15.

Rubinfeld, B., Robbins, P., El Gamil, M., Albert, I., Porfiri, E. and Polakis, P. (1997) Stabilization of beta-catenin by genetic defects in melanoma cell lines. *Science* **275**: 1790–1792.

Salghetti, S.E., Kim, S.Y. and Tansey, W.P. (1999) Destruction of Myc by ubiquitin-mediated proteolysis: cancer-associated and transforming mutations stabilize Myc. *EMBO J.* **18**: 717–726.

Schild, L., Lu, Y., Gautschi, I., Schneeberger, E., Lifton, R.P. and Rossier, B.C. (1996) Identification of a PY motif in the epithelial Na channel subunits as a target sequence for mutations causing channel activation found in Liddle syndrome. *EMBO J.* **15**: 2381–2387.

Schlesinger, M.J. and Bond, U. (1987) Ubiquitin genes. *Oxf. Surv. Eukaryot. Genes* **4**: 77–91.

Schoenfeld, A.R., Davidowitz, E.J. and Burk, R.D. (2000) Elongin BC complex prevents degradation of von Hippel-Lindau tumor suppressor gene products. *Proc. Natl Acad. Sci. USA* **97**: 8507–8512.

Skach, W.R. (2000) Defects in processing and trafficking of the cystic fibrosis transmembrane conductance regulator. *Kidney Int.* **57**: 825–831.

Solomon, V., Lecker, S.H. and Goldberg, A.L. (1998) The N-end rule pathway catalyzes a major fraction of the protein degradation in skeletal muscle. *J. Biol. Chem.* **273**: 25216–25222.

Staub, O., Abriel, H., Plant, P., Ishikawa, T., Kanelis, V., Saleki, R., Horisberger, J.D., Schild, L. and Rotin, D. (2000) Regulation of the epithelial Na^+ channel by Nedd4 and ubiquitination. *Kidney Int.* **57**: 809–815.

Staub, O., Gautschi, I., Ishikawa, T., Breitschopf, K., Ciechanover, A., Schild, L. and Rotin, D. (1997) Regulation of stability and function of the epithelial Na+ channel (ENaC) by ubiquitination. *EMBO J.* **16**: 6325–6336.

Suzuki, T., Ichiyama, A., Saitoh, H., Kawakami, T., Omata, M., Chung, C.H., Kimura, M., Shimbara, N. and Tanaka, K. (1999) A new 30-kDa ubiquitin-related SUMO-1 hydrolase from bovine brain. *J. Biol. Chem.* **274**: 31131–31134.

Tanaka, K., Suzuki, T. and Chiba, T. (1998) The ligation systems for ubiquitin and ubiquitin-like proteins. *Mol. Cells* **8**: 503–512.

van Leeuwen, F.W., de Kleijn, D.P., van den Hurk, H.H. *et al.* (1998) Frameshift mutants of beta amyloid precursor protein and ubiquitin-B in Alzheimer's and Down patients. *Science* **279**: 242–247.

Voges, D., Zwickl, P. and Baumeister, W. (1999) The 26S proteasome: a molecular machine designed for controlled proteolysis. *Annu. Rev. Biochem.* **68**: 1015–1068.

Vu, T.H. and Hoffman, A.R. (1997) Imprinting of the Angelman syndrome gene, UBE3A, is restricted to brain. *Nature Genet.* **17**: 12–13.

Wang, E.W., Kessler, B.M., Borodovsky, A., Cravatt, B.F., Bogyo, M., Ploegh, H.L. and Glas, R. (2000) Integration of the ubiquitin-proteasome pathway with a cytosolic oligopeptidase activity. *Proc. Natl Acad. Sci. USA* **97**: 9990–9995.

Ward, C.L., Omura, S. and Kopito, R.R. (1995) Degradation of CFTR by the ubiquitin–proteasome pathway. *Cell* **83**: 121–127.

Wilkinson, K.D. (2000) Ubiquitination and deubiquitination: targeting of proteins for degradation by the proteasome. *Semin. Cell Dev. Biol.* **11**: 141–148.

Wing, S.S. and Goldberg, A.L. (1993) Glucocorticoids activate the ATP-ubiquitin-dependent proteolytic system in skeletal muscle during fasting. *Am. J. Physiol.* **264**: E668–E676.

Zhu, H., Kavsak, P., Abdollah, S., Wrana, J.L. and Thomsen, G.H. (1999) A SMAD ubiquitin ligase targets the BMP pathway and affects embryonic pattern formation. *Nature* **400**: 687–693.

Index